ZULFIKAR ALI
BHUTTO
AND
PAKISTAN
1967–1977

ZULFIKAR ALI
BHUTTO
AND
PAKISTAN
1967–1977

RAFI RAZA

Karachi
Oxford University Press
Oxford   New York   Delhi
1997

Oxford University Press, Walton Street, Oxford OX2 6DP

Oxford   New York
Athens   Auckland   Bangkok   Bombay
Calcutta   Cape Town   Dar es Salaam   Delhi
Florence   Hong Kong   Istanbul   Karachi
Kuala Lumpur   Madras   Madrid   Melbourne
Mexico City   Nairobi   Paris   Singapore
Taipei   Tokyo   Toronto
and associated companies in
Berlin   Ibadan

Oxford is a trade mark of Oxford University Press

© Oxford University Press, 1997

ISBN 0 19 577697 6

Printed in Pakistan at
Mas Printers, Karachi.
Published by
Ameena Saiyid, Oxford University Press
5-Bangalore Town, Sharae Faisal
P.O. Box 13033, Karachi-75350, Pakistan.

CONTENTS

AUTHOR'S NOTE

I am grateful to numerous friends, particularly Mubashir Hasan, Mumtaz Ali Bhutto and Sardar Sherbaz Mazari, for encouraging me to publish this narrative. They have generously given their time and counsel, while providing me with additional insight on some important issues. I must also thank my wife for her help in editing the manuscript, and for her unfailing support over many years. My one regret is that my father is not alive to read this book; he encouraged me throughout my life in all my endeavours—though strongly advising me against politics.

I would like to draw the attention of the reader to the Notes to each Chapter. In several cases important and revealing facts have been confined to the Notes to ensure an uninterrupted flow in the narrative. I should also point out that the spellings of places and names are as officially in use today. Thus, it is Dhaka and not Dacca, Sindh rather than Sind, and similarly with Balochistan, Beijing, Mao Zedong and Zhou Enlai. They have been changed accordingly even in quoted passages.

24 October 1995
Karachi

LIST OF ILLUSTRATIONS
Between pp. 144 and 145

ABBREVIATIONS

BBC	British Broadcasting Corporation
BD	Basic Democracy
BPC	Basic Principles Committee
CEC	Chief Election Commissioner
CENTO	Central Treaty Organization
CML	Council Muslim League
CMLA	Chief Martial Law Administrator
DAC	Democratic Action Committee
DG, ISI	Director General, Inter Services Intelligence
DIB	Director, Intelligence Bureau
FSF	Federal Security Force
GHQ	General Headquarters (of the army)
GOP	Government of Pakistan
IAF	Indian Air Force
IMF	International Monetary Fund
JI	Jamaat-i-Islami
JUI	Jamiat-ul-Ulema-e-Islam
JUP	Jamiat-ul-Ulema-e-Pakistan
LFO	Legal Framework Order
ML(Q)	Muslim League (Qayyum)
MNA	Member of the National Assembly
MPA	Member of the Provincial Assembly
NAM	Non-Aligned Movement
NAP	National Awami Party
NWFP	North-West Frontier Province
OIC	Organization of the Islamic Conference
PAF	Pakistan Air Force
PLD	Pakistan Legal Decisions
PLO	Palestine Liberation Organization
PNA	Pakistan National Alliance

PPP	Pakistan People's Party
RCD	Regional Cooperation for Development
RTC	Round Table Conference
SEATO	South-East Asia Treaty Organization
UAE	United Arab Emirates
UDF	United Democratic Front
UN	United Nations
VOA	Voice of America
ZAB	Zulfikar Ali Bhutto

INTRODUCTION

'The Quaid-i-Azam[1] broke a country to wear three hats—that of Governor-General, President of the Muslim League and President of the first Constituent Assembly—so *Time* magazine wrote in 1947; I wonder what *Time* will say about me now that I have four hats—President of Pakistan, Chief Martial Law Administrator, Chairman of the PPP and President of the Constituent Assembly.' Zulfikar Ali Bhutto told me this half-jokingly in his chamber in the National Assembly after his election as President of the Assembly on 14 April 1972. This narrative sets out how ZAB[2] came to wear those four hats after founding the Pakistan People's Party (PPP) in December 1967, and traces his Government from December 1971 till his downfall in July 1977, all within a period of ten years.

This is not a biography, already undertaken by his childhood friend and admirer from Bombay, Piloo Mody, and an American academic, Stanley Wolpert. They and others have touched on the complexities and formative elements of his earlier life. He was born on 5 January 1928 to a Sindhi feudal and a Hindu lady of humble origin at a time when Hindu-Muslim differences were coming to the political fore. In contrast with his feudal background in rural Sindh, his boyhood was spent in Bombay where he experienced a cosmopolitan, urban life with his father, Sir Shahnawaz Bhutto, a senior official in British India. Higher education in California and at Oxford exposed him to further Westernizing influences. Another formative element was his involvement in politics on returning to Pakistan, when the dominance of the military under Ayub Khan convinced him of their inescapable role. In October 1958, at the age of thirty, he was selected as a Federal Minister under Martial Law, subsequently serving for eight years under President Ayub Khan, finally as Foreign Minister.

I do not intend to examine these aspects of his life further unless they bear specifically on events and decisions covered in this narrative. I have also avoided comment on his private life; this would be unbecoming in one who worked closely with him.

The ten-year period from 1967 to 1977 is of sufficient significance to merit undiluted attention from a political perspective. It witnessed the creation of the first political party with mass support in West Pakistan since 1947, the first ever general election based on adult franchise in Pakistan in 1970, civil war and the dismemberment of the country in 1971; the first democratically elected Government and its overthrow by Martial Law.

These years are illumined largely through the writings, words and actions of ZAB. Every event and aspect cannot be covered and there is an unavoidable element of selectivity; but nothing significant has been omitted. If frequent references are made to my role, this is both inevitable and justifiable.

I participated in much that happened over those ten years. I kept copious notes, and have looked back countless times over a period which was important not only in the life of Pakistan but my own. I have in the past hesitated to write because ZAB is now dead. His name lives on, but he cannot speak for himself. Any criticism could be misunderstood and attributed to poor taste on the part of a former colleague and lieutenant. I remain grateful for the opportunity he gave me to serve my country. On the other hand, merely writing a paean of praise would be a pointless exercise. To maintain a proper balance is difficult; nearer the event of his tragic execution it would have been almost impossible.

Many friends and colleagues from those days, and even some members of the Opposition, have urged me to publish an account of that critical decade. I owed it, they said, to Pakistan and to history, since no one had as yet written an accurate account of what happened, and certainly not from the PPP point of view. Indeed, a major tragedy in Pakistan is that the truth is neither told nor written. As a result, we have learnt little or nothing about ourselves.

Over the years, so many falsehoods and canards have been spread about ZAB that it is difficult to distinguish fact from fiction. He ceased to be a man and became a myth and martyr to those who believed in him, and evil incarnate to those who opposed him. The greatest casualty has been the truth, particularly as all efforts to destroy the man's image gave greater impetus to the myth. I have thus taken care to substantiate everything that is said. Only my account of personal conversations must stand on their own. Inevitably, however, in narrating events and developments I have been constrained by considerations of national interest and security.

NOTES

1. Quaid-i-Azam means 'Great Leader', by which appellation the founder of Pakistan, Mohammad Ali Jinnah, is generally addressed.
2. After Zulfikar Ali Bhutto became Chairman of the Pakistan People's Party, and then President and Prime Minister of Pakistan, he was not called 'Zulfi'; even his wife, Nusrat, referred to him formally. In this narrative, he is designated either by his title as Chairman, President or Prime Minister, or by the term 'ZAB', which came to be adopted by his colleagues.

Chapter 1

ORIGIN AND GROWTH OF THE PPP: 1967–9

Conceiving the PPP

'Welcome back,' we greeted Zulfikar Ali Bhutto at Karachi airport on his return from Europe in October 1966 as President Ayub Khan's former Foreign Minister. His cousin Mumtaz Ali Bhutto[1] and I were the only two present to meet him—far outnumbered by the police who, like ZAB, had expected a large gathering.

This was ZAB's period in the wilderness. Many felt his political career was finished and that opposition to a well-entrenched Ayub Khan was futile. Mumtaz, however, gave emotional support: 'Don't worry. We will form a party of our own; you can register me as the first member.'

A year before, ZAB had been at the height of his career as Foreign Minister and virtual Prime Minister under Ayub Khan. His path to prominence had been easy as the well-educated son of a wealthy land-owning family with a political background and good connections. He went to the University of California (1947-50), then Christ Church, Oxford (1950-2), before being called to the Bar in England. His father, Sir Shahnawaz Bhutto, nurtured this ambitious son on politics; letters to him in student days conveyed the political scene in Pakistan rather than family news and normal parental advice.[2]

His first official assignment was as chairman of the Pakistan delegation to the United Nations Conference on the Law of the Seas in April 1958. In appreciation, he wrote to President

Iskander Mirza from Geneva: 'For the greater good of my own country, I feel that your services to Pakistan are indispensible. When the history of our country is written by objective historians, your name will be placed even before that of Mr Jinnah.' In October 1958, within days of selecting ZAB as Commerce Minister, Iskander Mirza was exiled; historians have judged him differently. ZAB also bestowed praise on Ayub Khan, who followed as President, comparing him to Abraham Lincoln, Kemal Attaturk and Salahuddin. Such hyperbole was part of the language of politics in West Pakistan, particularly among those from a feudal background. Nine years later, he changed noticeably when he launched into an independent political career; his praise of others became markedly less fulsome.

President Ayub Khan came to value his intelligent and industrious young lieutenant, first as Commerce Minister and then as Minister for Fuel, Power and Natural Resources. ZAB realized his cherished ambition to become Foreign Minister on the sudden death in 1963 of Mohammad Ali Bogra, who had briefly been Ayub Khan's second Foreign Minister. In his new position, he enjoyed immense prestige and power. During the 1965 September War with India, his dramatic promise to wage a 'thousand-year war' made him an instant hero, particularly in the Punjab and among students. He promoted this image following the Tashkent peace talks between Ayub Khan and the Indian Prime Minister, Lal Bahadur Shastri,[3] held under the auspices of Soviet Premier Alexei Kosygin in January 1966.

On resigning as Foreign Minister—or as ZAB told Mumtaz Bhutto before meeting President Ayub Khan on 16 June 1966, 'I am getting the boot today'—he journeyed by train on 22 June from Rawalpindi to Lahore, where he was tumultuously received. The champion against India, however, disappointed the large, mainly student, gathering by failing to make a speech, or thank the crowd. He was not yet ready to challenge Ayub Khan publicly. After a brief stay in Larkana, he was met at Karachi by a group of cheering students led by Mairaj

Mohammad Khan. Thereafter he went abroad—into political oblivion, as many thought at the time.

His initial steps back into the political arena were slow and tentative. When he returned a few months later, in October, he was undecided about his future course. He started to meet politicians, and labour and student leaders in increasing numbers. The Government, who were watching him carefully, attempted to undermine his efforts and sought to dissuade him from condemning the cease-fire and Tashkent Declaration. Moves were made to prise away the support of his cousin, Mumtaz, when Governor Kalabagh of West Pakistan urged Mumtaz to establish an independent position in politics. The Government also pressured supporters; plain-clothes policemen harassed those visiting ZAB's residence at 70 Clifton, which became a Karachi landmark.

Direct pressure on ZAB was increased. The Government began by attacking his nationalist credentials and anti-Indian image. Ayub Khan's Minister of Information raised in the National Assembly an allegation already made earlier in the Indian Parliament after the 1965 War: ZAB had given up his claim to property in Bombay, based on his Indian nationality, only in November 1958 after becoming a Federal Minister. These charges were repeated in an editorial in a leading Karachi daily, *Dawn,* on 2 July 1967, under the caption 'Depressing Disclosures'. As the Minister's allegations were protected by parliamentary privilege, it was decided to refute the charges by issuing a detailed rejoinder and serving a legal notice on *Dawn* for defamation. Several lawyers refrained from involvement. At Mumtaz's request, I served a notice demanding retraction and damages on the grounds that 'Mr. Z. A. Bhutto's attitude towards Pakistan-India relationship, his acquiring a Pakistani passport, and his case in India and claim for property in Pakistan have been and are matters of record'.[4]

We decided against court action as it would have involved long drawn-out libel proceedings. The subsequent cancellation of the licences for his family collection of arms, designed to harass him, led to his first political case. Mahmud Ali Qasuri and I filed a writ petition on 11 October 1967 against this politically-motivated measure.

From his return till mid-1967, ZAB remained undecided on whether to join an established party or go it alone. He considered an arrangement with Khan Abdul Wali Khan's faction of the National Awami Party. He also held discussions, which were serious in intent, with the Council Muslim League about becoming its Secretary-General. He was dissuaded from this course by J. A. Rahim and Mairaj Mohammad Khan in Karachi, as well as by Mumtaz Bhutto and Ghulam Mustafa Khar, who flew down from Rawalpindi specifically for this purpose.

J. A. Rahim had been putting the case for a new party since meeting ZAB abroad in the autumn of 1966. Towards the middle of 1967, Mubashir Hasan and Rahim presented to him the principles they had formulated for a new socialist party. ZAB was conscious that no party with a socialist programme had made headway in Pakistan. He saw the merits of their views but was uncertain of the viability of socialism as a vehicle for change in West Pakistan's traditional society. World-wide, however, progressive forces in the 1960s were looking to socialism. In Pakistan, the disarray of the left offered opportunities and, moreover, it was mainly leftists who had responded to ZAB.

The original nucleus who helped him conceive the Pakistan People's Party were J. A. Rahim, Mubashir Hasan, Mumtaz Bhutto, Ghulam Mustafa Khar, Mairaj Mohammad Khan and Mohammad Hayat Sherpao. They were a curious mix, each talented in a different way. This was no elitist or distinct group as in other parties; they came from diverse classes, cultures and academic backgrounds. Rahim was a former officer of the Indian Civil Service and senior diplomat who had fallen out with Ayub Khan, living for a while in Paris. He helped ZAB, on leaving government, to write *The Myth of Independence*.[5] Most significant, he provided the intellectual framework and content for the proposed party. He was supported by Mubashir Hasan, a staunch leftist from Lahore, who had an American doctorate in engineering. Mumtaz Bhutto was a barrister and had been a Member of the National Assembly since 1965. Khar, a landlord from Muzaffargarh in the Punjab, who was elected to the National Assembly at the age of twenty-five in 1962, became an ardent follower of ZAB. Mairaj Mohammad was a Karachi student leader, an outstanding leftist orator and agitator. Hayat Sherpao came from a Muslim League family in the NWFP, a fiery young speaker who hero-worshipped his new leader. Differences were apparent, even in terms of language. Rahim and Mumtaz, both Oxbridge-educated, mainly spoke English; Mubashir was equally fluent in Urdu and English; the other three were more at ease in Urdu.

They were attracted by ZAB's undoubted dynamism, intelligence and charisma. A good team was formed, to be joined by others, though it contained no one of prominence in the early days. We were frequent visitors at 70 Clifton where free and frank discussion flowed. Serious talk apart, ZAB was amusing, a clever mimic, and entertained us with tales of his experiences with Ayub Khan and world leaders.

Once it was decided to form a new party, he planned ahead meticulously. He could see the broad picture and seldom forgot a detail. Rahim and he were a curious couple, essentially complementary in the early days. Rahim provided the ideology and ZAB the pragmatism and popular appeal. Their different approaches were exemplified in the Foundation Meeting.

Birth of the PPP

Amidst government-inspired rumours and disquiet, a convention was held in the compound of Mubashir Hasan's Lahore residence on 30 November and 1 December 1967. Most Pakistanis today are not familiar with the Foundation Documents. The documents provide a basis for understanding the forces unleashed by the PPP, its motivation, and ZAB's ideas and actions. Of the ten Foundation Meeting Documents,[6] ZAB wrote No. 3, *Why a New Party?* and Mubashir provided No. 7, *Declaration of the Unity of the People*. Rahim contributed the others. The Interim Constitution of the Party was finalized in a hurry by Rahim during the convention and set out its four-fold motto:

Islam is our faith;
Democracy is our polity;
Socialism is our economy;
All power to the people.

Rahim's ideological thrust was expressed in the opening words of the document, *Why Socialism is necessary for Pakistan*: 'To put it in one sentence, the aim of the party is the transformation of Pakistan into a socialist society'. The *Declaration of Principles* reiterated this as the 'objective', and called for fundamental changes: the nationalization of finance and key sectors of industry, the removal of the 'final remnants of the feudal system', and the strengthening of trade unions. It also sought the abolition of illiteracy, equal rights for women, the separation of the Executive from the Judiciary, and academic freedom.

The only question which did not involve socialism related to Jammu and Kashmir (Document No. 8):

This mission takes precedence over all other internal and external responsibilities of the party. The future of the people of Jammu and Kashmir forms a part of the future of Pakistan itself; the people of that State are as component to Pakistan as are the people of the Punjab or Bengal, Sindh or Balochistan, the Frontier or Mohajirs. Pakistan without Kashmir is as incomplete as is a body without a head.

The last of the documents, *Six Points Answered*, did not seem of much relevance at the time, certainly not in the West Wing. Later, it formed the basis of the PPP's critique against the Awami League when a Six-Point[7] constitution became the main issue, so it merits quoting:

The 6 points contain no proposal of economic and social reforms and are, for that reason, as a whole inadequate to solve the real problems of the relations between East Pakistan on one side and the Centre and West Pakistan on the other. The common people in West Pakistan suffer from exploitation just as their brethren in the East Wing. Pakistan is one nation and not two. The economic, social and political problems must be solved for the whole of Pakistan. It cannot be done in any other manner.

For Rahim and Mubashir, socialism was the answer to almost everything. In contrast, ZAB addressed the first of the four sessions on the problems facing the country: corruption, exploitation, fundamental rights, Jammu and Kashmir, Vietnam and American influence in Pakistan. Not once did he mention socialism.[8] Instead, he talked of a 'classless society such as has been conceived in the Faith of Islam'.

Similarly, the document he contributed, *Why a New Party?*, referred to the 'conservative' and 'progressive' forces finding their 'respective unities', followed by the national unity of all opposition parties in order to launch a constitutional struggle for the restoration of democracy. The new Party 'would form a bridge between the existing conflicting interests and give a lead in reconciling the historical dichotomies of the opposition'.[9] ZAB's object was to oppose Ayub Khan through such unity. For him, the struggle was not ideological.

In the fourth and concluding session on the afternoon of 1 December 1967, the Party was officially named and formed.[10] The convention rejected the three names proposed, the 'People's Progressive Party', 'People's Party', and 'Socialist Party of Pakistan', adopting instead 'Pakistan People's Party'. It also decided on the Party flag. ZAB was ready for battle.

The Rise of the PPP and Ayub Khan's Fall

ZAB had earlier started his 'meet the people' tours, and now there was no stopping him. He addressed every gathering prepared to listen, in the countryside or at roadside cafés, however small. He was a new phenomenon on the Pakistani political scene, well-groomed, young, articulate and charismatic. He went to the people. They were impressed and awed by this man in a Savile Row suit, Turnbull and Asser shirt and Sulka tie. His greatest assets were his unlimited energy, flair for politics and appetite for work. His driving force was above all his ambition, supported by his intellectual vigour.

The time was right for the PPP. Ayub Khan had been President for nine years and, if nothing else, longevity in power leads to decline in popularity, particularly in a Third World country. His authority had begun to diminish with the onslaught on him in the presidential campaign of 1964. The acceptance of the cease-fire in the 1965 September War followed by the Tashkent Declaration made him unpopular, and, most important, affected his standing with the army, hitherto his power base. He had also lost two major assets with the departure first of ZAB and then of his strong Governor of West Pakistan, Malik Amir Mohammad Khan of Kalabagh. He was sixty-one when, in January 1968, he suffered a serious pulmonary embolism followed by a relapse, and never fully recovered.

In East Pakistan, agitation for provincial autonomy had been gaining momentum, following years of protest against domination and exploitation by West Pakistan. Even recognition of Bengali as a national language had been won only after a struggle. In the West Wing, ZAB and his Party now provided a focal point for the expression of grievances and frustration. The regime's high-handedness had earlier angered students, who first demonstrated in 1961 and again in 1965-6. They were prepared to challenge Ayub Khan. The regime had allowed the 'twenty-two families' of industrial and business barons to flourish, but no trade unionism to develop. Labour was restless. In the rural areas, land reforms had not been implemented and the poverty of the peasant was extreme. 'Land to the landless' was an irresistible call. Young intellectuals felt stifled by press controls and authoritarian rule. To them ZAB was the 'Hero of Kashmir', made in the mould of Soekarno, Zhou Enlai and Nasser. It was the group around Mairaj that made *Roti, Kapra aur Makaan* the rallying call of the Party. It was a catchy abbreviation of a slogan among students in India:

Roti, Kapra aur Makaan
Mang raha hai har insan
(Bread, clothes and housing is the demand of all human beings).

In Pakistan, the slogan electrified the masses.

ZAB's first public meeting at Mochi Gate, Lahore, a test venue for all leaders, was reasonably well attended despite rain. He was wearing a Western suit, but in a typically dramatic gesture he threw off the jacket, saying he wanted to get properly wet like the crowd. He was learning how to rouse the masses and hold them in rapt attention, which became one of his major political assets. Although his Urdu was not good, the poor understood and welcomed him. The rally was discussed at a meeting in Mubashir Hasan's house. We expressed satisfaction at the outcome and also considered the appointment of various office-bearers.[11]

The year 1968 witnessed a whirlwind of tours to establish contact with the public. The Party was gaining strength and popularity despite Government obstacles and contrary forecasts by political pundits. ZAB's speeches through the year were barely reported in the newspapers. Mainly they were tape-recorded and replayed by devoted Partymen at various venues. The most comprehensive record was made by his ardent admirer, Hayat Sherpao, in the Frontier between 25 October and 5 November.[12] It was tragically ironic that he was assassinated in 1975 by an explosive device in a tape-recorder while speaking at Peshawar University.

The speeches showed an increasing build-up against Ayub Khan. They stressed the PPP programme, particularly how socialism was not in conflict with Islam. ZAB also felt the need to explain why he had left government, and his earlier silence on this subject. Towards the end of the year he began to lay the groundwork for his candidature in the forthcoming presidential elections. As with all his actions, he planned ahead with care and caution.

He appealed to the disadvantaged sections of society who, almost for the first time since 1948, had found a spokesman. Addressing Party women workers on 29 January 1968 at Lahore, he declared: 'We will offer the women of this country their proper place without prejudice. The place you are given today is by chauvinistic male courtesy.'[13] Later, to the Bar Association at Nawabshah, on 21 February, he pointed out, 'The student community is annoyed and up in arms' because the Government had introduced 'oppressive and obnoxious University Ordinances' and even had the 'audacity to take away degrees from them'. The legal profession was unhappy 'because the laws have been tampered with'. The labouring classes were being 'exploited under the system of plunder and loot which...exists nowhere in the world today. It is absolute highway robbery'. State lands should be given to peasants and tenants. He attacked the Government for 'throwing people into jails. How long can you fill the jails? Ideas cannot be imprisoned. Principles cannot be imprisoned. The entire population cannot be imprisoned'.[14]

Although his main focus was on the West Wing, he referred to East Pakistan while addressing the Khairpur Bar Association on 8 March:

> The Government says that East Pakistan want to go their own way. I cannot believe that. They are the majority. The seeds of Pakistan were sown in Bengal. The Muslim League was born in Bengal. Had there been no Bengal, there would have been no Pakistan. It is our majority province and it has made a very big contribution to Pakistan. How can a majority leave on its own.[15]

As the PPP gained momentum and more adherents, he held Party Conventions for each province, in Lahore for the Punjab and Balochistan on 22 August, for Sindh at Hyderabad on 21 September and, last, on 3 November, the Frontier Convention. At Hyderabad, publicly and for the first time, he threw 'some light on the circumstances that led to' his resignation from government: 'How these differences began is a long story. But during the September War of 1965 these differences erupted.' He went on to attack the cease-fire: 'The Government...fell at the feet of the great powers...I wanted to resist Pakistan's enemies, but my opponents dubbed my patriotic feelings as emotionalism.' Although he said he now felt 'obliged to put before the people all events', he did not in fact do so.

In his Hyderabad speech he also, for the first time, publicly expressed his readiness to contest against Ayub Khan. He would 'be happy to see a suitable candidate contesting from East Pakistan', but if no one was available, even from the West Wing, he would 'recite the *Kalima* (Article of Faith for a Muslim) and step forward'.[16] He had been building up to this announcement and now the time was ripe.

Events, he felt, were going his way, as he indicated at my wedding on 24 October, for which I had briefly returned to Karachi. His words are worth recalling: 'When are you coming back? I am going to Peshawar tomorrow and the real show will start. Pity you will not be there.'[17]

He rapidly stepped up the tempo against Ayub Khan and, of course, India. At Kohat on 25 October, he announced the PPP had 'decided to participate in the elections'. Four days later, at Mansehra, he was at his chauvinistic and anti-Ayub best: 'The conflict between Hindus and Muslims dates back one thousand years into history. Why can't it be carried into the future?...They (the Government) want to give up the historical struggle. All the time they are repeating that we cannot fight with India because it is a big country...They are trying to demoralize the country...how could the valiant people of Vietnam have fought against a power like America?...Gentlemen! Please mark my words. I am telling you that the days of this Government are numbered.'[18]

By the end of 1968, East Pakistan was in serious disorder. Sheikh Mujibur Rahman, the leader of the Awami League, had been arrested earlier and was being tried in the Agartala Conspiracy Case.[19] His party and the student leader, Tofail

Ahmed, spearheaded the opposition. ZAB by now had a solid following in West Pakistan where, from 1 November, trouble broke out. When he went to Dera Ismail Khan in the Frontier to address the local Bar Association, the police opened fire and used tear-gas to disperse the pre-arranged procession which violated Section 144.[20] This was the first time force had been used while he was present. At Hayat Sherpao's village, where he went for the Frontier Party Convention, he expected arrest.

When he addressed the Convention on 3 November, he underlined the strength of the Party and its rapid growth: 'When I announced the setting up of a party, the Government took it as a big joke...In the beginning, it was said that the party would not last long because it had the support of the students and a following in the Punjab alone...We can genuinely claim that the success of the party in eleven months' time has no parallel in the history of Pakistan.' He also tried to project a more acceptable face to the traditionally-minded, particularly in the military: 'in the socio-economic sector there is no difference between Islam and socialism; had these two systems been in conflict with each other, I would have given up socialism'. Again, he explained his earlier silence: 'When I came back from my European tour, following my resignation from the Government...I did not make political speeches for quite some time...I decided not to launch a political movement until we could stand on our feet and until our ideas became clear and unambiguous.'[21]

Serious agitation was sparked off on 7 November when the police shot dead a Polytechnic student in Rawalpindi as ZAB was arriving in the city. As he proceeded to Lahore, there were clashes with the police at every station on the way. In Lahore, on 10 November, ZAB made the first remark which could be termed 'incitement to agitation'. Asked at a press conference why he did not call on the students to refrain from burning buses and destroying public property, he replied, 'Why should I stop them!'[22] Three days later, he was arrested in the early hours from Mubashir Hasan's house, where he and Mumtaz were staying; within minutes Mubashir and Mumtaz were also arrested.

Besides the emergence of the PPP as an active force, the year 1968 also saw two major developments significantly affecting the course of events. The first involved the Judiciary. With Ayub Khan's weakness and increasing unpopularity, the Courts reasserted their authority, granting bail readily to PPP supporters arrested by the regime. More important, the army and the US administration became ambivalent towards Ayub Khan. The US arms embargo imposed during the 1965 War had not been lifted and the army lost confidence in Ayub Khan's ability to 'deliver the goods'. At a time when the US was still deeply hostile to China, President Ayub Khan had continued to develop close relations with China; and, in 1968, despite disagreement with the Commander-in-Chief of the Army, accepted Chinese assistance to build the Sinkiang Road. The US was equally upset by Ayub Khan's overtures to the Soviet Union. In April 1968, he served notice for the termination of the US communications base at Badaber near Peshawar, from where Gary Powers had flown his U2 reconnaissance aircraft which had been shot down in the Khrushchev era. This was done just prior to the visit of the Soviet Prime Minister, Alexei Kosygin, who, in support of Ayub Khan, reportedly commented that Pakistan was a strange country where the people agitated against a President who had done so much for the nation.

By the end of the year, all these developments came together to heighten the anti-Ayub campaign and strengthen the movement against him. Some prominent figures tried unsuccessfully to capture the leadership of this movement while ZAB was in jail. Retired Air Marshal Asghar Khan entered the political scene at this point; also, for the first time, ZAB's wife Nusrat came out to play a leading role. Agitation and the destruction of public property continued unabated.

When the establishment was truly threatened, the opposition forces split. The trigger was J. A. Rahim's announcement that ZAB would be a candidate in the forthcoming presidential elections against Ayub Khan under the hated Basic Democracy

(BD) system.[23] He was still in jail, and there are conflicting versions of whether or not he had consented to this; but on his release he denied it. However, Rahim had been acting Chairman during ZAB's imprisonment and maintained that he had been properly authorized to make the announcement.[24]

ZAB's denial was too late. Nawabzada Nasrullah Khan, an opposition leader, criticized the decision as tantamount to accepting the BD system. Asghar Khan joined the chorus. The Jamaat-i-Islami campaigned against socialists and communists, dubbing them anti-religious and so pro-Ayub. The Awami League pressed its own demands.

Ayub Khan tried to regain some lost ground. First his Minister, Khwaja Shahabuddin, and then the President himself, met Nawabzada Nasrullah Khan, the convener of the Democratic Action Committee (DAC), an eight-party opposition alliance. The opposition were invited to a Round Table Conference (RTC) on 17 February. However, the DAC refused to participate unless the Government withdrew the Emergency imposed since the 1965 War, lifted restrictions on political activities, and released all students who had been arrested. Ayub Khan withdrew the Emergency four days later, but conflicting demands increased, in particular, those relating to the dissolution of One Unit, 'parity'[25] and provincial autonomy. Sheikh Mujibur Rahman, who had earlier agreed to participate if paroled, now insisted he would only attend if the Government withdrew the Agartala Conspiracy Case against him. But even after Mujibur Rahman was released unconditionally, the situation, particularly in the East Wing, failed to improve. Condemnation of Ayub Khan continued, and on 21 February he announced that he would not be a candidate in the next elections. He also conceded the main constitutional demands of direct adult franchise and a reversion to the parliamentary system. However, the DAC still could not agree on the outstanding issues or their own future course, and lacked the requisite authority to make decisions.

The RTC finally met on 10 March but reached no conclusion after four days of deliberations. ZAB did not participate. Ayub Khan was powerless to accept the various demands of the as yet unelected leaders who composed the DAC. In his speech at the concluding session of the RTC, Ayub Khan said he would move the National Assembly 'to make the necessary amendments to the existing constitution to convert it into a federal parliamentary system without disturbing the basis of parity and the existing distribution of powers between the centre and the provinces until such time as the directly elected representatives of the people had had an opportunity to decide these matters.'[26] He called upon the political leaders 'jointly and manfully' to bring the present disorder under control and for 'collective resistance to the forces of agitation and disruption'. The RTC broke up in confusion. Sheikh Mujibur Rahman quit the DAC.

It remained for Ayub Khan to determine how to hand over power to the 'professionals'. The initiative passed to the army. When, a few days later, he tried to revert to a political option, the response of the army Commander-in-Chief, General Yahya Khan, was dismissive: 'I cannot worry about the niceties of law when the country is burning'; if Martial Law was not imposed immediately, 'the integrity of the armed forces will be severely undermined', allowing 'some madcap in the army' to intervene.[27] Previously, Yahya Khan had refused to allow the imposition of limited Martial Law in three principal cities and was unsupportive in suppressing the agitation.

Ayub Khan was left with little option. On 25 March he handed over power to Yahya Khan, who assumed charge as Chief Martial Law Administrator (CMLA) and abrogated the 1962 Constitution.

ZAB's role in these events has been criticized. He has been attacked for sabotaging the RTC by his non-attendance and agitation against it. Despite Ayub Khan's announcement that he would not stand for President, as earlier demanded by the PPP, ZAB continued his attack, returning the high award of *Hilal-i-Pakistan* conferred on him as Foreign Minister. He, in turn, claimed credit for the failure of the RTC and the toppling of Ayub Khan; indeed, he spent the next two years justifying his position. However, political observers feel his refusal to participate was at the behest of Yahya Khan, or at least in collaboration with the military who planned to take over. His apprenticeship in politics had taught him that, since 1954, real power in Pakistan was exercised only by or with the support of the army. His politics remained in tune with this fact. Certainly his interests converged with Yahya Khan in the ouster of Ayub Khan: 'It had to happen and on the whole it is a good thing. At least we are rid of Ayub Khan and the royal family.'[28] However, the charge of frustrating the RTC cannot be sustained. At the time ZAB was neither strong enough nor sufficiently relevant to achieve this. The RTC failed because of disagreement within the DAC, continued agitation in East Pakistan, Mujibur Rahman's demand for provincial autonomy, and the army's ambitions.

Unlike 1958, when Ayub Khan's Martial Law was generally acclaimed, Yahya Khan's was not. Nevertheless, there was a sense of relief when agitation ceased immediately and daily life returned to normal.

Yahya Khan: 25 March to 31 December 1969

Public relief was also mixed with an element of hope. The intensity of feeling against Ayub Khan had reached such a pitch that any alternative seemed preferable. General Yahya Khan at first appeared to fulfil public expectations. He assured the nation he would restore democracy as soon as possible, promising to discuss with political leaders a solution to the country's problems. The seriousness of the situation was accurately assessed by a journalist, Peter Hazelhurst, in *The Times* of London:

> ...if a revolt erupts in the eastern province again, nothing can prevent the tide of Bengali nationalism from swelling into a full-scale liberation movement.
>
> Only one course is open to General Yahya Khan; he must make an imaginative concession to the eastern province.[29]

Imagination was a quality which the military regime totally lacked. Yahya Khan also lost the resolve to settle the real issues. He soon began to gather around him a coterie of unrepresentative individuals as ministers and advisers who did not have the capacity to deal with the demands of the situation.

Yahya Khan held a few formal meetings with the main political leaders, but no serious discussions involving substantive issues took place. His declared goal was the holding of fair elections and, to that extent, he succeeded; but the wider issues at stake were more complex and beyond him. Whether a man of greater intellect and vision could have resolved the problems involved in the polarization of the two Wings remains moot.

Over the following nine-month period, President Yahya Khan's position changed perceptibly, as his public statements indicated. Initially, in his broadcast to the nation on 26 March 1969, he sought a 'smooth transfer of power to the representatives of the people elected freely and impartially on the basis of adult franchise', whose 'task' would be 'to give the country a workable constitution'. At this point, he envisaged a sovereign Constituent Assembly. Four months later, however, on 28 July, while referring to various constitutional issues, chief among them one-man-one-vote, One Unit, and adult franchise, he announced that 'the decision must lie with the people'. He added a proviso that he 'might have to go to the nation to obtain its verdict on the basis of a constitution' before elections were held, if the political leaders had not previously arrived at some agreement. With no political activity permitted and no forum for public debate or consensus, there was no possibility for such an elusive agreement before the elections. Even Ayub Khan had realized that the leaders would only emerge after the elections; meantime, there were many pretenders.

Yahya Khan revised his position precisely four months later, on 28 November. Without serious political debate, he settled three important constitutional issues by dissolving One Unit, abolishing parity between the two Wings, and allowing one-man-one-vote. These measures reflected public consensus and were generally accepted, but the Punjab remained apprehensive that the removal of parity, without any substitute mechanism, would result in political domination by the East Wing. He stated that his proposals on future constitutional arrangements 'would only be in the nature of a *provisional* (emphasis added) Legal Framework'. Yet again, he further altered his position four months later, on 28 March 1970, when he promulgated the Legal Framework Order which provided a detailed *constitutional* format.

Yahya Khan's arrangements for governing the country also changed significantly. Initially, they were firmly centred on the military and his position as Commander-in-Chief of the Army and Chief Martial Law Administrator. The CMLA's office was run from President House by Lt.-Gen. S.G.G.M. Peerzada and a small military staff responsible separately for civilian matters and martial law affairs. For general policy-making there was a Council of Administration, with the President in charge of Defence and Foreign Affairs and the remaining cabinet portfolios concentrated among his senior military colleagues.

Another important body was the National Security Council under the chairmanship of the CMLA, with the Directors of the Inter-Services Intelligence and Intelligence Bureau as members. Together with the Commanders-in-Chief's Committee, these bodies remained responsible for general policy even after the appointment of a civilian cabinet. The President was the link between all the institutions.

Yahya Khan was unable to fulfil such a demanding role. His financial probity was never questioned, and he had previously been well respected professionally; but by the time he became President the effects of alcohol had possibly got the better of him. His main channel of communication was his Principal Staff Officer, Lt.-Gen. Peerzada, who maintained a low profile but was very influential as the *de facto* chief of staff and virtual Prime Minister, although not a formal member of any of the governing bodies.

Power is neither readily released nor easily shared. A shift in Yahya Khan's arrangements for governing the country came in August 1969 with the induction of a civilian cabinet. Earlier, the nature of the purely military set-up had conveyed a transitional image. Now, those Generals around Yahya Khan who did not want to transfer power gained support from the members of this civilian cabinet, who were also interested in perpetuating their position, and favoured a move away from Yahya Khan's declared transitional role. Their interference exacerbated an already complex situation, contributing to mistrust of the regime among political parties, particularly the PPP.

ZAB's personal relations with Yahya Khan were good, but the coterie around the President were hostile, particularly Maj.-Gen. Ghulam Umar, such Federal Ministers as M. A. Qizilbash, Mahmood Haroon and Sher Ali, unofficial advisers like Z. A. Suleri, officials such as Sultan M. Khan and other bureaucrats. His access to Yahya Khan was thus gradually limited and his ability to influence the outcome of events reduced.

This ran counter to ZAB's expectations. Having welcomed the *coup,* he thought Yahya Khan would consult him, especially on foreign affairs. Instead, Peerzada asked him to confer with Sultan Khan, the Foreign Secretary. ZAB indignantly refused. On record, he declined because, 'it would not be advisable to be associated' with a Government which 'had come into power without the mandate of the people'. There was no let-up in the politically motivated inquiries and cases started against him in the Ayub period, involving the family collection of arms and his first wife's[30] rice-husking mills at Larkana. When the local Deputy Superintendent of Police continued harassment, ZAB protested by telegram on 14 May 1969 to the Martial Law Administrator for Karachi and Sindh Zone.[31] There was no response.

In August, the Deputy Commissioner, Larkana, complained that breaches of Martial Law Regulations prohibiting public meetings and processions of a political nature occurred during the Chairman's journey from Moenjodaro to Larkana on 2 August and at Jacobabad on 3 August. The Chairman asked to 'direct the members of the People's Party to desist from such displays in future'.[32] Already angered by the new civilian cabinet, he seized the opportunity to reply in strong terms on 12 September in a detailed nine-page letter. He stressed discrimination against him:

> In marked contrast Mr. Mujibur Rahman strides across the land of East Pakistan as a mighty colossus preaching his hymn of hate without let or hindrance. On the 10th of September 1969, *Dawn* reported that on his arrival in Chittagong, Mr. Mujibur Rahman was given a reception by one of the biggest crowds assembled on both sides of the road all over the 32 miles from the Airport to the City.[33]

The letter, copied to Lt.-Gen. Peerzada, to whom it was principally directed, referred to the arrest of several PPP leaders, among them Mubashir Hasan, two days after the declaration of Martial Law, Mumtaz Bhutto on 13 August, Mairaj Mohammad Khan a few weeks earlier, and Haq Nawaz Gandapur and others at Dera Ismail Khan on 30 August. He described how the enemies of the country 'both within and outside are active in their mischief to undo Pakistan', and condemned other political parties. The Awami League's 'Six Points which spell the destruction of Pakistan have gained support in Sindh and this is ominous. It can cast the shadow of secession over other regions of Pakistan.' He maintained, 'The Jamaat-i-Islami which wants to exploit our dear religion only to spread its baneful influence is not capable of overcoming the crisis... Pampering such parties led in the past to the Ahmedi riots and the imposition of Martial Law in 1953.' The National Awami Party 'is oriented to either Moscow or Peking...and in any event they want to impose systems alien to the concept of Pakistan.' As for the Convention Muslim League, it had limited support and was 'too obsolete to meet the grave challenge of our time'. The Pakistan Democratic Party was 'a collection of uninspiring fossils'. In contrast the 'Pakistan People's Party is the only national party in the country that inspires the confidence of the people and is both dynamic and vigorous enough to meet the test of our times.' Such efforts to convince Peerzada and the Government of the PPP's importance only increased the Government's distrust of ZAB's ambitions. The Yahya regime felt they could cope with West Pakistan without ZAB.

ZAB never allowed annoyance with the regime's behaviour to be reflected in criticism of the army. The fact that quiet descended on East Pakistan so quickly after the declaration of Martial Law led him to believe that the army was not only in control now but would remain so. He envisaged a military-cum-political arrangement. On the one hand, he stated on 10 June that, 'Pakistan cannot live forever under Martial Law...A political void for any society is harmful. But it is disastrous for a physically divided country like Pakistan'. Yet, at the same time, while talking about the Constitution, he said it was understandable that the Armed Forces should want to devise a system to prevent political chaos and the dismemberment of the country: 'The Armed Forces can stand behind the political structure to ensure that it is not subjected to strains that would break it and the country.'[34] He cited Turkey after the 1960 *coup d'état* by way of example. These views were to remain his guiding principles, particularly in the election year ahead and the period leading to the civil war.

From the declaration of Martial Law it took the Yahya regime eight months to announce the date of elections. On 28 November Yahya Khan stated that political campaigning would be permitted from the beginning of the new year, and elections would be held on 5 October 1970. With the existing polarization between the two Wings, the initial delay had already been damaging, but allowing a further nine months for electioneering, later extended to eleven months because of floods, was a prescription for disaster. The motivation for such extended delay is difficult to gauge. Possibly, Yahya Khan, like ZAB, could not visualize the nationalist forces that the Awami League would unleash. The delay might also have been a ploy to confuse the political scene and allow the military to continue holding the balance of power. There is, however, no evidence of such calculation; and no one at the time in West Pakistan pointed out the fatal consequences of delay.

This nine-month period contributed nothing constructive on the national level. Neither the Government nor the military consolidated their hold on the country, which remained in economic distress and without a sense of direction. The minor political parties sought government patronage. The Awami League was strengthening its position as its propaganda began to reverberate throughout East Pakistan. The PPP, too, had made gains but, busy with its own preparations, it did not take into account the developments in the East Wing. After several months of indoor meetings, the campaign ahead provided a unique opportunity that neither the Awami League nor the PPP could afford to miss.

NOTES

1. Mumtaz Ali Bhutto was educated at Christ Church, Oxford (1954-7), and called to the Bar in London in 1959. He and ZAB called each other 'adha', 'brother' in Sindhi, but Mumtaz's grandfather and ZAB's great-grandfather were brothers. Mumtaz's branch was the central line of the family and provided the *sardars* or chiefs of the Bhuttos, something ZAB never relished.

2. ZAB showed some of these letters to the author. Sir Shahnawaz Bhutto wrote a long note explaining his defeat by Sheikh Abdul Majeed Sindhi in the Sindh provincial elections in 1937 and his subsequent retirement from politics. He blamed members of Mumtaz Bhutto's branch of the family, who in turn maintained that Sir Shahnawaz lived in Bombay and did not visit his rural constituency.

3. The Tashkent Declaration was signed on 10 January 1966, following which Premier Shastri died of a heart attack at Tashkent. ZAB, as Foreign Minister, represented Pakistan at the state funeral in Delhi.

4. Text of the rejoinder is with the author. The notice was published by *Dawn,* Karachi. ZAB had, in fact, travelled to study abroad on an Indian passport issued on 11 August 1947, before Independence, and obtained Pakistani passport no. 0146 later in Karachi, on 12 July 1949.

5. Z. A. Bhutto, *The Myth of Independence,* Oxford University Press, 1969.

6. All references here to the convention proceedings and the Foundation Meeting Documents are contained in Dr Mubashir Hasan's *Foundation and Policy—Pakistan People's Party,* Masood Printers, Lahore, 1968.

7. Sheikh Mujibur Rahman, *6-Point Formula—Our Right To Live* (published by Tajuddin Ahmad, General Secretary, East Pakistan Awami League, Dhaka, March 1966). This became the basis of the Awami League election campaign in 1970 and led to the crisis in 1971. For a further analysis of Six Points, *see,* Chapter 3, pp. 52-5, 62-3, 87, and Appendix A for the text.

8. Mubashir Hasan pointed this out to the author.

9. ZAB prided himself on his term 'bridge', and told me to quote this sentence in his statement on the signing of the Accord of 6 March 1972. *See,* Chapter 5, p. 153.

10. Mustafa Khar and Mumtaz Bhutto could not be present. They were elected members of the National Assembly on Convention Muslim League tickets (Ayub Khan's party) and would have lost their parliamentary seats under the Political Parties Act, 1962, which ZAB did not want as the Assembly provided a useful platform. The author was away on professional work at the time.

11. Sheikh Rashid became the President of the Punjab PPP and Mubashir Hasan of Lahore. ZAB offered the Karachi Presidentship to the author, who had accepted an appointment as Counsel with the Asian Development Bank, Manila. Mubashir Hasan suggested the position be left vacant in case there was any change in the author's plans. However, the author left for Manila.

12. These recordings were collected and published in edited form after ZAB became President: *Politics of the People, Vol. 2—Awakening the People—Nineteen sixty-six to nineteen sixty-nine—Statements, Articles, Speeches (of) Zulfikar Ali Bhutto,* (Ferozsons, Karachi, 1972); quotations are mainly from this edition.

13. Ibid., p. 51.

14. Ibid., pp. 63-5.

15. Ibid., p. 75.

16. Ibid., p. 133.

17. While in Indonesia on a mission for the Asian Development Bank the author read about ZAB's arrest; he resigned and returned to Karachi when released by the Bank.

18. *Politics of the People, Vol. 2,* pp. 179-81.

19. The Agartala Conspiracy Case was named after the venue, Agartala, of the alleged conspiracy with India against Pakistan for which Mujibur Rahman and other Bengalis, including some military personnel, were brought to trial in 1967.

20. Section 144 is a provision in the Pakistan Penal Code by which the Government can ban any assembly of more than five persons in a public place.

21. *Politics of the People, Vol. 2,* pp. 179-81.

22. As narrated to the author by Mubashir Hasan.

23. For the Basic Democracy system, *see,* Chapter 6, p. 169.

24. J. A. Rahim's version is supported by Mubashir Hasan; the message from jail was conveyed to them at the latter's house in Lahore.

25. 'One Unit' was the expression used for the Province of West Pakistan on the merger of the provinces and other areas which comprised West Pakistan. 'Parity' described the equal representation, irrespective of population, given to the two Wings of Pakistan in the federal Parliament. This existed under the Constitutions of 1956 and 1962. For further details, *see,* Chapter 6, pp. 167-8.

26. *Dawn,* Karachi, 14 March 1969.

27. Altaf Gauhar, 'Pakistan—Ayub Khan's Abdication', in *Third World Quarterly* 7(1), January 1985, p. 123.

28. *The Sunday Times,* London, 28 March 1969.

29. Peter Hazelhurst, 'The Dilemma of Pakistan—A Nation in Danger', in *The Times,* London, 9 April 1969.

30. ZAB in his teens was married to a cousin who took no part in his public life.

31. Copies of the telegram of 14 May and the subsequent two letters of 17 May to the Martial Law Administrator, Khairpur Division, and the Superintendent of Police, Larkana, are with the author.

32. Copy of D.O.No. CB/272 of 5 August 1969 from the Deputy Commissioner is in the author's possession.

33. Copy of ZAB's letter of 12 September 1969 is in the author's possession.

34. *The Pakistan Times,* Rawalpindi, 11 June 1969.

CHAPTER 2

1970 GENERAL ELECTIONS

1970 witnessed the first general elections held throughout Pakistan since Independence in 1947, based on universal adult franchise and one-man-one-vote. The only previous national elections, held by President Ayub Khan in 1964, were indirect through an electoral college of 80,000 Basic Democrats. Even at the provincial level, no elections had been held since 1954 in East Pakistan, 1953 in Sindh and 1951 in the Punjab and the Frontier Province. In Balochistan, which had only been given the full status of a province with the dissolution of One Unit in 1970, they were the first ever elections. Everyone was eager. There was great excitement and expectation. The main political parties sprang into action with aggressive campaigns.

ZAB's Campaign Strategy

ZAB developed his strategy early, after careful planning and taking into account his experience of the past two years. He set the tone of his campaign and outlined the four main aspects of his platform at the PPP's first public rally at Nishtar Park, Karachi, on 4 January.[1]

First, he promoted the basic principles of the PPP, and attempted to give a more acceptable face to socialism by referring to it as 'Islamic Socialism', eliding the first and third lines of the PPP motto, 'Islam is our faith' and 'Socialism is our economy'. He sought to gain authority for the term 'Islamic Socialism' from Quaid-i-Azam Mohammad Ali Jinnah's use of it in a speech at Chittagong on 26 March 1948. If Algeria,

Egypt, Iraq, Libya and Syria among Muslim countries had made progress through Islamic Socialism, he maintained, the people of Pakistan should not be denied this opportunity.

Next, he bitterly attacked Ayub Khan and rival political parties. He condemned the participants in the Round Table Conference, which he described as a 'conspiracy against the people of Pakistan and a trick to break the people's movement'. At this rally, Mairaj Mohammad Khan, PPP General Secretary of Karachi, added to the denunciation of the past record of the Jamaat-i-Islami, challenging its leader, Maulana Abul A'la Maudoodi, on several issues.

Third, ZAB forcefully projected a nationalistic, anti-India position. He called for an independent foreign policy, withdrawal from CENTO and SEATO, the US-backed defence pacts, and the 'liquidation' of the Tashkent Declaration. He criticized Ayub Khan for halting the Rann of Kutch mini-war with India,[2] and for calling him 'emotional' for wanting to continue the operation. He accused Ayub Khan of jeopardizing national security, maintaining that, despite Premier Shastri's warning that India would open a front of her own choice, Ayub Khan and Finance Minister Mohammad Shoaib had refused the demand for two additional Pakistan Army divisions on the grounds that the country could neither afford them nor risk antagonizing the United States and prejudicing the supply of aid. ZAB claimed that, with these two divisions, Pakistan would have defeated India in the 1965 War. As intended, this appealed to national pride and the military establishment. The Tashkent Declaration was roundly condemned. He claimed that for three-and-a-half years he had avoided this matter for 'political expediencies'; now he would 'gradually unfold the Tashkent mystery'. He promised the next instalment at the Rawalpindi public meeting in a week's time—but no further disclosures were ever made.

Fourthly, he accused the Yahya civil administration of gross interference, despite President Yahya Khan's assurance of 'complete impartiality'. Information Minister Sher Ali was singled out for directing the hostile media campaign against the PPP. The bias of the civil administration was attacked in order to emphasize his position as underdog and the difficult task ahead. However, he avoided direct criticism of the President, and at no time challenged the military.

PPP rallies and meetings continued to be hard-hitting. He toured the Frontier and visited all areas of Sindh; but it was on the Punjab that he concentrated, realizing that success there was essential in his pursuit of power. He battled through eighteen to twenty hours a day of travel, speeches, processions and rallies, despite heat, dust, illness and other adversities. It was a remarkable performance.

The PPP bandwagon gained momentum. The slogan *Roti, Kapra aur Makaan* had caught the public imagination. Everywhere, Party flags were to be seen of different sizes and varying shades, demonstrating they were made individually by the poor at their own cost. The people saw and heard, for the first time, a leader whose words voiced their grievances and aspirations. A surge of support in the Punjab followed ZAB's Lahore Mochi Gate public meeting on 9 March, and this in turn had its impact on Sindh, particularly among the *waderas* (feudal landlords) who recognized the importance of the Punjab in the power game.

Economic, social, political and psychological factors were all present in his strategy. He carefully and successfully stressed the priorities of the people he was addressing. In the Punjab, he underlined confrontation with India. In Sindh, he promised to curb the *waderas* and improve the standard of living of the people. In line with Sindhi nationalist sentiments, he stressed the injustice of One Unit. He spoke about Islamic Socialism to the industrial workers, particularly in Karachi. To the Pathans, he recalled their glorious traditions and promised prosperity. He held out something for almost everyone, except capitalists, and endeavoured to envelop within his fold both peasants and landlords, workers and the well-to-do. He profited from the disarray of the leftists after the break-up of the NAP, and stole the leadership by calling for a socialist revolution. Subsequently, when he sought to appease religious sentiment, he talked about

Musawaat-i-Muhammadi (Equality of Islam) instead of Islamic Socialism.

In tune with the prevailing anti-American and anti-Indian mood, he became the standard bearer of nationalism. He even claimed that he had 'engineered' the 1965 War over Kashmir, which 'cowardly' Ayub Khan had failed to carry to its conclusion, and promised a *Shaukat-i-Islam* (Glory of Islam) Day in Delhi and Srinagar.

His personal and virulent attacks on opponents antagonized other politicians, the intelligentsia and the business community, but enthralled the common man. Never before had the people experienced such exhilaration. Above all, he projected confidence, asserting that a sweeping victory lay ahead. No single factor draws more adherents in an election campaign than confidence.

The Government became apprehensive about ZAB's rising tide of support, and pressure on him mounted. On 31 March, in Sanghar, Sindh, there was an attempt on his life. He called for an investigation, but the official Press note played down the incident. On returning to Karachi, he bitterly criticized the Pir of Pagaro, whose supporters he accused of carrying out the attack. He also condemned the Information Minister for authorizing a 'false distorted version of the facts' in an attempt to defend Pagaro, when the incident involved 'a pre-planned and pre-meditated conspiracy'.[3] With the campaign in full swing, the Sanghar incident was soon forgotten, except by ZAB.

Government intelligence agencies continued to focus their attention on the PPP, partly accounting for the absence of accurate information when the East Pakistan crisis overtook the country. At the beginning of the year, Yahya Khan appointed his brother, Agha Mohammad Ali, Director-General of the National Security Council with headquarters at Lahore, and, on 16 February, Maj-Gen. Umar became Secretary of the Council. It was generally recognized that, along with N. A. Razvi, the Director of the Intelligence Bureau, they deeply disliked ZAB and supported his rightist opponents. At the same time, Sher Ali orchestrated the media against the Party. Harassment of the

Chairman and Party members increased, about which ZAB lost
no opportunity to complain.[4]

The Legal Framework Order

Meanwhile, on 28 March 1970, Yahya Khan announced the
Legal Framework Order (LFO).[5] This provided a structure for
the future Constitution, containing such details as the contents
of the preamble, and fundamental and directive principles. On
the question of provincial autonomy, it prescribed that the
Provinces 'shall be so united in a Federation that the
independence, the territorial integrity and the national solidarity
of Pakistan are ensured and that the unity of the Federation is
not in any manner impaired.' And, more important, while
providing for maximum autonomy, it laid down that 'the Federal
Government shall also have adequate powers including
legislative, administrative and financial powers, to discharge its
responsibilities in relation to external and internal affairs and to
preserve the independence and territorial integrity of the
country.'[6] However, although it was evident from the outset
that Six Points could not be implemented within this framework,
the Government allowed the Awami League to propagate its
programme unchallenged.

Another problem arose because the LFO omitted to provide
for voting procedures. Without such a provision, particularly in
a Federation where one Province was numerically greater than
the others combined, the LFO could not prevent a one-sided
arrangement. It would have been more appropriate, in the
circumstances of Pakistan, to specify some minimum percentage
of votes in each Province, or at least each Wing, as Balochistan
had but five votes out of a total of 313[7] and East Pakistan 169.

The whole issue of autonomy and the balance between the
federating units ultimately proved the undoing of the LFO. This
was despite the fact that, under the LFO, the President was the
final authority in two important areas. Article 25 provided for
the Constitution Bill to be authenticated by the President, and

'the National Assembly shall stand dissolved in the event the authentication is refused', without specifying any reason. Moreover, under Article 27, interpretation of the Order 'shall be resolved by a decision of the President, and such decision shall be final and not liable to be questioned in any Court.'

ZAB was concerned that, under Schedule II of the LFO, the total number of general seats for the four West Wing Provincial Assemblies was the same as for East Pakistan: a form of parity of 300 each was being maintained. We discussed whether this was coincidental or of some significance and concluded it might be a fall-back position for Yahya Khan. If the National Assembly were dissolved, the five Provincial Assemblies could be used to elect him as President, or agree on a new constitution, with the equal say of the two Wings. Such were our limited calculations and concerns at the time. The analysis I prepared of the LFO was used to criticize its shortcomings and the restrictions placed on the sovereignty of the National Assembly.

He raised these misgivings publicly when he spoke on the 'Constitutional Problems of Pakistan' on 15 May. He spelt out how the Provincial Assemblies, in the event of dissolution of the National Assembly, could be manipulated by the Government through the grant of ministerships and other appointments to the new members. He therefore proposed that elections to the Provincial Assemblies should take place after the adoption of the Constitution. He was also critical of the way the LFO spelt out the five 'Fundamental Principles of the Constitution'. For example, he queried the need for the reference to 'Islamic Ideology' being 'preserved' in the Constitution, since Pakistan was in any case an Islamic State. Moreover, with seventy-two sects among Muslims, 'who will determine what is in conformity with the concept of Islam and according to which will it be decided?' On the principle enunciating Pakistan's solidarity, he declared, 'If the majority of people work against the solidarity of Pakistan the power of authentication will not save it.' For ZAB, the 'preliminary steps' for such solidarity were the liberation of Jammu and Kashmir and the 'liquidation' of the Tashkent Declaration. The one concrete proposal was for a bicameral legislature: 'No other way could create unity between the two Wings'.[8] Apart from the sovereignty of the Assembly, the main outstanding question was whether the PPP would participate in the elections. This was to be decided at a convention at Hala on 1 July.

He was the only political leader to attack the LFO, the Assembly's lack of sovereignty, the President's powers of veto, and the short period of 120 days for framing the Constitution. No other party in the West Wing questioned the LFO. In East Pakistan, the Awami League pursued its own campaign, disregarding the LFO provisions. Later, however, after the elections, ZAB largely reversed his stand, subordinating the sovereignty of the Assembly by calling on the President to give effect to the LFO in the face of the Awami League's over-whelming majority.

Progress of the PPP

ZAB withheld the announcement of the PPP's participation in the elections till July for two reasons. First, he sought to assuage the left, who believed in revolutionary, and not electoral, politics; second, he wanted to warn the Government that the elections could be frustrated by the PPP through a boycott. However, a boycott was an empty threat. The Government and the rightist parties would have welcomed the PPP's non-participation. Moreover, within the PPP, many landlords from Sindh and the Punjab, and others as well, eager for elected office for the first time, might have refused support for a boycott. Indeed ZAB, who had worked tirelessly to bring them into the fold, would not have countenanced seeing his efforts undone. The decision to participate in the elections announced at the Hala Convention on 1 July was in reality a foregone conclusion.

The Makhdoom of Hala formally joined the Party at the Convention. He was the first major *Pir* (religious leader) and landlord to join from Sindh. The tide had turned in the Punjab; it also became increasingly clear that the PPP would win the elections in Sindh. Other big landlords, including Ghulam Mustafa Jatoi and Pir Ghulam Rasool Shah, joined the Party. By October, there was no dearth of candidates for PPP tickets in rural Sindh.

The Party's Manifesto for the coming elections was prepared by Secretary-General Rahim. It was some fifty pages long, with a red cover symbolising revolution, and commenced and ended with the words from the Party's motto, 'All power to the People'. Outlining how the nation had been betrayed and a prey to neo-colonialism, it described the dispensation between the two Wings as an 'internal colonial structure' to exploit East Pakistan. The 'ultimate aim' of the Party was the 'attainment of a classless society, which is possible only through socialism in our time'. Its main thrust was economic. It called for the nationalization of twelve 'basic and key industries' including iron and steel, heavy engineering, motor assembly, machine tools, chemicals, cement and paper. It prescribed a ceiling on land ownership of between 50 and 150 acres of irrigated land, deliberately left vague on ZAB's insistence. It also advocated reforms relating to education, health, administration, Princely States and in other fields.

The Manifesto repeated the demand for an independent foreign policy, withdrawal from the SEATO and CENTO pacts, and a hard line against India: 'Confrontation will be maintained until the question of Kashmir' and other pending matters were settled, and 'the Tashkent Declaration will be repudiated, being a treaty extorted under duress'.

In elections in Pakistan, the selection of a symbol is important because the mainly illiterate population recognize and vote for it rather than the name of a particular candidate. While allowing J. A. Rahim leeway over the Manifesto, ZAB insisted on the Sword as the PPP election symbol. It had a religious connotation, representing *Zulfiqar-e-Ali* (the Sword of Hazrat Ali), but, according to Rahim, it was adopted to glorify ZAB's own name.

While ZAB carried the main burden of the campaign, others contributed significantly. Prominent in the Punjab were Sheikh Rashid, Mubashir Hasan, Haneef Ramay, Khar, Maulana Kausar

Niazi and Khurshid Hasan Mir. In the interior of Sindh, Rasool Bux Talpur made a notable contribution, as did Hayat Sherpao in the Frontier.

Parliamentary Boards were established for the National and Provincial Assemblies, but the award of tickets was finalized unofficially. In the key Province of the Punjab, where there was a lack of suitable candidates, the final scrutiny was carried out by Mustafa Khar, Mubashir Hasan and Mirza Tahir Ahmed.[9] After that it was formalized by the Central Parliamentary Board which included ZAB, Rahim, Mubashir, Sheikh Rashid, Khar, Khurshid Hasan Mir, Mahmud Ali Qasuri and Rasool Bux Talpur.

Meetings of the PPP Central Committee were held regularly and serious business was conducted. Characteristic of the frank discussion that occurred was Haneef Ramay's contention on one occasion that ZAB was not giving due importance to the *'fauji* belt', the districts from which the army was mainly recruited. ZAB was cautious and unsure to what extent he could criticize the Government in areas where the military dominated. He did not wish to raise the tempo of agitational politics, a necessary concomitant of such visits. The majority view prevailed after detailed discussions. Extensive tours of these areas proved very successful, to his delight.

During this period there was constant wrangling in the PPP, members criticizing both each other and ZAB. But, as support from the public grew after each rally he addressed, Party bickering temporarily died down. The growing mass support was in fact the principal strength of the Party, since it never acquired an organized structure, even during the campaign. J. A. Rahim attributed this to ZAB's desire to be a prima donna, seeking an audience and not a supporting cast. There was some truth in this, which was to have serious consequences later on.

The call for Islamic Socialism attracted concerted criticism from the religious parties, well-supported by the media. Socialism was described as *kufr* (anti-Islamic), and *fatwas* (religious edicts) were issued against the PPP. ZAB decided not to meet this challenge head on and, at Multan, dropped 'Islamic

Socialism' for *Musawaat-i-Muhammadi.* Mubashir Hasan criticized this growing tendency to cater to religious orthodoxy. ZAB replied: 'I look into the eyes of the people in public meetings; I know when they brighten up.'[10] He based his oratory on this. He was responding rather than leading, but nevertheless became *Quaid-i-Awam,* the Leader of the People.

The Government's efforts to frustrate the PPP now took a new form. In its attempt to unite the three Muslim League factions as its own support base, the Government pressurized them to come together. It seized the assets of the Convention Muslim League,[11] accumulated in the Ayub era, on 10 June and tried to force the resignation of Daultana, the President of the Council Muslim League. The effort failed. However, the harassment and arrest of PPP members continued, peaking in September. ZAB strongly condemned the interference of Federal Ministers when he addressed a public meeting at Memon Goth, about 20 miles from Karachi. He also repeated earlier demands for a sovereign Constituent Assembly, the dissolution of the Presidential Cabinet two months before the elections, and the release of all detained political workers, students and trade union officials. The people were tired of being badly ruled: 'Now it is the people's rule which will triumph.'[12]

The release of political prisoners became a major issue. ZAB asserted at the Peshawar Town Hall on 2 October that 'throwing Partymen in jail was an 'intrigue' intended to incapacitate the PPP. With over 150 top leaders of the PPP arrested before the elections, he charged that 'they have cut off my hands'. Still, he claimed that opposition propaganda about the PPP not having sufficient candidates would be proved wrong on 15 October, the new date for filing nominations. The PPP 'had strained every nerve and made such preparations' that others were frightened and 'running for election alliances, to form a united front' against him. The 'tremendous backing' of the masses for his Party had upset 'the scheme of the Government' for the election but it was too late to 'revise' it; the PPP would 'sweep the polls' in West Pakistan.[13]

At Lahore's Gol Bagh two days later, he again strongly condemned the Government's bias. He asserted that the PPP's determination to 'initiate revolutionary changes in the socio-economic pattern of the country' made the Yahya regime work from the outset against such economic emancipation. The rightists 'had no chance of winning' despite the arrest of PPP leaders; the Party would field a full list of candidates on 15 October, he defiantly claimed. For dramatic effect, he added that this might well be his last public appearance since he was 'unveiling' the Government's intrigues against his Party.[14]

East Pakistan

In the East Wing, the Awami League was working from a stronger base than the PPP in the West; it had long-standing support and a grass-roots organization which provided a natural platform for its campaign. The Six Points demand for regional autonomy received a massive endorsement. Mujibur Rahman was an outstanding organizer and his demagoguery captured the hearts and minds of his people.

As the original election date of 5 October approached, the Awami League appeared to be heading for a convincing victory. Tragedy then struck East Pakistan in the form of a cyclone and floods, and the Government used this opportunity to postpone the elections to 7 December. Possibly, they hoped that the PPP and Awami League would run out of steam in the additional two months. On the contrary, the PPP utilized this time to consolidate its progress. In East Pakistan, the delay, coupled with the even greater tragedy of another cyclone and tidal wave on the night of 12 November, ensured the overwhelming victory of the Awami League.

A few days after the cyclone, Yahya Khan touched down at Dhaka on his return from Beijing, but failed to visit the stricken areas. Rival political parties, expecting defeat in East Pakistan, demanded further delay in the elections because of the devastation. Governor S.M. Ahsan, who rightly anticipated a hostile reaction from the Awami League, persuaded the President to limit the postponement to the most seriously affected areas, covering nine National and eighteen Provincial Assembly seats.[15] Elections to these were rescheduled for 17 January 1971.

The indifference of the Government and West Wing leaders, who did not visit the area, vindicated Mujibur Rahman's accusations and claims. Maulana Bhashani, Ataur Rahman Khan and other non-Awami League Bengali nationalist leaders withdrew from the elections, using as an excuse their preoccupation with relief work. Nothing could now stop the Awami League.

Mujibur Rahman enjoyed an added bonus with the coverage he received in the international Press, which benefited him both in the elections and in the ensuing civil war. The number of foreign journalists covering the campaign increased dramatically because of the cyclone and tidal-wave disaster. They reported West Pakistani neglect and carried the complaints, mainly genuine, of the Awami League. The BBC, which had gained in prestige since the 1965 War, became the daily fare of Bengalis.

Throughout the campaign, even after the cyclone, Mujibur Rahman maintained that 'we are demanding regional autonomy and not independence'. He capitalized on the political impact of the disaster and wanted no further delay: 'If the polls are postponed, the people of Bangladesh will owe it to the millions who have died to make the supreme sacrifice of another million lives, if need be, so that we can live as free people and so that Bangladesh can be the master of its own destiny.'[16] His apparently conflicting use of such expressions as 'autonomy' and 'free people' contributed to some extent to, or gave an excuse for, confusion in the West Wing about the real intentions of the Awami League.

The campaign presaged the division to come; the election was conducted all but independently in each Wing. The PPP and JUI did not contest in East Pakistan. The Awami League fielded just a few candidates, none serious contenders, in the West Wing. Only two parties contested almost equally in both Wings, the Muslim League, and the Jamaat-i-Islami with 79 candidates in the West and 69 in East Pakistan; their chances were not considered good, except by the Government intelligence agencies. The Jamaat-i-Islami denounced the Awami League as a secessionist party, and the Muslim League (Qayyum) attacked Mujibur Rahman for his hate campaign against West Pakistan. Both were overwhelmingly defeated except for the Muslim League (Qayyum) in its traditional NWFP base. Events showed that the demands of each Wing were fundamentally different; parties which actively campaigned in both Wings merely divided their resources.

Another notable feature was that, unlike previous and subsequent elections in Pakistan, there were no electoral alliances in 1970. Without any elections based on direct adult franchise since 1954, there was no genuine measure of real strength. Attempts to form alliances were seen as signs of weakness, or a cover for ulterior designs. Efforts to achieve a grouping among the religious and rightist parties, including Jamaat-i-Islami, Nizam-i-Islam, JUI and PDP, failed early in November,[17] as did further endeavours to unite the Council Muslim League, the Muslim League (Qayyum) and the Convention Muslim League by mid-November.[18]

The Awami League concentrated on its Six-Point programme, ignoring the West Wing. Some critics, while overlooking this, have maintained that the PPP's lack of interest in East Pakistan was part of a deep-laid plot by ZAB to dismember the country. However, it must be remembered that the Party had only recently been formed, in December 1967, and that it had required a great effort in the Punjab and Sindh to secure a winning position in just three years. The Party's resources were limited and could not cover both Wings. Moreover, ZAB's call to the youth and new political forces had no appeal in East Pakistan, where Bengalis found satisfaction in the autonomy demands of the Awami League. ZAB's concentration on West Pakistan had no ulterior motive; it reflected the reality of the situation.

Outcome of the Elections

Throughout the country, 290 National Assembly seats were contested by 1570 candidates from 25 political parties and a further 315 independents. There were no elections in ten seats, nine in East Pakistan and one in the NWFP.

On 7 December 1970, nearly 57 million voters went to the polls, out of whom over 29 million were in East Pakistan. The turnout was 57% in East Pakistan and the Awami League won a predictable 75%. In the West Wing, the voter turnout was greater in the Punjab, with 63%, and in Sindh with 60%, while lower in the NWFP and Balochistan. The PPP's results are set out in the following table:

Province	Total seats	PPP candidates elected	% of seats	% of votes won
The Punjab	82	62	75.6	41.6
Sindh	27	18	66.6	44.9
NWFP	18	1	5.5	14.2
Balochistan	4	—	—	2.0

The Awami League's victory was complete, winning 151 out of 153 contested seats. Similar results had occurred in the 1954 provincial elections, which had also been fought on the issue of the rights of East Pakistan and exploitation by the West Wing. The rightist and religious parties were routed in both Wings.

The PPP's success, though smaller in numerical terms than the Awami League's, was in certain respects more significant. It crossed provincial boundaries. In Sindh, ZAB relied mainly on the feudals and landlords as candidates, as did the other parties, but his popular appeal contributed to substantial victories even against those with traditional local bases. In Karachi, however, the results were disappointing; despite ZAB's appeal to the proletariat, the PPP candidates were weak, and the main body of *mohajirs* (refugees from India after 1947) remained largely unconvinced about supporting a Sindhi leader with a feudal background.

The PPP's real success lay in the cities of the Punjab, particularly Lahore, Rawalpindi and Faisalabad,[19] which provided a clean sweep for previously unknown candidates; and even in the rural areas many big landlords were humbled. Some candidates, including Kausar Niazi, were elected while in jail. ZAB's main appeal in the Punjab was his anti-Indian stance. In the areas affected during the 1965 War, the vote was solidly for him. In personal terms too, his success was unmatched; with convincing majorities he won four seats from constituencies in Larkana, Lahore, Multan and Hyderabad-Badin.

In the NWFP, the performance of the PPP did not meet expectations. ZAB lost his own contest against Maulana Mufti Mahmood in Dera Ismail Khan, and the Party won just one seat. Hayat Sherpao's efforts secured a reasonable percentage of votes but this was not sufficient to defeat rivals from the NAP, the Muslim League (Qayyum) and the JUI, three strong parties in the Frontier. Although Sherpao himself lost the National Assembly seat to Qayyum Khan in the latter's Peshawar stronghold, he nevertheless defeated this major rival in the contest for the Provincial Assembly in Peshawar.

ZAB maintained that the PPP's victory was anticipated. In fact, he expected about forty seats,[20] and the Government even less. The victory might have been more resounding if, during the last month of the campaign, he had not confined himself mainly to Larkana. Although his cousin Mumtaz, who had nursed both their constituencies, assured him that their position was secure, he chose to listen to supporters who claimed that this was not so. As a result, some PPP candidates lost narrowly in constituencies where ZAB did not hold big rallies in those last critical days. His presence would almost certainly have ensured their victory. Taj Mohammad Langah lost by only a few hundred votes in Daultana's stronghold, because ZAB failed, despite promises, to attend a large public meeting arranged by Langah. His absence in this case was in response to Daultana's request. He had an odd relationship with Daultana;

while placating him in private, he showed contempt for him in public. Subsequently, in July 1972, he sent him as Ambassador to London, to avoid any possibility of Daultana gathering together the old forces in the Punjab to oppose ZAB.

The emergence of the PPP with a leftist programme, and the defeat of many feudals, prominent politicians and religious party candidates, heralded a major change in voting behaviour.[21] However, this important development was, at the time, overshadowed by the events that followed, which focused both national and international attention on East Pakistan.

ZAB was in a unique position. The PPP had not attacked Six Points during the campaign, except in so far as it undermined a feasible federal structure. The PPP sought to address the grievances of East Pakistan through a socialist programme rather than Six Points. Later, on 30 January, at the end of his first visit to Dhaka after the elections, he explained at a press conference:

> Throughout my election I did not mention anything about the Awami League or the Six-Point programme. I wanted all of you to understand our position, because I had come to understand your position. In West Pakistan much had been said about the Six-Point programme but always I had tried to avoid any discussion on it.[22]

Since ZAB was the main West Wing leader not to have condemned the Awami League outright, he was in a position to work with Mujibur Rahman to hold the country together. This was not to be; as events unfolded he became the object of hatred in East Pakistan.

NOTES

1. References to this public meeting are from *Dawn*, Karachi, 5 January 1970; it was also covered by other national newspapers of that date.
2. The Rann of Kutch is situated on the south-eastern border of Pakistan with India; the mini-war was fought in April 1965.
3. *Dawn*, Karachi, 4 April 1970.

4. Copies of ZAB's letters of 10 January 1970 to Afzal Agha, Chief Secretary, and to the Superintendent of Police, Larkana, are in the author's possession.
5. The Legal Framework Order, 1970, President's Order No. 2 of 1970, the *Gazette of Pakistan Extraordinary,* Part II of 30 March 1970.
6. The two provisions were contained in Article 20, clauses (1) and (4) respectively, of the Legal Framework Order, 1970.
7. The total number of 313 for the National Assembly included 13 seats reserved for women to be elected indirectly by the members from a Province in the Assembly. The total was composed as follows:

Provinces and Areas	General seats	Seats reserved for women
East Pakistan	162	7
Balochistan	4	1
The Punjab	82	3
Sindh	27	1
The North-West Frontier Province	18	1
Centrally Administered Tribal Areas	7	0
	300	13

8. *Morning News,* Karachi, 16 May 1970.
9. Mirza Tahir Ahmad was at that time a leading member of the Qadiani community (*see,* Chapter 8, pp. 291-6) and is at present their Chief.
10. As narrated to the author by Mubashir Hasan.
11. By Martial Law Order No. 14 of 1970.
12. *Dawn,* Karachi, 21 September 1970.
13. *The New Times,* Rawalpindi, 3 October 1970.
14. *Pakistan Observer,* Dhaka, 5 October 1970.
15. Former Governor Ahsan narrated this in 1988 to the author, when we discussed the elections and the East Pakistan situation.
16. *Morning News,* Dhaka, 27 November 1970.
17. *Dawn,* Karachi, 3 November 1970.
18. Ibid., 14 November 1970.
19. Faisalabad was called Lyallpur at that time.
20. Mumtaz Bhutto, who was with ZAB at Larkana when the results were being announced, said ZAB was 'utterly delighted' at this large number; in previous discussions and strategy sessions with ZAB, the PPP assessment had been forty seats.

21. This was generally not accepted at the time and the Yahya regime thought that the PPP would split. Even later, Richard Sisson and Leo E. Rose observed that a major concern of ZAB in 1971 was to keep the Party intact, which was not correct; *War and Secession—Pakistan, India, and the Creation of Bangladesh,* University of California Press, Berkeley, 1990, p. 56 ff.

22. *The People,* Dhaka, 31 January 1970.

CHAPTER 3

POST-ELECTIONS TO MILITARY ACTION

The outcome of the elections came as a shock to the Yahya regime, which had expected to remain as arbiter of a fragmented parliament. The military were confounded by the Awami League's absolute majority in the Assembly, which they had failed to anticipate. With the emergence of Sheikh Mujibur Rahman and ZAB as the elected leaders of each Wing, the regime lost political control.

Not only the Yahya regime, but most West Wing leaders, including ZAB, were also left confused and groping. However, as far as ZAB was concerned, the one sure factor in the situation was the power base he had acquired in the Punjab and Sindh. This, he believed, made him the spokesman for the whole of the West Wing and gave him the right to a major say in the constitution-making and government-formation that lay ahead. He was not going to be excluded from this process by the small West Wing parties combining with the Awami League. On the other hand, he was not certain about how to achieve what he regarded as his due.

Although the magnitude of the problems should have been evident to all at the time, particularly in the West Wing, it was surprisingly not fully comprehended. It took a foreign correspondent, Peter Hazelhurst of the *The Times*, London, to assess the situation accurately on 9 December: 'These two regional leaders are diametrically opposed on all the major issues in Pakistan, and the result of the election will, apparently, carve up the country into two separate political kingdoms.'

The contrasting approaches of the two leaders soon began to surface. In response to ZAB's felicitations, Mujibur Rahman sent a terse telegram, 'Congratulations on your greatest *(sic)* success in elections in Punjab and Sindh.' ZAB did not welcome his authority being confined to two Provinces and, exhilarated by his victory, declared at the conclusion of a victory procession in Lahore on 20 December:

> Punjab and Sindh are the bastions of power in Pakistan. Majority alone does not count in national politics. No government at the centre could be run without the cooperation of the PPP which controlled these two Provinces…I have the key of the Punjab Assembly in one pocket and that of Sindh in the other…The rightist press is saying I should sit in the opposition benches. I am no Clement Attlee.[1]

Six Points was obviously going to be the dominant issue following the elections and, for the first time, ZAB talked publicly about it on 15 December at Kotri in Sindh. He was prepared to make 'adjustment here and there' to achieve an agreed constitution, though not at the cost of Pakistan's unity, solidarity and integrity. The basis of his appeal was populist: 'We want a constitution which protects the rights of the labourers, workers, peasants and poor masses.'[2] He also denied differences with Mujibur Rahman. However, as speculation increased about him being side-lined, he countered that the PPP 'alone could voice the sentiments of the people of West Pakistan as it had emerged as the majority party…Any attempt to frame the constitution with the backing of splinter political groups alone could never meet with success'.[3] This theme dominated the next two months. It became increasingly difficult for the two major parties to achieve a viable accommodation within the existing federal structure.

As far as I was concerned, the differences crystallized at the end of the year when I met Rehman Sobhan,[4] an Awami League adviser. On a visit to Lahore he had talks with Mahmud Ali Qasuri, who suggested he should, in fact, pursue constitutional issues with me; in the course of 1970 I had prepared a PPP draft of the constitution, at ZAB's request. However, as I had not yet discussed the draft with anyone, I told Sobhan that I could not state the PPP position but would willingly listen and comment. During our long, friendly talk he stressed there could be no deviation from Six Points. When I posed the problems for the West Wing, he replied that the four Provinces could make whatever arrangements they chose, including a zonal sub-federation. Since the Awami League regarded the inter-Wing relations as settled, he suggested I should focus my attention on a framework for the West Wing. He attacked the 'bastions of power' speech: 'Bengalis are no longer prepared to accept the dictates of the military-bureaucratic establishment for whom Bhutto is the spokesman.' If the West Wing tried to obstruct Six Points, the whole of East Pakistan would 'stand up and resist to a man'. Such was the prevailing mood in East Pakistan, which he urged me to convey to ZAB. At the end of our meeting I remarked that tragedy lay ahead if the military were to intervene in East Pakistan.

Apart from Sobhan's visit, no contact was established by either side except when Mustafa Khar went to Dhaka in early January 1971 to prepare the ground for the meeting between the two leaders later that month. ZAB had previously indicated he would go to Dhaka earlier but, on 31 December, said the demands of by-elections in the West Wing did not permit it. Khar did not carry any specific proposals to Mujibur Rahman and nothing material transpired at their meeting.

December had passed without any significant developments. No progress was made on the constitution, nor was a date fixed for the convening of the National Assembly. When asked about the latter, ZAB expressed no objection to the Assembly meeting in February, though March was preferable.[5]

In the New Year, on 3 January, Sheikh Mujibur Rahman publicly affirmed his position on Six Points at a mammoth meeting at the Ramna Race Course, claiming that, 'None can stop it'. All the Awami League Members of the National and Provincial Assembly took an oath on Six Points. At the same time, he maintained that the people of Bangladesh believed in the integrity of Pakistan.[6]

We received news of this speech in Larkana, where ZAB went each year for his birthday on 5 January. I had been invited to discuss my draft of the constitution. However, he felt that, after Mujibur Rahman's unequivocal position on Six Points, there was little purpose in considering the draft containing our proposals. This was the first and last opportunity we had for a serious discussion of the text, which was then overtaken by events.

He telephoned Yahya Khan about the speech, only to learn that the President was shortly to visit Dhaka for talks with Mujibur Rahman. The President agreed to meet Khar first. ZAB was deeply concerned that Yahya Khan had not absorbed the significance of Six Points and its ramifications for the West Wing. Rahim and Khar, who had by then returned from Dhaka, were also present and concurred.

I narrated my meeting with Sobhan, but ZAB brushed this aside, setting out his own views on the prevailing situation. He then inquired why I looked apprehensive. I explained that, to me, it looked as though a Greek tragedy was unfolding in which the actions of each principal would lead inexorably to disaster. Mujibur Rahman was a prisoner of Six Points; even if he wanted to, he could not make adjustments because of his massive mandate. ZAB was in the invidious position of knowing that opposing Six Points would produce no positive results, while acceptance might have serious consequences leading to the dismemberment of the country. The military Government would lay responsibility on him for any consequences of acceptance; in any case, the Punjab, his support base, did not seem ready to accept Six Points. The third principal, Yahya Khan, could not accept Six Points which ran counter to the LFO provisions for the Federation. The constitutional hazards and financial implications for the West Wing, especially the question of how the Armed Forces would be financed, posed huge problems. Yahya Khan would either have to confront East Pakistan over Mujibur Rahman's insistence on Six Points, or, if ZAB accepted the Awami League programme, he would eventually turn on the PPP. Each of the principals appeared fixed in this triangle, and tragedy seemed inevitable. Such an outcome could only be avoided by an imaginative initiative. He agreed with my analysis, yet he also remarked, 'I will ensure Yahya remains on my side'.[7]

ZAB was right about Yahya Khan's lack of knowledge and understanding of the Six-Point programme. Khar found him unacquainted with its details when they met a few days later. We learnt that, on Yahya Khan's subsequent visit to Dhaka, he asked for a copy of Six Points to study them for the first time.[8]

In East Pakistan, the position on Six Points hardened. Maulana Bhashani called a meeting at Santosh on 8 January of all non-Awami League parties to reject any compromise on autonomy. Mujibur Rahman himself, the following day, declared Six Points 'a Magna Carta', and said that he had no right to make any amendment[9]—although the language of the Six-Point formula had in fact already been amended earlier by the Awami League. In the West Wing, ZAB was not to be easily outdone. He played his familiar anti-India card by attacking the proposed visit of the Indian team for the first World Cup Hockey Tournament, eventually held in Spain instead of Pakistan. He was still uncertain as to how to proceed, making tentative moves, testing the situation.

First Round of Meetings

Despite the position of East Pakistan on Six Points, Yahya Khan, on arriving in Dhaka on 11 January, expressed the hope that the 'second phase' of constitution-making, following the general elections, and the ensuing transfer of power, would proceed smoothly. Mujibur Rahman described their first two-hour meeting the next day as 'satisfactory', and, after their final talks, said he was 'fully satisfied'. The President maintained that he was 'not a separate power', and it was up to Mujibur Rahman and ZAB to settle the constitution, which had not been discussed.[10] Yahya Khan was reported as saying that Mujibur Rahman would be 'the next Prime Minister of Pakistan. I won't be there.' It was for Mujibur Rahman to talk of the future as the leader of the majority party.[11] Yahya Khan also indicated to Mujibur Rahman that the Assembly would be convened shortly after the *Eid* holiday, which was to be observed on 6 February.[12] Apart from these two statements, there was no evidence that Yahya Khan was prepared to withdraw from the political scene so readily. On the contrary, reports spread that Mujibur Rahman had assured him that he would remain as President.

Meanwhile, in the West Wing, ZAB continued to emphasize the popular nature of his base, which made him a force to be reckoned with. He promised at Tarbela on 12 January that he would revolutionize the economic conditions of the people and would refuse to join any government unless he could serve the masses in a real sense.

He spoke the next day at Rawalpindi on the subject of the constitution, offering 'wholehearted support' for the idea of a true federation of the Provinces having 'equal powers'. He sought a constitution based on the consensus of the federating units; if the Awami League framed the constitution without consensus, the responsibility would not be his. In these circumstances, he hinted, there would be no need to have a National Assembly session.[13] Throughout, he was cautious in not attacking the military regime and the Awami League simultaneously, and disapproved of some PPP members participating in a strike against the Government-controlled National Press Trust. He was a consummate politician.

The same day the Indian Government announced a ban on the Plebiscite Front in Kashmir, tightening its grip on the state. This move was viewed as anticipating increased trouble with Pakistan, and confirmed ZAB's concern about Indian intentions.

Shortly after returning from Dhaka, the President met ZAB in Larkana on 17 January. They were joined by Lt.-Gen. Peerzada for part of the talks; General Hamid Khan, Mustafa Khar and Mumtaz Bhutto were also present in Larkana. The meeting, it was later alleged, laid the ground for the 'conspiracy' against Mujibur Rahman. The details of what transpired between Yahya Khan and ZAB were never known. According to ZAB, Yahya Khan gave an outline of the Dhaka talks and the three alternatives he conveyed to the Awami League: to go it alone, to cooperate with the PPP, or to negotiate with the smaller parties of West Pakistan. In Yahya Khan's opinion, the best course was for the two majority parties to arrive at a suitable accommodation. Apparently, Mujibur Rahman was not particularly responsive to this course but did not rule it out. ZAB had serious misgivings about Six Points and handing over to Mujibur Rahman the government machinery both in the East Wing and the Centre. He said it would not work and would seriously damage the West Wing. However, he agreed to hold discussions with the Awami League leaders in Dhaka, and to seek a compromise on the future constitution.[14]

Publicly, at Larkana, Yahya Khan said he was hopeful that the constitution would be framed within the stipulated period of 120 days,[15] and that the two leaders would achieve the goal of constitution-making. It was not his but their task to negotiate the outcome.

On 22 January, the Awami League announced the finalization of the draft constitution on the basis of Six Points, claiming, nevertheless, that it would ensure 'the indivisible unity between the two Wings of Pakistan'. In the meantime, the smaller political parties of the West Wing began to get more actively involved. On 15 January, the Council Muslim League had announced its readiness to cooperate with Mujibur Rahman on constitution-making; five days later, Mufti Mahmood extended JUI's cooperation to the Awami League on condition that the constitution was 'Islamic'.[16] As the stage was being set for talks with Mujibur Rahman, ZAB was beginning to feel isolated.

On 27 January, a large PPP delegation arrived in Dhaka for discussions with the Awami League, 'to understand and comprehend the position'. The preliminary seventy-five minute talk between the two leaders made no headway; ZAB was concerned about both constitutional and governmental arrangements, namely, power-sharing, while Mujibur Rahman insisted on Six Points first being accepted before detailed discussions took place. The Press were told it was a courtesy call.

The following morning the negotiating teams of the two parties met to consider constitutional matters. The Awami League was represented by Syed Nazrul Islam, Tajuddin Ahmad, Qamaruzzaman, Khundkhar Mushtaq Ahmed, Captain Mansur Ali and Kamal Hussain. The PPP team consisted of J. A. Rahim, Sheikh Rashid, Hanif Ramay, Hafeez Pirzada and myself. Qasuri had been delayed in Lahore. We held two sessions of talks, both abortive. There was no common ground. The Awami League insisted on Six Points: 'This is our charter', and nothing short of it would suffice. The PPP team endeavoured in vain to explain the problems that would arise.

J. A. Rahim stressed that while Six Points might be the Awami League charter, it was not the PPP mandate; the real solution lay in a socialist government to alleviate the discontent in East Pakistan. However, he could make no headway with the Awami League leaders. In their minds there were two good reasons for dismissing him contemptuously: he was a former senior bureaucrat and of Bengali origin. No serious discussions on constitutional issues took place, contrary to what was believed outside.[17]

The one-hour meeting between ZAB and Mujibur Rahman the same day was equally unproductive. ZAB sought time to reach a broad settlement on the constitution and the government-to-be prior to the National Assembly session. Mujibur Rahman wanted no delay and was adamant that any talk about government formation must follow agreement on Six Points.

After these talks, Rehman Sobhan suggested I dine with him and Kamal Hussain, the Awami League's constitutional expert; I took along Mubashir Hasan to discuss economic matters. I told Kamal Hussain that since our formal meetings were clearly unsuccessful, it was all the more important that I learnt at least privately how the relevant provisions on the federation were expected to work in practice under the Six-Point dispensation. He said that the Awami League would be placing their draft before the National Assembly and he was not authorized to disclose it. I assured him that the contents would not be divulged; I wanted to understand and explain the position to ZAB and others. It was only a matter of weeks before the draft would, in any case, be made public, and I saw no harm. My request was to no avail.

In their final session on 29 January, the two leaders decided to meet again after ZAB had gone back to West Pakistan for consultations. There was no agreement, and J. A. Rahim was incorrect and over-optimistic in suggesting that there was an understanding on a 'socialistic future'.[18] A pleasant river-boat trip for the PPP delegation to meet senior members of the Awami League, and guarded statements conveying the impression of some success in the talks, could not disguise the failure to achieve any progress.

At the end of the visit on 30 January, ZAB asked the assembled Press to report fairly since 'the future of every citizen of Pakistan' was involved and it concerned everybody. Commenting on each of the Six Points, he said of the first and sixth points that there was 'no difference of opinion about a federation and a parliamentary form of government elected through adult franchise', or the right of the Provinces to have para-military forces. The remaining four points were 'controversial'; he would have to explain the views he had heard in Dhaka to the people of West Pakistan. He needed time to assess and influence public opinion and, if necessary, to prepare the ground for any compromise of a basic nature. However, it was not necessary to enter the Assembly with agreement on all issues, as 'negotiations would continue during the session'.

He added, 'Many mistakes were made in the past and there were callous decisions imposed on the people by the previous regimes. On our part we are quite aware of all the consequences. We had come here to find the areas of agreement. In the context of our national policy, we had gone as far as we could go.' When asked about the Assembly being summoned on 15 February, he replied that there was nothing wrong 'if we take time to the end of February at least' or the first week of

March, but added that he would not ask the President to delay the session.

Next morning, he again told the Press at the airport that he was 'not unsatisfied' with the talks, nor had they failed. Asked to elaborate on his earlier statement that comments on the four remaining points could not be expected in the foreseeable future, he replied, 'I have nothing further to add. If necessary I can make my comments before the National Assembly goes into session or after.'[19]

The Press gave full and fair coverage. ZAB had closed no door and left open several possibilities. At least on two important issues his position was clear at this stage. He had no objection to the National Assembly being convened in early March, and was prepared to present his submissions in the course of the Assembly session.

Deadlock over Six Points

By the time ZAB returned to the West Wing, a new scenario had emerged as a consequence of the hijacking of an Indian Airlines plane, the *Ganga,* to Lahore on 30 January 1971. As we shall see, he moved away from the position he had only recently taken in Dhaka on the two important issues relating to the National Assembly. Within weeks there was deadlock over Six Points. Before analysing ZAB's actions, it would be useful to set out the PPP's position on Six Points, the texts of which are reproduced in original and amended forms in Appendix A.

The Awami League's Six-Point constitutional proposal was for a unique federation in which the Central Government would exercise power only in relation to defence and foreign affairs, excluding foreign trade and aid. It failed to recognize that the foreign policy of Pakistan, as with most Third World countries, was largely concerned with international trade and aid, and, if these were excluded, foreign policy would be severely curtailed. Without full control over foreign policy it would be difficult to determine and implement an effective defence policy, in practical terms leaving only confrontation and war within the competence of the Central Government.

The Six-Point programme was based on the resolution passed by the All-India Muslim League at Lahore on 23 March 1940, which provided that 'geographically contiguous units' be demarcated into 'regions which should be so constituted...that areas in which the Muslims are numerically in a majority, as in the North-Western and Eastern zones of India, should be grouped to constitute *Independent States* in which the constituent units shall be *autonomous and sovereign'* [20] (emphasis added). It is commonly referred to as the 'Pakistan Resolution' although Pakistan was not named. Indeed, the use of such imprecise and undefined terms as 'units', 'regions', 'zones' and 'autonomous and sovereign' can only be accounted for by the fact that Pakistan itself was yet a distant dream in 1940. It was not until the All-India Muslim League Legislators' Convention was held in Delhi on 7-9 April 1946 that a resolution gave this idea definite shape: 'That the zones comprising Bengal and Assam in the North-East and the Punjab, North-West Frontier Province, Sindh and Balochistan in the North-West of India, namely Pakistan zones where the Muslims are in a dominant majority, be constituted into a sovereign independent *State...of Pakistan'* [21](emphasis added). From 1946 onwards, and certainly after the founding of Pakistan, the terms 'States' and 'autonomous and sovereign' were not given serious consideration. However, from 1966 the Awami League returned to the text of the 1940 Resolution, when its literal implications were put forward by, first, Sheikh Mujibur Rahman and then Maulana Bhashani.

Six Points meant an overnight transformation from a quasi-unitary system to an extremely loose federation or, in reality, a confederal arrangement. It was the PPP's understanding that the West Wing would have to bear the financial burden of Rs 38,000 million out of the existing Rs 40,000 million external debt, and the entire internal debt of Rs 31,000 million. The contribution of East Pakistan towards the cost of the Central Government would be only 24%, though their population was 56% of the total, and, moreover, this could be set off over the next few years against 'reparations' due from the West Wing for past exploitation. As the representative of the West Wing, ZAB could not accept this position.[22]

The day after the hijacking of the Indian plane, the two Kashmiri 'commandos' involved were granted asylum by the Pakistan Government. On his arrival from Dhaka, ZAB hailed them as heroes. The passengers returned to India by road, and on 2 February the commandos blew up the aircraft. Initially, the Indian Government retaliated by cancelling 'military' flights over Indian territory, disrupting the supply of matériel to East Pakistan. In contrasting reactions, Mujibur Rahman deplored the destruction of the aircraft and urged an inquiry, while ZAB called the hijackers 'freedom fighters'. Three days later, Delhi banned all flights over India. Now all Pakistani planes had to fly from Karachi to Dhaka via Colombo, a distance of over three thousand miles. ZAB declared it a war-like measure, triggering thoughts of the 1965 War and an uprising in Kashmir. No regard was paid to the present military imbalance between the two countries, Pakistan's domestic problems, nor, a month later, to Indira Gandhi's massive election victory. There seemed little realization that, by distancing the two Wings, India had virtually won the first battle for East Pakistan.

Caught up in the aftermath of the *Ganga* incident and the anti-India demonstrations, ZAB seemed to be waiting for something to happen which would obviate a decision on Six Points. Nothing did. At the same time, in Lahore, he met Party members of the Assemblies from Lahore and Rawalpindi Divisions.[23] He described his talks with Mujibur Rahman as 'neither successful nor unsuccessful', but only exploratory. Those present gave him a mandate to seek 'adjustment' in the Awami League's programme.[24] He then met Party representatives from Sindh on 4 February in Karachi and, after the *Eid* holiday, those from Multan and Bahawalpur Divisions at Multan on 10 February.

The general view of the PPP was that Six Points could not be accepted as a whole and required modification. There was little understanding of the Awami League's rigidity and unwillingness to compromise. It was felt, erroneously, that Bengalis were 'weak and inferior' and could not resist the power of the West Wing's military-bureaucracy. The Punjab was not prepared for any major accommodation and ZAB did not elaborate on the consequences of non-acceptance, which he felt was the responsibility of the President under the LFO. Meanwhile, in Dhaka, Mujibur Rahman reiterated that the constitution would be based on Six Points: 'if anyone refuses to cooperate it will be his responsibility.'[25]

ZAB set about trying to achieve an agreed position for the West Wing in the negotiations on the constitution with East Pakistan. First, he met the President in Rawalpindi on 11 February and explained that as the Awami League had already dictated the constitution, which ZAB refused to rubber-stamp, little or nothing could be achieved by attending the Assembly. He emphasized the need for making a last attempt at a broad understanding with the Awami League before the Assembly was convened, and for six weeks' additional time to complete discussions. The President gave no commitment but appeared to understand the position.[26] The Press, however, reported that the National Assembly was expected to be convened on 18 February.

He also met West Wing leaders including M.M.K. Daultana of the Council Muslim League, Khan Abdul Qayyum Khan, the PML chief, Khan Abdul Wali Khan, the NAP President, and Maulana Mufti Mahmood and Maulana Hazarvi of the JUI. His efforts to achieve some consensus among them were unsuccessful. Nawab Akbar Bugti agreed with Mujibur Rahman on the constitution, and Daultana was reported as saying that Six Points should be debated in the National Assembly and its utility explained to the members.[27] Earlier, Daultana had supported ZAB in the constitution-making process while his colleagues in the Council Muslim League, including Shaukat Hayat Khan, had supported the Awami League.[28]

ZAB maintained that a broad understanding on constitutional and political issues was necessary before the Assembly met; an acceptable settlement could not be reached in the charged atmosphere of public debate within the Assembly. If efforts prior to the session failed, at least the Assembly would remain intact. A deadlock within the Assembly would result in its breakdown. Equally, if the President refused to authenticate a constitution passed by a sovereign body, his decision would not be accepted in East Pakistan.

From Rawalpindi he proceeded to Peshawar to meet Party members and other leaders, including Qayyum Khan, Mufti Mahmood and Wali Khan. On 13 February the President announced that the Assembly would meet on 3 March at Dhaka. ZAB telephoned Lt.-Gen. Peerzada, Principal Staff Officer to the President, to say that the PPP would not attend as he had sought extra time only two days earlier.[29] He told the Press he would comment later. The following day, Hayat Sherpao in Peshawar read out a statement that ZAB would hold a 'historic' meeting on 'constitutional matters' in Lahore after the PPP leaders' convention in Karachi on 20-21 February.[30]

The crisis was clearly deepening. After the Executive Committee of the Awami League met in Dhaka on 14 February, Tajuddin Ahmed announced that the basic postulates of Six Points admitted of no possible readjustment. However, he assured West Pakistani leaders that interests peculiar to their region could be accommodated. East Pakistan would not dictate the arrangements for the West Wing Provinces; nor should the West Wing interfere with the East.

The general situation, particularly the economy, was now deteriorating rapidly. In Sindh, there was renewed agitation over the language issue. The ban on overflights and the threatening statements of the Indian Prime Minister compelled Pakistan's Permanent Representative at the UN, Agha Shahi, to give notice to the Security Council of the 'serious situation which had currently developed between Pakistan and India'.[31] But, for the political leaders of Pakistan, the power struggle continued to dominate their perceptions.

In Peshawar, on 15 February, ZAB abruptly announced that the PPP would not attend the Assembly session as the Awami

League had framed a Six-Point constitution on a 'take it or leave it' basis. Referring to the 'alarming and threatening' deterioration in Pakistan's relations with India after the hijacking incident, he said he would not take the PPP representatives to Dhaka to be in 'jeopardy' as 'double hostages', to both Indian hostility and the non-acceptance of Six Points.

On the points at issue with the Awami League, the 'two-subject Centre' was not acceptable. On currency, he felt something could be worked out: 'I am not despondent on that.' However, on taxation he was less sanguine, 'But I am not without hope.' He maintained, 'We have really gone to that precipice' beyond which there was a fall: 'I want a transfer of power but not a transfer of Pakistan.' He was not prepared to accept a constitution imposed as a 'vendetta on Pakistan'. Emphasizing that the PPP had been silent on Six Points during the elections, he called upon those West Pakistan leaders, who had first attacked and now were praising the programme, to explain their views on this controversial subject.

He sought an assurance from Mujibur Rahman, even a private one, on the question of 'give and take': 'I think we can work out something which will satisfy both of us. There is hope for understanding. But if we are asked to go to Dhaka only to endorse the constitution which has already been prepared by the Awami League and which is not to be altered an inch here and an inch there, then you will not find us in Dhaka.' Within the Party, he said, some took an extreme position, wanting to fight it out, a small number accepted the Awami League's programme, but the majority supported a compromise based on reasonable adjustments. He promised that, 'We will take a formal and final position' at the Karachi meeting of the PPP on 21 February.[32]

The abrupt announcement in Peshawar that the PPP would not attend the Assembly session in Dhaka proved to be the first definite move in the tragedy that began to unfold. Several PPP leaders were then meeting in Mahmud Ali Qasuri's Lahore house to consider, for the first time, our position on the constitution. The announcement, without any prior consultation with the PPP leadership, took us by surprise. ZAB came directly from

Peshawar to Qasuri's house and assured us that he had no alternative, and it would 'work out for the best'. Our misgivings remained.

Why such a vital decision was taken in Peshawar when ZAB knew many PPP leaders were meeting in Lahore, and an important PPP convention was to take place in Karachi in less than a week's time, is difficult to understand. Ordinarily, he would consult on important matters, even if he had already made up his mind, in order to associate others with his decisions. There were two exceptions: when he sought to take the sole credit; or, as it appeared in this case, when he wanted to avoid discussion of decisions he had already taken, the reasons for which he did not wish to disclose. Some critics of ZAB maintain that his announcement was made to force the President's hand, others that it was done in consultation with Yahya Khan. However, the issue of his involvement with the President will be considered subsequently.

Significant developments were also taking place elsewhere. In Dhaka, Mujibur Rahman told Awami League leaders on 15 February that there could be no adjustments on Six Points: consensus was a much abused word designed to subordinate the majority. The following day, India recalled her envoy from Pakistan. The fears which ZAB had already expressed now seemed justified. As he had anticipated, Mujibur Rahman had disavowed consensus, while India was adopting an increasingly belligerent stance.

On the Awami League side, there was strong objection to such expressions as 'double hostage' and 'transfer of Pakistan', but no hasty decision was taken. Meanwhile, in Lahore, Maulana Maudoodi, whose Jamaat-i-Islami had fared badly in the elections, called for avoiding extreme positions: the majority party should present its draft of the constitution, others could object, and, 'if the majority party still insisted on the basis of its numerical strength, it should be made clear that even if such a constitution was passed, it would not be a success and the majority party would be fully responsible for the results'.[33] This was a reasonable position, but at the time seemed directed against the PPP.

ZAB became increasingly strident and inflexible, telling a news agency on his return to Karachi that his decision was 'unshakable and irrevocable...Anyone who goes to Dhaka from West Pakistan whether in khaki or in black and white does so at his own cost.' When warned that his Party members might lose their seats by non-attendance, he replied it would be the 'finest thing if that happens'.[34] The next day, he told the Press that if a 'viable' constitution was to be framed, 'all of us must have a hand in that'. The Assembly would be a 'slaughterhouse' in the present circumstances.[35]

Despite this tough talk, he was still worried that the smaller parties from the West Wing would combine in the Assembly with the Awami League to marginalize him. By this time, the Hazarvi Group of the JUI had decided to attend. Mufti Mahmood's JUI also declared they would participate, but would pull out if the Awami League were not prepared for reasonable adjustments, or if the genuine interests of West Pakistan were bypassed. Mufti Mahmood was critical of the foreign trade and taxation aspects of Six Points, which he considered in direct conflict with the LFO. Wali Khan's NAP had decided to attend the Assembly, believing that all issues should be discussed in this forum. The majority of the representatives from Balochistan and the Frontier had thus agreed to participate.

On 18 February, the Council Muslim League MNAs announced that they would go to Dhaka. The following day the MNAs from the Tribal Areas followed suit. As the Tribal MNAs invariably acted according to the Government's wishes, it was assumed they were doing so now. ZAB was also pressured to accept the argument that Yahya Khan's consent would be withheld if the constitution conflicted with the LFO. If ZAB's announcement of 15 February was made with Yahya Khan's approval, the President was clearly trying to retain all options.

ZAB was summoned to Rawalpindi by Yahya Khan. Accompanied by Mairaj Mohammad Khan, he left Karachi on 18 February. Talking to the Karachi Press, he rejected the idea of any arbitration or mediation. For the first time, he publicly acknowledged the importance of the army; there were three 'forces' in the country, the Awami League, the PPP and the Armed Forces. Stopping briefly at Lahore, he reiterated the decision not to attend the Assembly; no Party member would dare go. He was dismissive of the significance of 'small parties' attending the session.[36]

The President met him for five hours on 19 February. No other member of the PPP was present and no details were disclosed. ZAB told the Press that the President had 'spelled out the magnitude of current issues', stressing, 'of course we are now facing a crisis, rather a crisis of extreme nature, which is not of our making'. He modified his position on Six Points: 'I think, on currency, we might be able to evolve some kind of settlement, agreeable to East Pakistan. On taxation I am not without hope...but...we are terribly stuck up on foreign trade and foreign aid.' Still, his decision not to attend was 'irrevocable and unshakable', unless some 'substantive adjustment' was made: 'We must have a part in the framing of the constitution. We must be given a chance to make our contribution. We are half of the country and this half must be heard.' With his Party absent, the constitution-making exercise would be 'like staging Hamlet without the Prince of Denmark'—a futile, barren, negative and counter-productive effort, he declared.[37] He talked about the threat from India. Hinting at his removal from the scene, he nominated Mustafa Khar and Mairaj Mohammad Khan as his two successors; according to him, the PPP had a parliamentary and a revolutionary face, which these two represented.

On 20 February, the President made a curious amendment to the LFO, permitting those elected to resign before the first meeting of the National Assembly, even prior to being sworn in as members. This was seen as accommodating ZAB's position, since elected representatives of the PPP were thereby enabled to threaten immediate resignation, which they did at the PPP Convention in Karachi. The Yahya-ZAB relationship was further reflected by the presence at the convention of Z. A. Suleri, a leading pro-government journalist hostile to the PPP who, it was believed, would report to the President on the conduct of the proceedings. J. A. Rahim, Mairaj Mohammad Khan and others objected strongly, but the Chairman overruled us, insisting Suleri should stay.

Despite any possible 'understanding' between the two, Yahya Khan played his own game. The JUI of Maulana Hazarvi and Mufti Mahmood met the President, as did Qayyum Khan, after which they announced they would go to Dhaka. The JUI said they would seek clarification on the questions of foreign trade and loans and of taxation, but not the issues relating to currency as these were not being pressed by the PPP. In Dhaka, the Awami League spokesman said ZAB's recent Rawalpindi press conference helped clear the air a little by specifying the areas of disagreement. After meeting Mujibur Rahman, Hazarvi and Mufti Mahmood vouched for his belief in a strong Pakistan.

The two-day PPP Convention concluded on 21 February with a vote of confidence in the Chairman, empowering him to decide all matters. A resolution called for 'a democratic Pakistan' and rejected 'the authoritarian principle of elected despotism'. It explained: 'We are ourselves most anxious that a normal civilian regime should be installed at the earliest possible moment, especially as the nation is facing a crisis imperilling its very existence. It is better, however, to undergo tribulation now rather than risk the certain dissolution of the country a little later.'[38] Little did we realize at the time that the country was soon to suffer both tribulation and dissolution.

Simultaneously, on 21 February, the President removed the Cabinet 'in view of the political situation obtaining in the country'. This eliminated civilian encumbrances before the military crackdown. Next day, at a high-powered meeting, he told the Provincial Governors and Martial Law Administrators that the National Assembly session was to be postponed without a date fixed for its meeting. Governor Ahsan argued strenuously against the decision, supported by General Rakhman Gul, the Sindh Governor. But Yahya Khan insisted, instructing Ahsan to inform Mujibur Rahman about it the night prior to the announcement. Before returning to Dhaka, Ahsan tried again from Karachi to contact Yahya Khan but was not 'allowed' to reach him directly. He asked Peerzada to convey his view that at least a date should be given for the session.[39] He also met ZAB. No one else in the PPP or the country was aware of the critically important decision taken on 22 February.

The next few days were devoted mainly to the question of participation in the Assembly proceedings. The Muslim League (Qayyum) now decided against attending the Dhaka session, which was believed to reflect the Government's latest position. The PPP said its MNAs should not resign. The Council Muslim League prevaricated about attending as they could see 'no fruitful outcome'. On 24 February, the JUP decided not to attend, but two days later their leader, Maulana Shah Ahmed Noorani, said they would review this decision, and on 27 February he said they would participate. Meanwhile, the JUI confirmed they would attend, but warned of a boycott if the Awami League acted unilaterally.

Mujibur Rahman issued a lengthy statement on 24 February condemning what he described as a 'conspiracy' to prevent the transfer of power to the elected representatives. He insisted that foreign trade and aid must be provincial subjects, and accused opponents of seeking to perpetuate the exploitation of the East Wing and the past colonial structure. While insisting on Six Points, he said this need not apply to West Pakistan; if the federating units in the West Wing did not wish to have precisely the same degree of autonomy as Bangladesh, they could cede certain additional powers to the Centre, or establish regional institutions. Mujibur Rahman's statement served a dual purpose. He was placating concern about units in the West Wing pulling apart under the Six-Point arrangement, while also putting forward a completely new inter-Wing dispensation. He did not yet call it confederal, but that was indeed what it was.

The eagerly awaited 'historic' PPP public meeting was held at Lahore on 28 February. Before a huge gathering, ZAB declared that adjustments could be made concerning currency and federal taxation, but foreign trade and aid could not be entrusted to any provincial government. He went on to say that if the Assembly session was held as scheduled, there would 'be a general strike from Peshawar to Karachi' and he would launch a movement. Members from the West Wing were sternly warned not to attend the Dhaka session.[40] He gave three alternative scenarios. First, in the event of deadlock, the President might dissolve the Assembly, which was not acceptable at any cost as it would mean the continuation of Martial Law and a dangerous crisis for the nation. Second, the Assembly might be postponed to allow time for discussion of constitutional issues with the Awami League; for that he would go immediately to meet 'elder brother' Mujibur Rahman. Third, the limit of 120 days prescribed for preparing the constitution might be removed, permitting his Party to debate fully the issues involved. If one of the last two alternatives were not accepted, the crisis would continue, resulting in the end of democracy in the country.[41] The speech was rapturously applauded, reflecting the sentiments of the Punjab.

For the first time, he had spelt out a tough position. Speculation that the speech was made after consulting Yahya Khan increased when, on the same day, the Election Commission deferred the preliminaries for the convening of the Assembly, involving indirect elections for women members in the West Wing Provinces.

The following day, the Assembly session was postponed *sine die*. In explanation, Yahya Khan stated, 'A majority party, namely the PPP, as well as certain other political parties, have declared their intention not to attend the National Assembly session'. Tension created by India had further complicated the position. With so many staying away, 'if we were to go ahead with the inaugural session on 3 March, the Assembly itself would have disintegrated and the entire effort made for the smooth transfer of power that has been outlined earlier would have been wasted. It was, therefore, imperative to give more time to the political leaders to arrive at a reasonable understanding on the issues of constitution-making.'[42] The

President had accepted one of the alternatives demanded by ZAB without any reference whatever to Mujibur Rahman.

Beginning of the End

This postponement by the President changed the position dramatically. East Pakistan, already highly tense, now exploded. The previous evening Ahsan performed his last function as Governor of East Pakistan; he read out the President's postponement statement to Mujibur Rahman, who listened in stony silence and then left the room.[43] Ahsan resigned. He was and remained a gentleman to the last. Yahya Khan lost a sober adviser, one whom the Awami League respected. He was replaced as Governor by Lt.-Gen. Sahibzada Yakub, the Martial Law Administrator of East Pakistan, who lasted only a few days.

The Press were informed that ZAB would give a 'considered statement' in due course. Qayyum Khan hailed the post-ponement, calling on all political leaders to work out a constitutional formula. Daultana called for dialogue. Few in West Pakistan realized the dire consequences of the decision.

The postponement *sine die* was a grave error for which no satisfactory explanation can be found: it was the result of either ignorance or complete disregard for the position in East Pakistan. It confirmed in the minds of Bengalis their worst suspicions about 'conspiracy' and the intentions of the West Wing. It signalled, for all practical purposes, the beginning of the end.

Resentment and resistance in the East Wing was total. Addressing a huge rally at Dhaka's Paltan Maidan on 1 March, Mujibur Rahman was brief, saying he would very soon unfold a programme of action to achieve self-determination. The crowd abused Yahya Khan and ZAB.

The Awami League called a general strike in Dhaka on 2 March and throughout East Pakistan the following day, both completely successful. The army under Yakub took some hesitant steps the first day but then did not intervene. Mujibur

Rahman assumed *de facto* control of the Province; the movement would continue 'till the people of Bangladesh realize their emancipation'.[44]

On 2 March, the leaders of the NAP (Bhashani), the Bangla National League and other nationalist elements in the East Wing joined the movement, and called for unity. A shooting incident created considerable confusion but, by next morning, everything outside the control of the Awami League was at a standstill. Communications between the two Wings were confined to army and air force telephones. The curfew, which had been reimposed on 3 March, was withdrawn the next day. Mujibur Rahman threatened that if the troops did not return to their barracks he would ask the Awami League volunteers to resist them.

After meeting Yahya Khan on 2 March, ZAB held a 'hurriedly-called' press conference. There had been a 'disproportionate reaction' in East Pakistan, he explained: 'Surely, nothing is lost, if the premise of a united Pakistan is accepted, by the delay of a few days to enable the major parties of the whole of Pakistan to come to an agreement upon the nature of [the] constitution that ought to last for years and years.' He elaborated: 'When we say we are trying to come as close to the Six Points as possible, it obviously means a loose federation because you can call it a federation under the Six Points, but it has another name too.'[45] For the first time, he publicly spoke of a 'loose' arrangement. He also repudiated the idea of any prior consultation on the postponement; later, on 14 March, he took 'oath on his children' about this at Karachi.[46]

On 3 March, Yahya Khan invited the leaders of all the political parties to meet him in Dhaka a week later to discuss the situation and settle an early date for the National Assembly session. West Wing leaders accepted, but Mujibur Rahman described the invitation to sit in a meeting while people were being killed as 'a cruel joke'—the more so 'to sit with certain elements whose devious machinations are responsible for the deaths'. The invitation was made at 'gun point' and had to be declined. Surprisingly, Nurul Amin, the only other East

Pakistani leader to be invited, declined, though throughout he was pro-Pakistan.

Mujibur Rahman's public condemnation of ZAB's 'machinations' pulled the two parties further apart. He maintained that if he were 'prevented' from being present at the public meeting called on 7 March, implying arrest or assassination, others would make his announcements. By now the Awami League had virtually taken over the administration while the Government did nothing, remaining a silent spectator.

In Karachi, a visibly shaken ZAB held a two-day PPP Central Committee meeting to consider the political crisis. He told the Press on 4 March that, 'if a brief postponement can spark such an unfortunate and tragic situation, you can imagine what would have happened at the end of the day, had the Assembly met without some sort of agreement between the two parties. Whenever the Assembly meets you will realize why I took this decision. Time will tell you, and, when I say time, I do not mean months or years'. As the Awami League had already drawn up the constitution based on Six Points, without any prior understanding, his going to the Assembly would merely mean endorsing its constitution.[47]

That morning Yahya Khan announced he would be leaving for Dhaka, but at night he changed his mind and held a lengthy meeting with ZAB at Rawalpindi the following day. There is no full account of what transpired. According to ZAB, the President felt the LFO contained sufficient safeguards for a united federal Pakistan, and no separate assurance was necessary from the Awami League. ZAB pointed out that the Government had made no pronouncement that Six Points was in conflict with the LFO provisions pertaining to provincial autonomy; on the contrary, the programme had been permitted to gain ground during and following the elections. Mujibur Rahman could not be prevented from using his majority to push his constitution through the Assembly. If the President, under the LFO, refused to authenticate the constitution, the Assembly could declare itself a sovereign body, not subject to the LFO. Yahya Khan, however, prevailed and ZAB agreed to attend the session provided the

President publicly held out the assurance about the LFO, even though ZAB considered it a 'dead letter'.[48] However, what he failed to disclose, and we in the PPP did not know, was that Mujibur Rahman's proposal for 'two Committees'—one each for the elected MNAs of East and West Pakistan meeting separately—had been conveyed to the President that very morning[49] and was no doubt also discussed between Yahya Khan and ZAB.

To defuse the tension building up before Mujibur Rahman's speech of 7 March, when many expected a declaration of independence, it was announced that the President would address the nation at noon on 6 March. In Dhaka, the army was withdrawn while a complete *hartal* (strike) continued in the city. The Jamaat-i-Islami and other parties demanded the Assembly be summoned immediately.

On 6 March, the President summoned the session for 25 March. He explained the two objectives of the previous postponement, to save the Assembly and to allow time for negotiations. The first succeeded but the second failed. His decision had been completely misunderstood. Mujibur Rahman had earlier not been averse to the idea of a conference on the issues, and his rejection was a surprise: 'Instead of accepting the decision in the spirit in which it was taken, our East Pakistan leadership reacted in a manner which resulted in destructive elements coming out in the streets and destroying life and property.' Having laid the main responsibility on the Awami League, he concluded with these strong words: 'I will not allow a handful of people to destroy the homeland of millions of innocent Pakistanis. It is the duty of the Pakistan Armed Forces to ensure the integrity, solidarity and security of Pakistan—a duty in which they have never failed.'[50] Having throughout the election campaign, and even afterwards, allowed the LFO to be flouted by the Awami League, Yahya Khan belatedly held out this public assurance, which by implication meant the enforcement of the LFO, as requested by ZAB.

Yahya Khan's firm stand was reinforced by the appointment the same day of Lt.-Gen. Tikka Khan as Governor of East

Pakistan. Tikka Khan had earlier earned the reputation of 'Butcher of Balochistan' and was expected to implement the hardline approach which, according to the President, Lt.-Gen. Yakub had failed to carry out.

ZAB immediately responded that the PPP would 'proceed to the National Assembly' on 25 March, and expressed the hope that discussions with the Awami League on constitutional issues would take place before the session. He was trying to 'show willing', though a meeting of the Assembly in the high tension in Dhaka was quite unrealistic. He explained to the Press why he had earlier decided to stay away: the PPP had done its 'best' to achieve a 'broad consensus', seeking to reconcile differences of a fundamental nature, and to avoid jeopardizing the Assembly.

By the time Mujibur Rahman addressed the public meeting on 7 March 1971, feelings were running very high. Reports of a unilateral declaration of independence gained ground, especially since, at the time, there were insufficient troops in East Pakistan to prevent it. However, despite the circumstances, Mujibur Rahman was not belligerent.[51] He announced that the Awami League would 'consider' attending the session on 25 March if the Government met four demands: the immediate withdrawal of Martial Law, the transfer of power to the elected representatives of the people, an enquiry into the army killings, and the return of the troops to their barracks. Countering Yahya Khan's statement that the postponement had been 'misunderstood', he claimed it had been 'effected solely in response to the machinations of a single party—constituting a minority of the total members—against the declared wishes of the majority party and also those of numerous West Pakistani members'. The PPP had obstructed the transfer of power and, he predicted, 'military confrontation' would follow 'political confrontation', because the majority would not submit to such minority dictation. Recognizing, possibly, that his speech was less than his audience expected, he raised his fist and concluded: 'Our struggle this time is a struggle for independence. *Joi Bangla.*'[52]

The Awami League issued ten directives which effectively gave them control of the entire provincial administration, cutting it off from the West Wing. Never has an opposition, let alone under Martial Law, asserted such total control within a state. So successful was civil disobedience that Wali Khan reportedly observed, 'Even [Mahatama] Gandhi would have marvelled'.

Much killing and brutality had occurred in this early period. But, contrary to general belief, of an estimated 3000 killed, only 300 were victims of the army.[53] Relevant reports showed that the Awami League magnified their casualties, publicized through the foreign media. Despite provocation and hostility, the military authorities did not reveal the numbers of those supporting Pakistan who were killed by the Awami League, for fear of exacerbating opinion in the West Wing and within the army itself.[54] This changed dramatically from 25 March 1971.

After an extended stay in the capital to keep in touch with developments, ZAB returned to Karachi on 9 March. In the changed situation, criticism against him mounted. It was alleged that the President's postponement of the Assembly was a direct consequence of ZAB's earlier decision not to participate in its proceedings, and his latest Lahore speech. Isolated, he asked me to prepare a message for Mujibur Rahman. Although he did not anticipate receiving any reply, he thought it was necessary to placate public opinion. The telegram, simultaneously released to the Press, expressed readiness to visit Dhaka immediately 'to devise a common solution to end the crisis'; and ended with the words, 'Let not people say, nor history record, that we have failed them.'[55] Mujibur Rahman made no response; the General Secretary of the Awami League, Tajuddin Ahmed, contemptuously commented that they were 'not even prepared to consider the message'.

There was no over-all coordinated response to the crisis by the parties in the West Wing. The PPP, after eleven-hour discussions, expressed great concern—and awaited developments. Asghar Khan, who was then in Dhaka, warned that only a few days remained to save the situation; mentally, the two Wings had already separated and the last link, through Mujibur

Rahman, was about to be broken. He urged the acceptance of Mujibur Rahman's four preconditions. However, his words went unheeded as his recent electoral defeat made him politically irrelevant. Meanwhile, the smaller West Wing parties resented the PPP being the sole spokesman for West Pakistan and decided to meet on 13 March to form a common front. Qayyum Khan, however, refused to attend because ZAB was not invited. Wali Khan was not able to participate as he was *en route* to Dhaka but said he would accept the group's decision. Following the meeting, the group declared that Mujibur Rahman should form the government, 'interim to the framing and promulgation of a new constitution', before the session of the National Assembly on 25 March. They recognized Mujibur Rahman's 'firm commitment to the solidarity of Pakistan by putting in the present crisis four demands that were not in the least parochial or regional but exclusively based on a national approach'.[56]

The President decided to visit Dhaka for talks with Mujibur Rahman and stopped on the way in Karachi to discuss the situation with ZAB. No one else was present at this meeting and the precise nature of their talks remains unknown. According to ZAB, he told the President that in principle there could be no objection to the four demands put forward by Mujibur Rahman on 7 March; but any interim or final settlement, relating to the lifting of Martial Law and the transfer of power to the elected representatives, required the consent of the PPP. He was prepared to visit Dhaka provided Mujibur Rahman would hold meaningful discussions.[57] An indication that future political arrangements might also have been discussed came during ZAB's important speech to a large PPP rally at Karachi on 14 March. He urged that since the Awami League and the PPP were in the majority in each Wing, if power were to be transferred, it should be to both these parties. Normal democratic principles of simple majority rule could not apply to a country divided into two parts. Bearing in mind the background of Six Points, the majority parties from both Wings should together reach a settlement. The PPP was willing to sit with the majority party to frame Pakistan's 'comprehensive' constitution.[58] He was reported in the Urdu Press as saying, *'Idhar ham, udhar tum'*, (We here, you there).

The speech created a furore in the West Wing since it was interpreted to mean a demand for two Pakistans. Overlooked in the bitter condemnation was ZAB's plea for one Pakistan, and Mujibur Rahman's earlier proposal for separate arrangements for each Wing. Critics also ignored the fact that Mujibur Rahman was now imposing conditions for attending the Assembly, whereas earlier he had maintained that all issues should be discussed within it. The situation had indeed changed.

Some of us met ZAB that night. We were surprised that he should have gone so far in his speech. Later, it appeared that the statement on the transfer of power in the two Wings was probably a trial balloon to test public opinion. He was agitated and taken aback by the adverse reaction and the numerous hostile telephone calls. He felt compelled to clarify his position and asked me to prepare a statement with great care, repairing the damage but not altering the purport of his speech. He said he did not 'want to deter Yahya Khan'. This comment presumably related to progress on the 'two Committees' proposal, about which most of us knew nothing at the time. In the circumstances, the clarification could not be convincing. It referred to the geographical peculiarities of Pakistan which required that the two majority parties should come together in a 'broad coalition'. This would help avoid further polarization and safeguard against a federal arrangement in which the West Wing would, in the foreseeable future, remain under the majority rule of East Pakistan. The statement called for the 'united power' of both majorities at the Centre 'in the interest of a united Pakistan'.

ZAB was grilled on this statement at his press conference. When asked 'what sort of constitutional arrangement' he visualized in power-sharing at the Centre, he replied, 'I don't want to go into that now; I only want to say what is absolutely necessary for the present. I am mindful of the discussions that President Yahya Khan will be having in Dhaka. I don't want even to comment unnecessarily because that might foul up the atmosphere.' He explained that, if power were transferred to the majority party in the West Wing, the PPP 'would certainly take into government the representation of Balochistan and Frontier'.[59] In this comment too he was anticipating separate arrangements for each Wing.

Final Round of Meetings

The ten days from 16 to 25 March 1971 witnessed discussions in Dhaka which were to prove fateful. Talks were held first between Yahya Khan and Mujibur Rahman, and their respective negotiating teams; for the last five days, ZAB too was involved. What actually transpired during those ten days is difficult to verify. What is certain is that the negotiations ended with military action. The following is an outline of the sequence of events from the perspective of one who participated in the PPP team.

The President arrived on 15 March and, the following day, held a three-hour meeting with Mujibur Rahman. The Awami League leader told the Press that the 'country's political situation and other matters' were discussed, but declined to comment further. He emphasized that the non-cooperation movement would continue till 'emancipation'.[60] The meeting the next day between Yahya Khan and Mujibur Rahman did not move forward on a compromise formula, although the Government ordered a probe under a High Court Judge into the circumstances of the army being called out in aid of civil power on 2 March. Mujibur Rahman rejected this concession as the terms of reference 'prejudged the main issue' of whether the deployment of force was in aid of ulterior political purposes or of civil power. Meanwhile, Yahya Khan ordered Tikka Khan, the newly-appointed military commander of East Pakistan, to prepare for action.[61]

Over the next few days, however, there appeared to be progress on an arrangement with East Pakistan, and it was agreed that experts from both sides would discuss an appropriate constitutional framework. Mujibur Rahman told the Press, 'Let us hope for the best and prepare for the worst'.[62] Newspapers carried optimistic reports of the talks, and no one denied there had been some advance. The prospects for a settlement brightened following Mujibur Rahman's directive that 23 March, Pakistan National Day, should be observed as a holiday and not as the usual day of non-violent non-cooperation. The experts were next scheduled to meet on 22 March.

During the first five days ZAB was in Karachi in conclave with several Party members and held a Central Committee meeting. He was agitated at being excluded from the talks, but had to remain in Karachi as Mujibur Rahman would not meet him. On 16 March, the President sent a message asking him to come to Dhaka. ZAB declined since no useful purpose would be served by talks in which Mujibur Rahman refused to meet the PPP. The following day he rejected a similar request from Peerzada.[63] However, concerned at being sidelined, he decided to inform the President that the developments in Dhaka were being carefully watched by the PPP; any settlement without the PPP's consent would break down as it would not be acceptable to the people of the West Wing. We anxiously awaited an answer.

A peculiar feature of Pakistani politics is that very little dampens the urge of leaders to jockey for position, and this was true even in the crisis then facing the nation. Mufti Mahmood questioned ZAB's right to speak for the West Wing since he had a majority only in the Punjab and Sindh, and urged the formation of a National Government representing all five Provinces. On the other hand, the Muslim League (Qayyum) and the PPP claimed the joint support of twenty-one out of forty members in the NWFP Assembly and the right to form the Provincial Government. The PPP Secretary-General accused the other parties of being unpatriotic, warning on 19 March against a 'sell-out' of the West Wing, and the implementation of the 'London Plan'.[64]

In the evening of 19 March, ZAB was asked to come imme-diately, along with his advisers, for discussions with the

Government and Mujibur Rahman. Our suspense was over. He told the Press he was proceeding to Dhaka and stressed that he would not allow the 'London Plan' to break Pakistan.

We arrived on 21 March in a bitterly hostile city. The PPP were considered even more the enemy than the army, and ZAB was given a rough reception. The Awami League had undertaken to make arrangements for our stay as they controlled the city, but, after being manhandled at the hotel, he sought military protection. Tension was running so high that we were deterred from leaving the hotel.

Several potentially significant developments occurred that day. *The Pakistan Times,* a National Press Trust paper, speculated about an interim Coalition Government, with Nazrul Islam as Prime Minister, constituted on the basis of provincial strength. Yahya Khan held an unscheduled hour-long meeting with Mujibur Rahman, who declared there would be no relaxation in the Awami League movement. In addition, the leaders of the smaller West Wing parties met in Dhaka; this time Qayyum Khan was not invited.

Late that evening Yahya Khan briefed ZAB on the meetings he had already held with Mujibur Rahman, and outlined the proposals discussed between their advisers. According to ZAB,[65] the main features of the arrangement, to be implemented by Presidential Proclamation, were the immediate withdrawal of Martial Law and the transfer of power to the elected representatives in the five Provinces. The President would continue to run the Central Government with the assistance of advisers who would not be elected representatives. The National Assembly would be divided *ab initio* into two Committees representing each Wing, which would sit respectively in Dhaka and Islamabad to prepare 'separate reports' within a specified period to be submitted to an Assembly whose function would be restricted to discussing these proposals and devising ways for the two Wings to live together. Until the constitution had been framed, East Pakistan would have provincial autonomy on the basis of Six Points. The West Wing units would have powers under the 1962 Constitution and would settle their future autonomy, subject only to the President's approval.

The President said ZAB's agreement was necessary for the proposal to be put into effect, though it would also be desirable to have the consent of other West Wing leaders. He wanted all the parties to arm him with written authority for the Proclamation. ZAB asked for time till next morning to consult his Party members present in Dhaka. After the two-hour meeting, he told a news agency that everything would turn out all right: 'This much I can tell you now.'[66]

Most members of the PPP team were surprised to learn how far the discussions on the 'two Committees' proposal had proceeded. All were of the view that such a far-reaching settlement could not be made in personal talks between the leaders, to be merely confirmed by letter; the National Assembly could not be ignored. Moreover, sending a letter to the President might have led the PPP into a trap, so ZAB decided against it. We also considered a compromise formula to be presented to the President. This provided for the National Assembly to meet briefly to ratify the Presidential Proclamation, after which it would separate into two Committees, one for each Wing, before sitting again as a single body to approve a final constitution for the two Wings, with a specified number of votes from each Wing.

Next morning, the three leaders met for the first and last time. ZAB recorded[67] that Mujibur Rahman asked Yahya Khan to approve the proposal; it was up to the President to obtain ZAB's consent, if required. Formal discussions could only proceed after agreement, and Mujibur Rahman said that in the meantime he would merely tell the Press that he had met the President and ZAB had also been present. It is difficult to reconcile this with ZAB's account of the sudden change which then took place in Mujibur Rahman's attitude when they went out of the room.

According to ZAB, the Awami League leader took this opportunity to seek his help in overcoming the crisis. Mujibur Rahman apparently claimed that things had gone too far, there was no turning back; the best course was to accept the Awami League's proposal as there was no workable alternative. He urged ZAB to agree with the Awami League and, as the West Wing's leader, become Prime Minister of Pakistan, leaving Bangladesh to Mujibur Rahman. Otherwise, he warned, the army would first destroy him and then ZAB. The National Assembly should be adjourned *sine die,* allowing the two Committees to function. He wanted to meet again and would keep in touch by sending an emissary the next evening to fetch Khar.

ZAB pointed out that the earlier postponement, sought in good faith, had unfortunately become the pivotal issue. The new proposal would be considered carefully, but had to be placed before the Assembly. They agreed that no letter to the President was necessary, but Mujibur Rahman remained adamant that the Assembly should not meet initially.

After this brief discussion, ZAB returned to the President and told him that the proposal would inevitably result in two Pakistans. Unless the Assembly established a new source of sovereign power at the national level, there would be a vacuum; Martial Law was the source of law and the basis of the President's authority, and had to be properly replaced. In the absence of this, nothing could stop East Pakistan from seceding. The Assembly should sit, even briefly, as a single body to provide a minimum umbrella, after which the two Committees could meet separately. They agreed that these issues required detailed discussion between their advisers.

At the conclusion of these talks, Yahya Khan announced that, 'In consultation with the leaders of both the Wings of Pakistan and with a view to facilitating the process of enlarging areas of agreement among the political parties, the President has decided to postpone the meeting of the National Assembly called on March 25.' No new date was set for the session; this time the postponement *sine die* was what Mujibur Rahman had wanted.

When newsmen asked Mujibur Rahman whether the postponement meant progress, he replied, 'You can see for yourself'. He said that he had met the President and that 'Mr Bhutto was also present there. The President informed me in front of Mr Bhutto that whatever I had discussed with the President, he had conveyed to and discussed with Bhutto.' He expected to meet Yahya Khan again, 'tomorrow or the day after'.

For ZAB's part, he told the Press after leaving the President that he had no formula to solve the present crisis. However, his Party was examining the terms of the broad agreement between Yahya Khan and Mujibur Rahman. This was the only public mention of any understanding.

The representatives of the smaller parties, Wali Khan, Mufti Mahmood and Daultana, expressed disapproval of the postponement. Qayyum Khan and Maulana Noorani were at the time on their way to Dhaka.

As had already been agreed, discussions then took place between the President's advisers and the PPP negotiating team. The Awami League proposal and our compromise were considered. ZAB was accompanied by Secretary-General Rahim, Mahmud Ali Qasuri, Abdul Hafeez Pirzada and myself; the President's team consisted of Lt.-Gen. Peerzada, Justice A. R. Cornelius, M. M. Ahmad and Col. M. A. Hassan. The PPP Chairman reiterated the need for the Assembly to meet initially as the PPP formula proposed. He pointed out that once Martial Law was lifted, and in the absence of any legal sanction for the Presidential Proclamation, the unilateral declaration of independence, 'legal UDI', by the Awami League would be possible. He also referred to such matters as the appointment of Governors. The Government team proposed to meet us again after the PPP's points were discussed with the Awami League.

In retrospect, such discussions were meaningless. The situation changed significantly on 23 March, when the new flag of Bangladesh was formally hoisted on all buildings in Dhaka and the recently created Bangladesh Militia paraded. Mujibur Rahman also raised the new flag at his residence. We were as if on foreign territory. A three-man PPP team led by Mahmud Ali Qasuri informed other West Wing leaders in Dhaka about our position. They accepted the PPP's stand on the Proclamation; they too had informed the President the previous day that there could be no permanent settlement without a meeting of the Assembly.

On the evening of 23 March, when the Chairman called some PPP members for discussions, it was agreed by virtually everyone that military action was necessary. J. A. Rahim described Mujibur Rahman as a fascist, who could only be countered by the army. I said this would inevitably spell the end of Pakistan and cautioned against it. To my surprise, the only person who supported my view was Mustafa Khar. Next morning, ZAB met the President and Lt.-Gen. Peerzada and told them a decision had to be made. Many of his Party leaders had gone back to Karachi that morning; other West Wing leaders had also returned after meeting Yahya Khan and Mujibur Rahman. The Awami League's inflexible attitude left little reason for him to remain in Dhaka. However, as Mujibur Rahman was sending an emissary for Khar that night, ZAB had stayed back with a few of us.[68]

After this meeting, ZAB told the Press that matters could be expedited if the three sides talked together, but the Awami League had unfortunately refused to sit across the table with the PPP. However, he still hoped that something tangible would emerge and a system satisfactory to the people of both Wings devised. But the chance of this seemed negligible following Mujibur Rahman's meeting with Khar that night. The Awami League leader merely told him that the situation had reached breaking point; ZAB must accept the Awami League proposal. He feared the situation was getting out of control, particularly in Chittagong. He promised to send an emissary to fetch Khar the following night for further talks.

By this time the Government had conveyed to ZAB the proposals put forward by the Awami League leaders during the meetings between them on 23 and 24 March. Instead of two Committees, the Awami League now demanded two 'Constitutional Conventions' to submit two constitutions in the National Assembly—not reports containing proposals as previously suggested. The Assembly would meet only to tie up the two constitutions for a 'Confederation of Pakistan'. At the end of these meetings, Tajuddin Ahmad, the Awami League General Secretary, informed the Press that they had given their 'complete plan', and 'from our side there is no need of any further meeting'.

On the morning of 25 March, ZAB, Rahim and Khar had a final meeting with the President and Peerzada. Rahim was specifically taken to spell out his strong views on how to deal with the 'fascist Mujib'. They considered, in outline, the latest revised proposals of the Awami League. Subsequently, in the afternoon, a PPP team consisting of Rahim, Qasuri and myself met the President's advisers, who provided us with details of these proposals:

- The Presidential Proclamation was to be issued without reference to the National Assembly.
- The Centre would only have powers over defence and foreign affairs, excluding foreign trade and aid, for the 'State of Bangladesh'. The Awami League insisted that Bangladesh would negotiate foreign loans directly, although the Government had suggested as a compromise that foreign affairs should include 'policy aspects of foreign aid and foreign trade'.
- The State Bank at Dhaka should be redesignated as the Reserve Bank of Bangladesh, and placed under the Provincial legislature, while the State Bank of Pakistan would only issue currency notes and otherwise act as required by the Reserve Bank.
- The term 'Confederation of Pakistan' was used for the first time. The two Constituent Conventions would be sworn in separately to frame Constitutions for each Wing, and then would make the Constitution for the Confederation.
- The Awami League wanted Martial Law to be lifted and the appointment of Governors of each Province within seven days from the Proclamation. The President would have no control over a Governor after appointment.
- The President had to authenticate the Constitution within seven days of its presentation, after which it would be deemed authenticated.

- The President was to have no powers of interference in emergency.

It was difficult to understand the purpose of the meeting. We were merely informed about the Awami League proposals and the talks. We were not asked for our views, nor did anything appear to be expected of us. However, in view of the military crackdown which took place that night, the purpose of the meeting was probably to associate the PPP with the Government's planned action.

In the late afternoon we learnt that the Awami League had called for a strike on 27 March to protest against the heavy firing on the civilian population in Saidpur, Rangpur and Joydetpur. With the whole East Wing under Awami League control, we considered this call curiously weak. Some felt it was a ruse.

When Mujibur Rahman's emissary came for Mustafa Khar at 8 p.m., as earlier arranged, we were taken aback to learn that the President had left the city at 7 p.m. If ZAB knew anything about this in advance, he certainly gave us no inkling. There were only six PPP members in Dhaka at the time and, when Khar returned, we tried without success to ascertain the position about Yahya Khan's departure from President House.

Just before midnight we were disturbed by the sound of gunfire outside the hotel, and assembled in ZAB's room. We witnessed the military operations for about three hours. We saw the nearby office of the newspaper, *The People,* on fire, and flames in other parts of the city. The moment for which many Party members were waiting, and a few dreaded, had come.

The next morning, a colonel came to take us to the airport. On the way we saw the flags of Bangladesh, only recently hoisted, being taken or pulled down from the housetops, and street barricades removed. To my surprise, Qasuri clapped his approval of the troops along the streets. It was strange coming from a man who always upheld the rule of law, but then this was civil war. We were informed that the Awami League leaders had escaped; only Mujibur Rahman had remained at home where he was arrested at 1.30 a.m. and then lodged in a school in the cantonment. ZAB expressed surprise that military intelligence had permitted their escape; now, they would create trouble. In fact, there was to be more trouble than anyone in the West Wing anticipated.

Some Conclusions.

I have set out in this Chapter the sequence of events. These have been explained and interpreted by some of the participants. ZAB stated his position in *The Great Tragedy*. Both Yahya Khan and ZAB testified before the Hamoodur Rahman Commission[69] whose report remains classified confidential and, one may add, questioned in both its scope and findings. Without discussing the contents of the report, one can still analyse events to find answers to three of the main questions posed by this period. First, did any or all of the three parties have a preconceived plan? Second, was there collusion between Yahya Khan and ZAB? Third, why did the final round of talks break down?

Sheikh Mujibur Rahman throughout remained adamant about the Six Points programme. We have already considered how this conflicted with a feasible federal structure. Following his overwhelming mandate he could not, nor did he want to, retract from it in any way. Despite the fact that the Awami League knew their programme was not acceptable in the West Wing, particularly in the Punjab and among the military, they did little to assuage the fears it created; they used force to subdue opponents in East Pakistan. In addition, although they stressed the separate identity of each Wing, they refused during the last days to negotiate directly with the PPP, which was the majority party in the West Wing. On his release in January 1972, Mujibur Rahman declared he had been striving for a separate Bangladesh for many years, an aim which was shared by most of the top leadership of the Awami League. All this suggests Mujibur Rahman worked to a plan.

Yahya Khan had decided on the military option as early as 22 February. Why then did he engage in detailed discussions on the 'two Committees' arrangement during the ten days prior to the military crackdown on the night of 25 March? His critics maintain that the talks were intended for three purposes: to gain time in order to build up the strength of the troops in East Pakistan; to demonstrate that he had tried the path of negotiation; and to involve the West Wing parties, particularly the PPP, in his decision. More likely, however, is the possibility that although the military option was settled by 22 February, the massive upsurge in East Pakistan, following the postponement of the Assembly session, was of such unexpectedly disturbing proportions that it necessitated some re-thinking of strategy. Yahya Khan re-summoned the Assembly to meet on 25 March. At the time it must be remembered that Mujibur Rahman's 'two Committees' proposal had also been conveyed to Yahya Khan and, no doubt, had been discussed by Yahya Khan with ZAB both in Rawalpindi and Karachi before the President left for Dhaka on 15 March. ZAB felt the proposal had merit and sought to gauge public opinion in the West Wing through his *'idhar hum, udhar tum'* speech on 14 March. At this stage Yahya Khan too probably felt the viability of the proposal should be explored as an alternative to military action, and thus pursued it in talks with the Awami League. However, Yahya Khan finally reverted to a military solution for reasons we will shortly consider.

So far as ZAB was concerned, he had no preconceived plan apart from the fact that he intended to stay on the same side as the military in confronting the Awami League. On the whole, he was reactive rather than proactive in his efforts to preserve what he termed a 'united Pakistan', and at the same time to safeguard the interests of the West Wing, which included his own. However, his announcement of 15 February that the PPP would not participate in the Assembly proceedings precipitated events. By the time he again agreed to attend, when the Assembly was summoned afresh, it was too late. The Awami League position had hardened. It must be kept in mind that in his view the army would be able to deal with the Awami League and so prevent the ultimate break-up of Pakistan. For this reason, as we shall see in the next Chapter, he at all times supported the military action in East Pakistan.

He would have attracted less blame if he had attended the Assembly in the first place. He could there have objected to the Awami League constitution, and the majority party would then have been 'fully responsible for the results'.[70] In that event, it would have been the President's responsibility to refuse to authenticate a Six-Point constitution which was contrary to the provisions of the LFO. On the other hand, if ZAB had followed such a course from the outset, he might have forfeited the confidence of the Punjab and the army, on which he relied.

Above all, ZAB has been accused of deliberately seeking to embroil the army in East Pakistan, in the belief that its defeat and the dismemberment of the country would leave him undisputed master of the West Wing. This is to cast him in a more Machiavellian mould than can be supported by his actions. The involvement of the army for its own self-destruction did not appear to me as part of his calculation. Events proved that he over-estimated military power, under-estimated Bengali nationalism, and, most surprisingly, did not properly anticipate India's determination to seize this opportunity to break Pakistan.

On the question of collusion, there can be little doubt that in the phase of confrontation with Mujibur Rahman, from mid January to 25 March, the President acted in concert with ZAB. This was inevitable since, unlike other periods of Yahya Khan's Presidency, they had a common goal, the preservation of the interests of the West Wing and the army, though each had his own interpretation. At the time, most of us in the PPP did not know what actually passed during some of their meetings, particularly in connection with the military option and the 'two Committees' proposal. The facts that emerged, when correlated with subsequent events, inevitably suggest collusion. It cannot have been merely coincidental that the two held lengthy discussions before each major decision or change of direction during this period, as is evident from the following sequence of events.

After the December 1970 elections, their first meeting took place in Larkana in mid January 1971, soon after Yahya Khan's discussions with Mujibur Rahman in Dhaka. While Yahya Khan had, in Dhaka, referred to Mujibur Rahman as the future Prime Minister of Pakistan, there was no similar utterance following the Larkana meeting. Moreover, the President did not summon the Assembly to meet soon after *Eid* as he had previously indicated to Mujibur Rahman. Following ZAB's announcement on 15 February that he would not participate in the Assembly proceedings, he met the President four days later, on 19 February, for five hours at Rawalpindi. We now know that the decision at a high-powered Government meeting to pursue the military option in East Pakistan was taken on 22 February; Governor Ahsan was told to inform Mujibur Rahman on 28 February about the postponement of the Assembly session. That very day, ZAB held a major public meeting at Lahore to make a 'historic' statement on 'constitutional matters'. The next day, Yahya Khan formally announced the postponement *sine die* of the National Assembly session he himself had summoned a mere two weeks earlier.

Again, after talks with ZAB, the President on 3 March called a conference of all leaders on 10 March to defuse the agitation in East Pakistan. When Mujibur Rahman rejected the invitation, another Yahya-ZAB meeting was held before the address of the President on 6 March when, at ZAB's request, he gave a public assurance that the army would do its duty. They had, as we have already noted, also discussed the alternative strategy presented by the 'two Committees' proposal which ZAB tried to test publicly on 14 March. Even his subsequent 'clarification' did not fully negate the purport of his speech. When we arrived at Dhaka on 21 March, the 'two Committees' arrangement was almost settled between the Government and the Awami League, as ZAB clearly expected.

We come to the last of the three questions, why did the 'two Committees' arrangement not materialize? President Yahya

Khan made only two pronouncements during this period. One was on 6 March prior to the final round of negotiations; the second was on 26 March following the military crackdown, when he gave an account of his readiness to accept in principle the plan for the Assembly to sit *ab initio* in two Committees, 'despite its flaws', on condition that it received the 'unequivocal agreement of all political leaders'.[71] He went on to state how the leaders of the West Wing had objected that this would increase division, and had suggested that the Assembly should meet and pass an interim Constitution Bill for the President's consent which would permit the transfer of power without creating a vacuum.

The Awami League too felt that by 22 March substantial progress had been made in the discussions with the Government, and only 'a few loose ends needed to be tied up'.[72] This view was supported by ZAB himself who, after his first talk with the President, on 21 March, stated that everything would be all right. Then, following the only meeting between the three leaders, on 22 March, ZAB told the Press that the PPP was examining the terms of the broad agreement between the Government and the Awami League. In fact, he accepted the proposal, with the minor amendment that the Assembly should meet initially and not be bypassed.

The turning point in the talks came with the introduction of the term 'Confederation' by the Awami League in discussions with the Government team on 23 and 24 March. Subsequently, no serious talks on the constitution were held. The President met ZAB and they considered what action was to be taken. The answer was military intervention.

In retrospect, the differences between the 'two Committees' arrangement and the final Awami League proposal for 'Confederation' involved questions of emphasis and terminology rather than matters of such fundamental significance as to warrant an abrupt end to the negotiations by army action. However, the term 'Confederation' had an emotive quality which seemed to have struck home with Yahya Khan and the army in a way that 'Six Points' and the 'two Committees' proposal did not.

If the military crackdown of 26 March was not merely the implementation of the earlier 22 February decision, a combination of further factors must have determined the outcome. The parties of the West Wing were indecisive. None of them wanted to risk bypassing the Assembly *ab initio*; but equally, no one had a viable alternative acceptable to the Awami League. ZAB was cautious on the 'two Committees' proposal following the very adverse reaction to his 14 March speech. Military force seemed to be the most ready solution for the problem, and the final trigger for action was the Awami League's demand for 'Confederation', which came as the proverbial last straw. Subsequently, on 6 May, a government spokesman claimed the Awami League had been planning to make a unilateral declaration of independence on 26 March, though the President, it should be noted, made no mention of this in his speech of 26 March.

Critics blame ZAB for allowing his all-consuming pursuit of power to override national considerations. They maintain that his collusion with the President, and the deadlock on the constitution for which he was responsible, ultimately resulted in the break-up of Pakistan. In this, they ignore the role of Mujibur Rahman and the ultimate responsibility of Yahya Khan, who had the power of the gun and used it. No amount of persuasion or precipitation of events by ZAB could have resulted in Yahya Khan's resort to military action, if that was not in accord with the President's own plans.

Yahya Khan, Mujibur Rahman and ZAB were the principal protagonists at the time, but it must be borne in mind that the crisis in the East Wing had long been in the making. Starting as early as 1948, the Bengalis had protested against domination by the West Wing. There can be no doubt that by the time these three had emerged as the leading players, it would have been extremely difficult to arrest the momentum of events which had acquired an inevitability of their own. These facts were largely overlooked in the narrow perspective of the West Wing. Against the background of Martial Law since March 1969, the military option seemed the safest and easiest. No West Wing leader or military officer condemned it, not even those who declined to participate. Rejection would have been considered treason. The action was condemned only when it failed.

The situation following the December 1970 elections demanded a degree of statesmanship, imagination and courage which was not present in Pakistan. Throughout our sad history, the power game played by our leaders has been the dominant feature of Pakistani politics. It was no different in 1971. It would be inaccurate to apportion all, or even most, of the blame to ZAB; equally, he was not blameless.

Ironically, the 'Confederation' proposal, which was considered treason at the time, might ultimately well have been the best. The 'two Committees' arrangement on which it was based was more appropriate for a Confederation than Six Points. It would have established the West Wing as an equal partner, and safeguarded its interests, especially in the event of breakdown. This would not have been the case with Six Points, under which the Awami League, or at least the East Wing representatives, would have controlled not only East Pakistan but also the Centre in the foreseeable future. The confederal solution would certainly have been preferable to civil war, defeat and the dismemberment of the country.

NOTES

1. *Pakistan Observer,* Dhaka, 21 December 1970.
2. *Dawn,* Karachi, 16 December 1970.
3. *Dawn,* Karachi, 26 December 1970.
4. Rehman Sobhan was a schoolfriend of the author.
5. *The Pakistan Times,* Lahore, 1 January 1971.
6. *The Pakistan Times,* Lahore, 4 January 1971.
7. Some years later, at a small dinner in Peshawar, when the Air Force Chief inquired about East Pakistan, ZAB repeated this discussion and said that if the army thought they knew how to ensnare him, we had anticipated it and made a precise analysis early in January 1971.

8. Admiral Ahsan, Governor of East Pakistan at the time, confirmed this to the author when we were comparing notes in 1988.

9. *The Pakistan Times,* Lahore, 10 January 1971.

10. *The Pakistan Times,* Lahore, 14 January 1971.

11. *The Pakistan Times,* Lahore, 18 January 1971.

12. Confirmed to the author by former Governor Ahsan in 1988.

13. *Dawn,* Karachi, and *The Pakistan Times,* Lahore, 14 January 1971.

14. Z. A. Bhutto, *The Great Tragedy,* Karachi, 1971, p. 20, and as narrated by ZAB to the author.

15. *The Pakistan Times,* Lahore, 18 January 1971.

16. *The Pakistan Times,* Lahore, 21 January 1971.

17. Sisson and Rose, op. cit., p. 69-70, say 'substantive' issues were discussed.

18. *The Pakistan Times,* Lahore, 30 January 1971.

19. *The People,* Dhaka, 31 January 1971, *The Pakistan Times,* Lahore and Rawalpindi, 1 February 1971, and other national newspapers on those two dates reported the main features of ZAB's statements and answers, some of which are quoted. *See also,* Bhutto, op. cit., p. 24.

20. Syed Sharifuddin Pirzada, ed., *Foundations of Pakistan—All-India Muslim League Documents: 1906-47,* National Publishing House, Karachi, 1970, Vol. II, p. 341.

21. Ibid., p. 513.

22. Bhutto, op. cit., pp. 11-3 and 22.

23. A 'Division' is the largest administrative unit in a Province.

24. *The Pakistan Times,* Lahore, 3 February 1971.

25. *The Pakistan Times,* Lahore, 10 February 1971.

26. Bhutto, op. cit., p. 25.

27. *The Pakistan Times,* Lahore, 13 February 1971.

28. *The Pakistan Times,* Lahore, 2 February 1971.

29. Bhutto, op. cit., p. 26.

30. *The Pakistan Times,* Lahore, 15 February 1971.

31. Ibid.

32. *See, Dawn,* Karachi, *The Pakistan Times,* Lahore and Rawalpindi, of 16 February 1971, for details of this press conference and these quotations.

33. *The Pakistan Times,* Lahore, 17 February 1971.

34. Ibid.

35. *The Pakistan Times,* Lahore, 18 February 1971.

36. *The Pakistan Times,* Lahore, 19 February 1971.

37. *The Pakistan Times,* Lahore and Rawalpindi, 20 February 1971.

38. *The Pakistan Times,* Lahore, 22 February 1971.

39. Former Governor Ahsan narrated this to the author in 1988. He confirmed that the decision to take military action, if necessary, was made at that time. Sisson and Rose also confirm this, as did ZAB subsequently to the author.

40. He was heard to say crudely that he would pull their legs apart, though reported as saying 'break their legs'.
41. *The Pakistan Times,* Lahore and Rawalpindi, 1 March 1971.
42. *The Pakistan Times,* Lahore, 2 March 1971.
43. Narrated to the author in 1988 by former Governor Ahsan, who said it was quite the worst duty he had ever performed. *See also,* Sisson and Rose op. cit., p. 89.
44. *The Pakistan Times,* Lahore, 3 March 1971.
45. *Morning News,* Karachi, 3 March 1971.
46. *Jang,* Rawalpindi, 3 March 1971, and *Nawa-i-Waqt,* Lahore, 15 March 1971.
47. *Morning News,* Karachi, and *The Pakistan Times,* Lahore, 5 March 1971.
48. Bhutto, op. cit., pp. 31-2.
49. Sisson and Rose, op. cit., p. 98.
50. *The Pakistan Times,* Rawalpindi, 7 March 1971.
51. ZAB told the author that Yahya Khan had sent a message to Mujibur Rahman that an SSG sharp-shooter would kill him on the spot if he declared UDI. However, no proof of this came to light subsequently. ZAB at all times had a poor opinion of Mujibur Rahman, referring to him as a coward and a mere rabble-rouser. Others were of the view that Mujibur Rahman still hoped for 'one Pakistan' and wanted to avoid civil war.
52. *The Pakistan Times,* Lahore, 8 March 1971.
53. David Loshak, *Pakistan Crisis,* William Heinemann, London, 1971, p. 67.
54. Confirmed to the author by Lt.-Gen. Jilani in 1972, and relevant documents examined by the author himself.
55. Bhutto, op. cit., p. 33.
56. *The Pakistan Times,* Lahore, 14 March 1971.
57. Bhutto, op. cit., p. 36
58. *The Pakistan Times,* Rawalpindi and Lahore, and other national dailies of 15 March 1971.
59. *The Pakistan Times,* Lahore, and *Morning News,* Karachi, 16 March 1971, and other national papers of the same date.
60. *The Pakistan Times,* Lahore, 17 March 1971.
61. Sisson and Rose, op. cit., p. 132.
62. *The Pakistan Times,* Lahore, 20 March 1971.
63. Bhutto, op. cit., p. 37; the author was present.
64. *The Pakistan Times,* Lahore, 20 March 1971.
65. Bhutto, op. cit., p. 40, and as conveyed by ZAB to the author.
66. *The Pakistan Times,* Lahore, 22 March 1971.
67. Bhutto, op. cit., pp. 42-6.
68. Ibid., p. 48.

69. The Commission of Inquiry into the 1971 December War appointed by ZAB on 24 December 1971 came to be known as the Hamoodur Rahman Commission. *See,* Chapter 5, p. 158.

70. *See,* p. 58–9 and Note (33) above.

71. *The Pakistan Times,* Lahore, 27 March 1971.

72. Conveyed subsequently to the author by Rehman Sobhan.

CHAPTER 4

CIVIL WAR TO
TRANSFER OF POWER
(26 March to 20 December 1971)

Civil war is one of the most brutalizing experiences any nation can suffer. In Pakistan, divided as it already was by over a thousand miles of hostile India, 'Hindu intervention' further exacerbated the callous attitude in the West Wing to the happenings in East Pakistan. This narrative will bring out the unreal, almost surreal, nature of concerns in the West Wing. While civil war raged in East Pakistan and the country was being dismembered, the West Wing was engaged in the esoteric niceties of constitutional issues and the raw realities of political power. Even the involvement of India failed to alert the top echelons to the wider ramifications of the situation.

ZAB himself was in a dilemma. He supported the military action to counter secession in East Pakistan; at the same time he realized that the real solution lay within a political framework—which also alone provided a role for him. However, the nature of the military intervention precluded, as far as the East Wing was concerned, a political settlement. Although the military acted to save the integrity of the country, their very intervention led inexorably to its dismemberment.

Early Suppression of Insurrection

After a long flight via Colombo, we arrived in Karachi in the late afternoon of 26 March. At the airport, ZAB told journalists, 'By the Grace of Almighty God, Pakistan has at last been saved'.[1] Other reports carried, 'Thank God, Pakistan has been saved'. This spontaneous comment was later to earn him severe criticism. At the time, most West Pakistanis shared his view.

We rushed home to hear the President's broadcast to the nation. He announced a ban on the Awami League, and the tightening of Martial Law until the situation was under control. He explained the Awami League's proposal and why it had not been accepted. The President also condemned Mujibur Rahman's action as 'treason', describing the Awami League leaders as 'enemies of Pakistan': 'This crime will not go unpunished'.[2]

At a press conference the following day, ZAB explained that in Dhaka he had called for the Assembly to meet first, elect a Speaker, and determine the central subjects, 'before, if at all, it was found necessary to divide into two committees'.[3] This last proviso was a new position adopted in an attempt to distance himself from the 'two Committees' proposal, which was not viewed favourably in the West Wing. He also pointed out that he wanted the Presidential Proclamation to be adopted by a resolution of the National Assembly. He stressed that the Awami League was bent on secession, maintaining that Mujibur Rahman had struggled for a separate Bangladesh since 1966. Next day, he repeated that Yahya Khan's military measures 'were in the interest of the country' and accused Mujibur Rahman of wanting 'an independent fascist and racist State in East Pakistan'.[4]

The forced restraint exercised by the military in early March was now replaced by fury,[5] and the suppression of the insurrection was reported very unfavourably by the international Press. Foreign journalists were herded out of Dhaka after 26 March, and, in the absence of eyewitness accounts, stories of terror and horror abounded. The Foreign Office summoned the American Ambassador and the British High Commissioner to complain about 'misreporting' by VOA and the BBC.

In the West Wing, people either were, or pretended to be, ignorant of what was happening in East Pakistan; worse, some expressed satisfaction at the 'Bingos' getting what they deserved. Except within limited PPP circles, there was no criticism of the treatment meted out to Hindu minorities and progressive elements. When I telephoned Rehman Sobhan's wife to ask after him, she said I was the only West Pakistani friend to make inquiries; no further proof of the separateness of the two Wings was necessary, she remarked.

The brute impact of military action succeeded initially. By 29 March, curfew was lifted during daylight hours in Dhaka, and all government employees were ordered back to work. The secretariat of the Organization of the Islamic Conference stated that the situation in East Pakistan was an internal crisis and it was 'not right for outsiders to interfere'.[6]

This early success of the military caused India to internationalize the issue. On 31 March, Prime Minister Indira Gandhi obtained a unanimous resolution of the Indian Parliament expressing 'profound sympathy and solidarity with the people of *East Bengal* (emphasis added) in their struggle for a democratic way of life'; it also asserted that 'their struggle and sacrifices will receive the wholehearted sympathy and support of the people of India'. The resolution called for an immediate end to force and the 'massacre of defenceless people', and for all governments to prevail upon Pakistan to halt genocide.[7] The same day the people of West Bengal (in India) observed a total strike to express 'solidarity with East Bengalis fighting for independence from West Pakistan'. These references to East Bengal, and not East Pakistan, were significant. It had been East Pakistan since 1955. It was not yet Bangladesh.

Yahya Khan met ZAB on 1 and 2 April to discuss the Indian reaction. Following this meeting, ZAB lashed out against India at a press conference. Her crude interference had exposed her to the world; she should look at her own 'shameful record with Kashmir, the Mizos and the Nagas'.[8] He was equally scathing about Six Points, which he described as a scheme for the country's 'dismemberment' through 'constitutional secession step by step'. East Pakistan had voted to end exploitation and not for separation. If he were power hungry, he would have accepted Mujibur Rahman's offer of Premiership and Six Points. He declared that history would judge his actions from 15 February in the light of whether Six Points was 'a sugar-coated pill for secession', or a demand for autonomy within the federation.

At the time, the situation in East Pakistan appeared to be coming under control. On 4 April, Nurul Amin and other non-Awami Leaguers extended their cooperation, offering to form peace committees. Four days later, M. Sabur, Mahmud Ali and other East Pakistani politicians condemned India. Kamal Hussain, who was hiding in the American Consulate at Dhaka, surrendered. On 9 April, Tikka Khan was formally sworn in as Governor; during the previous month, the Chief Justice of East Pakistan had refused to perform the oath-taking ceremony. The next day, Mujibur Rahman was brought as a prisoner to the West Wing.

India continued to step up pressure on the international front, calling for 'suitable action' by the world community at the United Nations against 'increasing repression' by Pakistan. The Indian High Commissioner in Islamabad was told that the Indian Parliament's resolution was totally unwarranted: 'Delhi has chosen a path that could only lead to serious consequences'.[9] Prime Minister Indira Gandhi justified India's response, claiming it was neither proper nor possible for India to keep quiet in the present situation; it was not 'interference' as India always raised her voice against 'colonialism and repression, and for freedom'.[10]

At this point, the Soviet Union became involved for the first time. President Podgorny in his message of 3 April, which, along with Yahya Khan's reply, was published three days later, expressed concern at the 'arrest and persecution of M. Rahman and other political leaders'[11] who had obtained such an overwhelming majority. He made an 'appeal to take most immediate measures so as to put an end to bloodshed and repression against the population of East Pakistan and take measures for peaceful political settlement'. For the present, the Soviets' attitude was not unhelpful. Later, their position changed; for example, they began to refer, like India, to 'East Bengal'. At the time, the public generally acclaimed what they considered a strong reply by Yahya Khan.

It is important to note that at this stage the President was not acting in consultation with ZAB. In fact, having secured his full support for the military action and his condemnation of the Awami League and India, the President did not meet him again till the end of the month. Between 2 and 27 April, the Government did not consult him on international issues nor on East Pakistan. He was disturbed at this and, to assert his importance, told a gathering of Party workers at Multan that the Government alone could not resolve the crisis; the PPP had a 'historic role' to maintain the solidarity and integrity of the country. As always, he sought to underline his knowledge of international issues and his relevance in the field of foreign affairs. He referred to his speech as Foreign Minister in July 1964, and the Chinese ultimatum in the September 1965 War, suggesting the role he could play in persuading China to prevent India from starting an expansionist war in the region.

China, in fact, sent a stern note on 7 April protesting against India 'flagrantly interfering in the internal affairs of Pakistan'.[12] This was occasioned by Indians demonstrating outside the Chinese Embassy in Delhi. A few days later, in a message to Yahya Khan, Premier Zhou Enlai stressed that the 'unification of Pakistan and the unity of peoples' are basic guarantees to attain prosperity and strength; 'should the Indian expansionists dare to launch aggression against Pakistan, the Chinese Government and people will, as always, firmly support the Pakistan Government and people in their just struggle to safeguard State sovereignty and national independence'.[13] But that was as far China was prepared to go.

At this time, a small but significant news item reported the arrival of the first ever US sports team in Beijing. Premier Zhou described the visit as 'a new page' in Sino-US relations.[14] President Nixon reciprocated by relaxing trade and travel restrictions with China, imposed since the Korean War. The new pages of 'ping-pong diplomacy' turned fast and, as it transpired, to Pakistan's detriment.

Meanwhile, in East Pakistan, despite ZAB's warnings, Government activity remained exclusively military. This was plain from the principal features of Lt.-Gen. Tikka Khan's orders to his commanders:

- to disarm the battalions of the East Bengal Regiment, the East Pakistan Rifles and the Police;
- to secure the military cantonments;
- to secure the naval base at Chittagong;
- to control airfields including Lalmunirhat and Ishurdi; and
- to secure all major towns.

It was like the conquest of an alien territory and people.

The Awami League response, with the backing of India, was equally military. On 11 April, Tajuddin Ahmad, now Prime Minister of the Provisional Government of Bangladesh, stated in Calcutta that 'a mighty army is being formed around the nucleus of professional soldiers from the East Bengal Regiment and East Pakistan Rifles'. He announced the names of commanders for five designated military sectors, including, for Chittagong-Noakhali, Major Zia-ur-Rahman[15], who shot to prominence as the first to announce over Chittagong Radio an independent Bangladesh. His forces in and around the port of Chittagong put up the fiercest resistance to the Pakistani military. The Mukti Bahini, the term used to describe the Bengali forces, were thus formally established and, on 14 April, Col. N.A.G. Osmany became their Commander-in-Chief.

By now, ZAB felt increasingly marginalized. Cut off from the President, he sent a memorandum on the current situation to Yahya Khan on 14 April through Khar, who discussed it with Peerzada. This document, and other letters[16] addressed by him subsequently to Yahya Khan and Peerzada, are quoted in the following pages because they reveal his thinking at the time. However, it must be pointed out that in preparing these papers

ZAB asked me to bear in mind that they would become 'historical documents'.

The memorandum of 14 April warned the Government that the people of East Pakistan must understand

> to what degree they have been deceived by the Awami League and how the conspiracy would have deprived East Pakistani Muslims of their freedom, thus bringing them again under the domination, political, economic and cultural, of the Hindus, if not directly under that of India.

It also described economic and relief measures as 'imperative to save the situation', and proposed specific steps. On the subject of propaganda, both at home and abroad, it pressed that:

> The justification of the Government's action must be immediately and properly projected. Full facts about the conspiracy with India should be revealed along with evidence of misdeeds of the Awami League. A critical analysis of Six Points must be published.

On the political front, the document urged that:

> The present vacuum should not be allowed to continue...The people have to be taken into confidence, both by the Government and the political party which will support the Government's action taken in East Pakistan...In this situation, the Pakistan People's Party has both a moral and political obligation.

The same day, ZAB addressed a press conference, mainly on international matters. He slated the Soviet Union for its role, protesting 'on behalf of the People of Pakistan'. He praised Yahya Khan's reply to Podgorny, and condemned Indian interference. Once again, he attacked political opponents for their 'London Plan', calling for a ban on all parties opposed to Pakistan.[17]

By this time, East Pakistan had gained major international attention and news coverage. On 16 April, the American Ambassador to India, Kenneth Keating, was reported as saying the situation was of concern to the world, and should not be treated as an 'internal affair' of Pakistan. Maulana Kausar Niazi was directed by ZAB to demand publicly that the Government should take cognizance of the US envoy's statement, and to praise Yahya Khan: 'He is no longer a soldier President, but a popular President'.[18] The Ambassador denied making the statement. ZAB was clearly still keeping Yahya Khan on his side. The military remained all-important.

Concerns in the West Wing

An air of unreality hung over the West Wing which was reflected in the priorities established at the time. The magnitude of the crisis in the East Wing, and the ramifications of India's involvement, were not understood. These issues were relegated to a subordinate position and the main debate centred on the transfer of power and a new constitution. The Government also failed to concentrate on the most important problems. Instead of unifying support in the West Wing in a search for solutions, it attempted to undermine the position of the largest party, the PPP. In its efforts to weaken the Party, it gave prominence in the controlled Press to reports of major internal differences and a so-called 'Forward Bloc' within the PPP. Such rumours were picked up by political commentators,[19] and the PPP was compelled to waste time in refuting allegations which had little basis in fact.

On 1 April, ZAB denied reports of a body being established to probe into the question of discipline in the Party, particularly in the Punjab. There had indeed been some differences between the Punjab President, Sheikh Rashid, and Khar, the General Secretary, motivated by a combination of ambition for the Province's leadership and objections to the control exercised by feudals such as Khar. The leadership question was settled on 25 April by ZAB, supported the next day by six senior Party members from the Punjab. He then dealt directly with those he termed 'trouble-makers'. At Kasur, he castigated the local MNA,

Ahmed Raza Kasuri, which led to a physical attack by PPP
workers on the MNA, who described ZAB as a 'fascist'.[20]
Handled firmly, dissensions within the Party died down.

The predominant issue of this period was the transfer of
power. Mustafa Khar was asked by ZAB to raise the question at
a press conference held on 17 April. Khar proposed that the
transfer should take place early in the Provinces of the West
Wing, to be followed, when conditions permitted, in East
Pakistan under an interim Constitution to be promulgated by the
President. ZAB did not raise the issue himself as it was contrary
to his earlier assertion that elections to the Provincial Assemblies
should not precede the framing of a constitution, obviating the
necessity for any elected Provincial Governments. Khar's call
was severely criticized as not feasible politically or
constitutionally. This question gave rise to serious debate in the
West Wing. Qayyum Khan maintained that democracy had to
be established in both Wings simultaneously. Two days later,
on 25 April, the Council Muslim League also opposed any
piecemeal transfer of power. The fact that ZAB's opponents did
not want him in power certainly contributed to these objections.

The debate resulted in Yahya Khan meeting him in
Rawalpindi on 27 April. ZAB declined to comment or disclose
details to newsmen; the President had told him to exercise
restraint. Two days later, he urged patience: 'We are going
through a difficult period; we hope this period will be behind us
as we march forward to strengthen national integrity to bring
progress and prosperity for the people.'[21] He was conciliatory,
not challenging the establishment. He did not comment adversely
on measures taken in East Pakistan; in fact, Mahmud Ali Qasuri
was to be sent there and ZAB would go 'at the right time'. He
was adamant only on the point that the PPP would not contest
successive elections, since new elections might affect his recent
victory.

Although not in a position to challenge Yahya Khan, ZAB
could not accept the criticism of the other parties; he condemned
their lack of patriotism in dealing with the Awami League. In
contrast, as he stressed before a gathering of Party workers at

Sheikhupura on 30 April, it was the PPP that tackled the 'nefarious designs' of the outlawed Awami League; Pakistan would have been in pieces if the PPP had not emerged as the majority party in the West Wing. Soon, he suspected that the President and some of these parties were conspiring to keep him out of power. As a result, he changed his tactics. He gave up his call for patience and, instead, claimed at Lahore on 4 May that an elected government should make the June Budget: 'Only an elected and representative government could succeed under the circumstances.'[22] He referred to the deepening economic crisis and the ill-advised unilateral moratorium on external debt declared on 30 April. Defending the proposal to transfer power in the West Wing, he described it as a 'magnetic incentive' for East Pakistan to create conditions to make democratic rule possible.

Five days later at Peshawar he put forward a practical scheme for the transfer of power, pointing out that constitutions 'are made by the people but interim arrangements can be made by individuals'. So far as the elected members of the outlawed Awami League were concerned, 'If they can be sifted—and in my opinion they can be—those who are for Pakistan should retain their membership of the assemblies; for the remaining by-elections can be held'.[23]

His opponents remained deeply hostile to an early transfer of power, particularly in the West Wing. Nurul Amin, after meeting the President, said it was not possible in East Pakistan; 'Whom will you transfer power to?'[24] Mufti Mahmood also said it should not be done 'yet', as it would create further disillusion in East Pakistan.

We discussed this issue as the main item at a meeting of the Party Central Committee, and also considered Mahmud Ali Qasuri's report on his visit to East Pakistan, which said the situation was serious but not hopeless. It was decided to deflect the growing criticism of the PPP, so ZAB explained to the Presi that 'transfer of power' incorrectly described his Party's demand which stood for democratization. He warned that if a 'bold plan of action' was not put into operation, especially on the economic front in East Pakistan, the situation could get out of hand; 'the people's problems need the participation of the people's accredited representatives', and could not be solved by 'bureaucrats'.[25] He concentrated on the economy because he could not attack the army action in East Pakistan.

Meanwhile, relations with India continued to deteriorate. On 20 April, the Inquiry Commission established by the Government to investigate the hijacking incident described it as a 'culmination of a series of actions taken by the Indian Government to bring about a situation of confrontation between Pakistan and India'.[26] The two 'commandos', Hashim and Ashraf, were held to be Indian agents. The motive behind the incident, according to the Commission, was to disrupt communications within Pakistan and to dislocate the movement of people and supplies between the two Wings with a view to strengthening separatist tendencies.[27] However, when the Pakistan Mission at Calcutta was forcibly occupied, the Indian High Commissioner assured the Foreign Office in Islamabad on 22 April that it would be vacated. More important, he said that India did not recognize Bangladesh, although the Awami League leaders in exile had five days previously issued a declaration of independence and formed the People's Republic of Bangladesh. Despite this, the following day the Government decided to close the Calcutta Mission. This diplomatic escalation proved a mistake. Four days later, India banned the movement of all Pakistani embassy staff, and refused to respond to protests against this restriction. Pakistan took reciprocal measures on 29 April. In India, government-backed committees began raising funds for Bangladesh, and the former Pakistan Mission in Calcutta was allowed to be used as a Bangladesh Mission.

In the course of May, the army gained the upper hand in the East Wing and, for the first time since the military action and Yahya Khan's speech of 26 March, the Government on 6 May felt sufficiently confident to comment publicly about events in East Pakistan. A spokesman announced that the early hours of 26 March had been set for an armed uprising by the Awami League and for the formal launching of the independent Republic of Bangladesh.[28] The news was poorly presented and received with general scepticism. If this were so, why did the President not mention it in his speech of 26 March? The second week of May saw further proof of the army's success when the entire south-eastern region was cleared of rebel forces and port facilities restored at Chittagong, Chalna and Khulna. On 15 May, Lt.-Gen. Tikka Khan announced that 'organized armed resistance' had been 'crushed'.[29] The next day, reports about the massacre of university professors were also contradicted personally by some of those rumoured dead.

The Government's silence about the situation in East Pakistan between 26 March and 6 May proved costly. India's propaganda machine was very active. Abroad, horror stories about the civil war were widely reported. They were given greater credence by Anthony Mascarenhas' report which appeared in *The Sunday Times* of London on 13 June. Ironically, he had been taken from Karachi with other West Pakistani journalists to see how East Pakistan was returning to normal. He proceeded to London and published a damning exposé, which focused world attention on East Pakistan. This eyewitness account by a West Pakistani led to heightened international condemnation and a hardening of attitudes. Whereas, on 20 April, Prime Minister Edward Heath had told the House of Commons that the British Government sought an end to the strife and a political solution to the problem, soon after the publication of the Mascarenhas story, the British Foreign Secretary announced that approval of new aid would be withheld till Pakistan took concrete steps to solve the problem. The World Bank Consortium at its meeting of 21 June in Paris also decided to postpone aid to Pakistan.

The military seemed unconcerned about international repercussions. They were satisfied that the Mukti Bahini had been pushed out of most parts of East Pakistan except a few small pockets in difficult areas north of Rajshahi. However, the Mukti Bahini had not been fully defeated, having withdrawn in part to India to regroup and receive training and weapons. In India, anti-Pakistan activity increased, and even leading pacifists such as Jay Prakash Narayan spoke of the necessity of war:

'The country, the government, and the people are unworthy if they are not prepared for a war...Defeat of Bangladesh would be the defeat of India.'[30]

The ascendancy of the Pakistan Army in the East Wing had caused an escalation in Indian efforts. Instead of consolidating support in the West Wing, the Government's attitude towards the PPP hardened. ZAB repeatedly complained to the Government[31] about the hostility of the controlled media, particularly *The Pakistan Times,* which denounced as 'treason' the call for an early transfer of power.

From May till the dramatic events at the end of the year, there was little change in the political situation in the West Wing. Activity centred on formulating a constitution and the related question of the transfer of power to the elected representatives, which led to the announcement by the President of a new plan on 28 June. A day before the broadcast, ZAB had written to Lt.-Gen. Peerzada:

> ...The Indians seem to have succeeded in winning over world opinion to their point of view. This is not so much their achievement as our failure.
>
> ...India may launch a series of mischievous actions falling short of classical armed aggression although that possibility cannot be ruled out either. Never before has a country become so dependent for its sovereignty and survival as Pakistan is on China.
>
> ...Immediately on my return from Dhaka I...urged you to remain firm on an early transfer of power...Among the factors that influenced me to take this position was the apprehension of foreign interference...the Government must prepare immediately to hand over effective power to the elected representatives of the people...
>
> In view of the extraordinary conditions prevailing in the country, Martial Law might have to remain in the background for some time...We are prepared favourably to consider a realistic arrangement...We are prepared to offer sincere cooperation in the national interest on a reciprocal basis...
>
> ...Once we have heard the broadcast, we shall avail ourselves of the opportunity to contact the Government...[32]

The broadcast announced that, since the elected representatives had failed to provide a constitution, the President had set up a committee to draft one, after which he would put it into effect. Although this amounted to imposing a constitution, the President's departure from his earlier Legal Framework Order had been encouraged and then endorsed by opponents of the PPP. Fearing isolation and confrontation with the military Government, ZAB restrained Party members from commenting till full consideration had been given to the President's plan by the PPP Central Committee.

These were difficult times for the PPP. East Pakistan was in turmoil, despite the apparent calm which led the Yahya regime and others in the West Wing to believe the situation was under control. The regime was not prepared to transfer power in the West Wing, and the PPP lacked the necessary street power to wrest it. Worse, any attempt to stir up trouble would undoubtedly have resulted in the accusation that the PPP's hunger for power jeopardized the army's position in East Pakistan.

ZAB felt the need for support to convince Yahya Khan about the importance of cooperation in the West Wing to overcome the crisis. He knew that the Shah of Iran wielded influence in Pakistan, particularly on Yahya Khan; and the Shah would soon be seeing the President at the Iranian monarchy's 2500th anniversary celebrations, and at the Islamic Summit Conference at Rabat. ZAB went on a private visit to Tehran as a guest of his friend Ardeshir Zahedi, Iran's Foreign Minister, together with J. A. Rahim, Khar and Mumtaz Bhutto.[33] He had a lengthy meeting with the Shah from which Rahim was excluded. Rahim deeply resented this and later maintained that it was a part of a conspiracy to dismember Pakistan, as the Shah shared ZAB's view, and had always felt, that a geographically divided country was not viable.[34]

While in Tehran, ZAB told *Kayhan International* during an interview that the military could not solve Pakistan's problems. He wanted a 'speedy return to civilian rule', and the elected representatives to draft the constitution. There should be 'no attempt at installing a civilian puppet government; I shall destroy such a government through mass movement in less than a month'.[35]

Before leaving for Tehran, ZAB had asked me to work on a formula that would be acceptable to Yahya Khan and avoid loss of face by either side. I suggested that the President could announce a *draft* constitution for the National Assembly to consider. Amendments would have to be proposed within a specified time; if the President did not accept them, they would have to be passed again by the Assembly with a greater majority. Thus, Yahya Khan could still present his constitution but the Assembly would have the final say.

On ZAB's return, we considered these matters in a meeting of the PPP Central Committee on 15 July at Rawalpindi. He also discussed with some of us the recent visit of Henry Kissinger to Islamabad and his 'disappearance' to Nathiagali, a hill station forty-five miles north of Islamabad. We were unaware of Kissinger's secret visit to Beijing, and rumours were rife that he was meeting Kamal Hussain in Nathiagali in order to reach a settlement on East Pakistan and to revive the Awami League. ZAB warned about this in a lengthy letter to the President written that night: 'If there is any truth in such news... we should then really be discussing the future structure of the country itself and not the government'. He went on to state:

The elected representatives have a mandate to frame the Constitution within the concept of One Pakistan...Here, after nearly 10 years, we arrive at a situation where not even the names of the so-called group of experts are announced. And then you have said that you will consult various leaders regarding the provisions of the Constitution. Who are these leaders to be—defeated and discredited politicians who have been rejected by the people at the polls?...

In my letter of 27 June I dealt with the Government's efforts to combine the three Muslim Leagues. You have now publicly announced the desirability of their combining and the future role they may be asked to play...

Finally, I need hardly add that if I refused, in the national interest, to put my thumb-impression on Sheikh Mujibur Rahman's Constitution—and that would have been a Constitution drawn up by an elected Assembly—how can the Pakistan People's Party be expected to rubber-stamp a dictated Constitution?...If we are to be cheated of our mandate, we must seriously consider whether we should resign our seats; and then you may put forward yet another plan of action.[36]

This threat to frustrate Yahya Khan's plan was as impractical as the plan itself to impose a constitution. Both ZAB and the President knew it would not be easy to persuade MNAs to resign hard-won PPP seats in the Assembly. Both fully realized that the East Wing crisis did not permit further confrontation. Yet their meetings over the next two days started ominously with the President referring to the 15 July letter as an 'ultimatum'. It was only after ZAB's assurance to the contrary, and his conceding on some basic issues, that the President agreed to hold further detailed talks in Karachi. Following this meeting with the President, he informed newsmen that there were features of the plan with which the PPP agreed, but that we had our own point of view on some issues. He emphasized that, 'It will be in the national interest if we continue to hold these discussions a little more frequently...In the spirit of under- standing and cooperation I do not see why points of view cannot be reconciled in the national interest'.[37]

After the PPP Central Committee meeting, Mubashir Hasan and I stayed on briefly in Nathiagali. On long walks we discussed the situation, concluding that East Pakistan was lost for all practical purposes, and considered the problems that lay ahead for any PPP government. As events unfolded, much of our combined thinking proved accurate. We have often looked back on those talks and referred to situations when we were proved right in one word, 'Nathiagali'.

The President discussed his plan with ZAB in Karachi on 29 July together with their advisers.[38] ZAB was conciliatory, pointing out that the PPP did not seek confrontation with the Government because, in the present circumstances, this would prove disastrous for the country. He explained that previous constitutions had failed for lack of public support and, for this reason, it was necessary to reconcile the President's announcement with the aspirations of the people represented by the PPP. He dwelt on the procedure we had evolved for the constitution. Secretary-General Rahim talked on the economic causes of the crisis; he attacked the capitalists for supporting the secessionist movement, saying their industries should be confiscated and placed under public control in the East Wing to encourage the people.

The President listened patiently without interruption. In reply, he was accommodating towards ZAB but dismissed Rahim's suggestions. He accepted that nothing would be gained by confrontation, and maintained that he did not intend to impose a constitution. The PPP formula had merit, he agreed in principle, provided details could be satisfactorily settled. It was decided that the two teams of advisers should meet the following month for this purpose. There was, remarkably, no sense of urgency; the Government clearly felt that East Pakistan was under control.

The optimism generated at Karachi was quickly dispelled when the advisers met in Islamabad on 25 August. The first encounter lasted less than an hour. We were taken aback by A. R. Cornelius, a retired Chief Justice of Pakistan, who behaved as if we were pleading a hopeless case rather than negotiating details of a settlement already agreed in principle. Indeed, when the PPP team[39] tried to elaborate on the formula for constitution-making, Cornelius denied there had been any earlier agreement in principle, maintaining that the President would 'give' a constitution after consulting all the political parties. We decided to withdraw and await the outcome of the separate Yahya-ZAB talks.

ZAB informed Yahya Khan that the two teams had reached an impasse since the Government had gone back on the earlier understanding. They agreed to settle the constitution-making formula between them, while the advisers considered other details. They also discussed the situation in East Pakistan, where Yahya Khan wanted to install a civilian cabinet with A. M. Malik, a senior Bengali educationist and politician, as Governor. Considering that the crisis in East Pakistan was the reason earlier advanced for delaying the transfer of power, it was strange to propose a civilian government in that war-torn Province while there was no mention of similar governments in the West Wing.

Next morning the advisers met again, and M. M. Ahmed joined the Government team as a moderating presence. I had prepared several briefs setting out the PPP position on certain constitutional issues which the President had requested in the earlier meeting of 29 July. These dealt with the question of joint electorates, the necessity of having a bicameral federal legislature, the amendment procedure to be incorporated in the constitution, and our views on fundamental rights and directive principles.[40] While M. M. Ahmed was ready to discuss these matters, Cornelius made it clear that the Government would only record the PPP's views. He gave us a questionnaire relating to almost all aspects of the constitution, and we agreed to meet the next day after considering it. However, since no meeting had been fixed by Yahya Khan with ZAB to consider the formula for the Assembly, we felt there was nothing to be gained by restating the PPP's views. We merely stressed the need for the Assembly to have a constitution-making role, and declined to discuss the questionnaire. The talks finished within an hour, much to the consternation of the Government team.

The next weeks saw a rapid deterioration in relations between Yahya Khan and ZAB. The Government issued a *White Paper on the Crisis in East Pakistan* on 5 August 1971, attempting to explain the position. Within a month, in an interview with *Al-Fatah,* a pro-PPP Karachi weekly,[41] ZAB called for the lifting of press censorship and a return to political activity to pave the way for a democratic government. He emphasized the need for accurate information rather than rumours about recent events in East Pakistan in order to create public confidence. He criticized bureaucratic control of foreign policy, maintaining that the recent Indo-Soviet Treaty was a 'war pact' against Pakistan and China. However, he claimed that any attempt by India to undo Pakistan would be thwarted.

Indeed, while we in the West Wing were wrangling over issues concerning a constitution and the transfer of power, a major new international dimension had been introduced in the subcontinent by the Indo-Soviet Treaty of Friendship of 9 August 1971. India had previously avoided a formal alignment with the Soviet Union; now, in her view, Kissinger's secret mission to Beijing, via Islamabad, had undermined India's position in the region. Indian aims in East Pakistan could no longer be achieved without positive Soviet support. The Kissinger mission also annoyed the Soviet Union. However, after pulling off the 'coup' of Kissinger's visit, the Government felt secure in the belief that the friendship and gratitude of the United States and China would see Pakistan through the domestic crisis. This proved a naive assessment.

The Government's complacency also extended to the domestic scene. Although the PPP had asserted the sovereignty of the Assembly as early as 15 July, with which the Government agreed in principle on 29 July, the regime failed to take the matter forward and initiate a transfer of power. More serious was the absence of any political progress in East Pakistan. ZAB was scheduled to meet the President on 9 September. A week earlier he wrote to Lt.-Gen. Peerzada, sending the letter through Mustafa Khar, as ZAB also wanted him 'to discuss certain important developments' and to convey the 'need for urgency in arriving at a decision regarding the character of the Assembly in relation to the Constitution'.[42] The letter pointed out that the appointment of a civilian Governor in East Pakistan was 'not the same thing as the introduction of the democratic process', and warned that 'if force, and not reform' was for a 'protracted period the main instrument of policy then East Pakistan would be pushed beyond the pale and there would also be serious repercussions in the West Wing'.

In the absence of any response, he postponed his meeting with the President on a plea of illness. A further letter was sent through Khar on 8 September 'to facilitate our future discussion'. This letter recapitulated the past two years from ZAB's standpoint, emphasizing the Government's errors and hostility towards the PPP during this period.[43]

It was apparent that differences with the Government were coming to a head. ZAB decided to state his position publicly. An appropriate occasion presented itself on the Quaid-i-Azam's death anniversary on 11 September. At the Quaid's *mazar* (mausoleum), only two days after being too ill to travel, he vehemently condemned the regime for failing to transfer power to the elected representatives. The country was 'virtually on the brink of collapse', and if the military were unwilling to leave office they should clearly tell the people and end this game of 'hide and seek'.[44]

For a brief period, relations between the President and ZAB improved. Yahya Khan was about to go to Tehran for the 2500th anniversary celebrations, which were to be attended by many world dignitaries. It was important to him to show the world that he faced no problems in the West Wing. He therefore met ZAB on 13 September for a friendly exchange of views, though one without positive outcome. On returning from Tehran, the President held further talks with ZAB on 18 September. He pointed out that other political parties and leaders had agreed to his 28 June plan; but, because of the PPP position, he had now evolved a mechanism which would reconcile the 'exigencies of the crisis with the expression of the people's will'. He was prepared to make no further concession. That evening, in a national broadcast, the President announced he would publish a provisional constitution which would be given final shape after consideration by the Assembly.

Soon after returning from Dhaka at the end of March, ZAB had dictated his version of events leading to the military crackdown. When he decided to publish it, he asked me to put the text into shape after considering 'the implications and ramifications for the future'. It entailed much redrafting and numerous discussions, in the course of which I got to know him closely and gained some insight into his motivation. *The Great Tragedy* was finally published on 29 September. A day earlier he autographed the first copy to me: 'Thank you for all the trouble you took to lessen the tragedy in more ways than one.

For this and other reasons, I have much pleasure in presenting the first copy of our effort to you.'

Simultaneously with the release of the book he issued a thirteen-page statement at a press conference. The statement set out the PPP's earlier negotiations with the Awami League, and touched on the PPP-Government meetings on constitutional matters between July and September. It described the President's 18 September announcement as 'a step forward' in the right direction. It also pointed out that certain prior requirements still had to be fulfilled, for example, the LFO needed amendment to provide a form of consensus among the Provinces without derogating completely from the principle of simple majority. ZAB spoke out for reforms, and the early formation of constitutional governments at the Centre and in the Provinces. He concluded emotionally and extempore: 'The long night of terror must end. The rule of the generals must end. The people of Pakistan must take their destiny in their own hands.' This part was carried by the *New York Times* [45] but not in the national Press.

The Government reacted to ZAB's 29 September statement by banning the PPP daily, *Musawaat,* and arresting prominent Party workers. ZAB complained in a letter to Peerzada on 4 October: 'The prejudice of your regime which appeared blatantly before the elections has come to the fore again. It seems that disaster is in the stars of Pakistan.'[46]

Earlier, on 19 September, further to the President's broadcast, the Government announced the schedule for by-elections in East Pakistan. However, the game of hide-and-seek and pro-crastination was soon to restart. The 'numbers game' had already been put into play. After banning the Awami League, the Government had been left with the problem of what to do with the party's 167 members in the National Assembly. Finally 79 were disqualified, while 88 were allowed to retain seats in their individual capacity. To disqualify all would have amounted to an admission that every elected Awami Leaguer was against Pakistan. The figure of 88 was the same as the PPP's membership at the time; there was no doubt that through by-elections the military regime would ensure control of any future central government.

The Government proceeded without reference to the PPP. A civilian cabinet was inducted in East Pakistan with the representatives of the defeated parties, while the PPP was not even consulted. Martial Law arrangements continued in the West Wing. ZAB condemned as 'dishonest' such 'double standards'. He emphasized that having an elected government at the Centre would not be to the detriment of the army: 'In this change, our Armed Forces would be further strengthened.'[47] He praised the army for saving Pakistan, and pointed out that if he were against the army he would have opposed the military action in East Pakistan.

Along with the Government, such parties as the Jamaat-i-Islami stepped up their activities against the PPP. Mian Tufail, its acting *Amir* (Chief), urged party workers to spread out in the villages to counter the PPP; he also expressed satisfaction that Yahya Khan's new constitution would be Islamic. ZAB retorted, 'It was heartening that Mian Tufail had found the constitution written by a non-Muslim as duly Islamic, even before seeing it.'[48]

Following his 29 September outburst, he had become despondent about the transfer of power. When Mubashir Hasan and I discussed the situation with him at 70 Clifton, he abruptly pointed out that it was all right for Mubashir, who could return to his engineering consultancy, and to me, he added, 'You have your legal consultancy, but what can I do?' I replied he had already made history in Pakistan by giving the people a voice. Mubashir agreed, but ZAB did not. It was pointless without power. Surprisingly, till the very end, he never quite understood the change he had introduced in the popular mind.

The Situation Deteriorates

Throughout September and much of October 1971, it was not evident to many either within or outside the Government that we were experiencing the lull before the storm. The recent failure of the Mukti Bahini's monsoon offensive encouraged the Government's belief in its ability to control East Pakistan, obviating the need to effect an early transfer of power. ZAB was kept at a distance and, with nothing to do in the West Wing, he left for Cairo on holiday in October.

He was called back from Egypt to lead a military mission to Beijing in November. At last, the Government was emerging from its complacency and realizing the difficulties of the situation. India had by now increased pressure on the international front. With the Soviet Union firmly on her side, Prime Minister Indira Gandhi visited European capitals and the United States, and by the end of October the world seemed convinced not only of the rights of East Pakistanis but also, to some extent, the right of India to support them, particularly in relation to the several million refugees from East Pakistan. The United States was still maintaining its arms embargo and so it was to China that Pakistan had once again to turn.

Although holding no official position, ZAB led a high-level military mission, which included the Air Force Commander-in-Chief, Rahim Khan, and the Army Chief of the General Staff, Gul Hassan Khan. The mission was advised by the Chinese leadership to prevent the situation from developing into war with India. China was not in a position to confront the Soviet Union or even India following the Indo-Soviet Treaty of Friendship. Secretary-General Rahim and some others in the PPP disapproved of ZAB leading such a mission at a time when there were no signs of a civilian government being introduced. However, ZAB felt it was an opportunity 'to get in' with the Armed Forces,[49] and, in any case, it was incumbent on him to do whatever he could for the country.

On his return, he was asked to mobilize public opinion in the West Wing as part of the Government's efforts against the Indian threat. Here too, the regime's ambivalent attitude towards him became apparent; while wanting to use him, it hesitated in extending him support and publicity. Mubashir Hasan, the PPP

Lahore President, pointed this out in strong terms in a letter of 13 November to the Information Secretary.[50]

Despite adverse developments on the international front, the President remained preoccupied with his constitution. On 13 October, he announced that it would be published on 20 December, by-elections to vacant seats would be completed in East Pakistan by 23 December, and the Assembly would meet four days later. The Government's world was truly unreal. To further its aims, the regime encouraged an amorphous group of seven rightist parties known as the United Coalition Party, which the controlled Press heralded as the future civilian government. It was to be buttressed through 'by-elections' in East Pakistan. These were indeed a travesty of democracy. There were no elections; 'results' were merely announced by the military authorities. The Yahya regime was determined to secure an over-all majority for opponents of the PPP by declaring the results of 58 of the 79 seats without contest, of which 52 were awarded to rightist parties and 6 to the PPP. Thus the PPP's strength in the Assembly now increased to 94 while the rightist parties went from 37 to 89. With 21 by-elections outstanding, and the presence of many Awami Leaguers who had retained their seats, the PPP could not acquire power.

The PPP leadership had held long and heated discussions on whether to boycott these 'by-elections' or to claim a share of allotted seats in East Pakistan along with other parties. Finally, it was decided to participate in what we subsequently described as a 'legal and political fiction' of by-elections to the National Assembly as 'a necessary evil' and a 'bitter pill'[51] on the grounds that there could be no transfer of power without a completed Assembly. This was the highest form of sophistry. On the other hand, there was no similar need to contest the Provincial Assembly seats, yet the PPP team visiting East Pakistan under Mahmud Ali Qasuri were persuaded by the Government to nominate candidates for by-elections to the East Pakistan Assembly. Although this decision was severely criticized by several of us, including Sheikh Rashid, Haneef Ramay and

Mairaj Mohammad Khan, ZAB did not reverse it as he felt this would annoy the military.

With the regime's efforts to install a government of the United Coalition Party gaining ground, and power eluding ZAB, he attacked the Government for what he termed this 'fraud' on the people of Pakistan. Addressing Party workers at Karachi on 7 November, he also criticized the 'anti-people renegades and reactionaries' comprising the Coalition, especially Daultana and Qayyum Khan, for calling him a fascist: 'Actually those two are fascists who have slaughtered democracy and ridden roughshod over the people's rights'. He held Daultana, along with Governor-General Ghulam Mohammad, responsible for scuttling the first Constituent Assembly just as Hitler had burnt the Reichstag'.[52]

The politics of posture and manoeuvre dominated the domestic scene, distracting attention from the dangers of the situation in East Pakistan. In the meantime, Indira Gandhi had prepared her ground internationally and, on 22 November, Indian infantry supported by tanks advanced on several fronts in East Pakistan. To our surprise, we learnt that this did not affect the programme for by-elections scheduled for 28 November in Chatta and Gujrat in the West Wing. Indeed, the Government remained remarkably unhurried; Yahya Khan declared a National Emergency on 24 November but made no significant move in the United Nations except for urging the Secretary-General to use his 'personal influence' in the matter. U Thant replied, understandably in the circumstances, that he had done what he could within the powers granted under the UN Charter. By failing to take more positive steps to rouse international opinion against India's action, the Government conceded that the country was already divided; the attack on the East Wing was not regarded as war or aggression against Pakistan as a whole.

ZAB flew to Lahore on 23 November, where he declared at the airport, 'The hour has struck; we must be given charge otherwise we will not be responsible for anything that might happen'. The situation was very grave indeed: 'look to what a pass we have come as a result' of a government which was run 'by three people'. The next day he told a large by-election meeting in Gujrat that only a truly representative government could pull the nation out of its present difficulties. He also elaborated on the Indian strategy of opening four fronts in East Pakistan, which aimed to weaken Pakistani forces by dividing them thinly. The Indians would capture some area in East Pakistan and, after negotiations, 'trick Pakistan' into signing another Tashkent-type Declaration.[53] To rapturous applause, he asserted that the people of Pakistan would not accept this under any circumstances.

His assessment that India would seize a part of East Pakistan install a government of Bangladesh and then recognize it as an independent state was, in essence, what the President also believed. Both were proved wrong. No serious thought was given by GHQ to the establishment of a 'Fortress Dhaka'; the Government could not comprehend a situation which, in three weeks, would result in the surrender of the whole of East Pakistan.[54] Most people, including ZAB, envisaged the same scenario as in the earlier two wars with India, which ended inconclusively with UN-imposed cease-fires. They overlooked the fact that in 1965 no one, least of all a superpower, contemplated a change in the map of South Asia. Since then India's strength had grown considerably and she now enjoyed full Soviet support, while the position in East Pakistan had also changed radically. By December 1971, the situation in South Asia was very different, but few in West Pakistan realized the implications of this.

ZAB was to meet the President on 26 November and, on arriving at Islamabad airport, told waiting newsmen that the right course was not to go on 'bended knees' to seek the intervention of the Big Powers because 'if you show helplessness' they 'will jump upon you'.[55] A solution had to be found in the subcontinent itself. This realistic gesture towards India, like everything else in 1971, came too late in the day.

With virtually full-scale war in the East Wing, it should have seemed very strange for the President and his advisers to settle

down with the PPP team to discuss the salient features of the
proposed constitution; but such was the unreal atmosphere
prevailing at the time that we felt we were actually fulfilling a
need. The President was assisted by M. M. Ahmed, Lt.-Gen.
Peerzada and Syed Sharifuddin Pirzada, while the PPP team
consisted of Chairman Bhutto, Mian Mahmud Ali Qasuri,
Hafeez Pirzada, Khurshid Hasan Mir (replacing J. A. Rahim,
who was ill) and myself. Although Pakistan was being
dismembered, the situation in the East Wing was not discussed.
The President merely said it was under control: 'My boys can
handle it.' For two hours we talked instead about the
constitutional scheme proposed by the President. He was to
remain President, Supreme Commander, C.-in-C. of the Army,
and also retain some Martial Law powers. Apart from such
extensive authority, an interesting new feature gave the
President 'special responsibilities for the preservation of the
integrity and ideology of Pakistan and for the protection of
fundamental rights'. The Executive was to consist of a President
and Vice President, either one or the other from each Wing,
and similarly a Prime Minister and Vice Premier. There was to
be a bicameral legislature.

On provincial autonomy, the proposal was to restrict the
Centre's field and allow the Provinces residuary powers. There
were to be new arrangements for dealing with such subjects as
foreign exchange, trade and aid, as also with income tax,
customs and excise duty, which M. M. Ahmed explained. The
proposals[56] would have been salutary and relevant several years
earlier. The Government talked as if the situation in East
Pakistan had remained the same. ZAB was called by the
President two days later for a further brief meeting on
constitutional matters.

The PPP was acting as if in a void, with no relevance or
effective say. In these circumstances, ZAB decided to set out
the PPP position on the proposed constitution in a letter to
Yahya Khan.[57] The next few days witnessed the same
inconsequential activities, with a PPP Central Committee
meeting in Peshawar on 1 December to pass a detailed

resolution, a further eight-page letter on 2 December from ZAB to Lt.-Gen. Peerzada, which remained undelivered,[58] and finally a long letter of 3 December from ZAB to Yahya Khan and a public meeting at Liaquat Bagh, Rawalpindi, the same day. A major concern, apart from preparing these documents, was to secure an accurate and confidential typist, and we had to fly one in from Karachi. If only all problems were so readily remedied.

ZAB's letters are relevant as they reflect his position at the time, even though their preparation kept in mind the perspective of 'history'. Neither the President nor Peerzada replied. The letter of 28 November dealt with constitutional issues. In fact the PPP was, for the second time in 1971, discussing a proposed constitution without seeing its text. He pointed out that a 'Presidential system in the name of a Parliamentary system' was being introduced by Yahya Khan. ZAB criticized the special powers which would make the President 'a virtual dictator': 'With so much power to be vested in the President, who could have so wrongly advised you to top it all by conferring upon the President the disproportionate and provocative special responsibilities for the preservation of fundamental rights.' While Yahya Khan had initially wanted the C.-in-C. of the Pakistan Army to be the President for 'quite some time', ZAB confirmed their agreement that this should be automatic for only the first fixed term. He had two reasons for placing emphasis on provincial autonomy in his letter. First, it would obviate criticism that 'the main aim of the military action in March was not to stop secession but to prevent genuine autonomy'. The second and real reason was that, at the time, the PPP expected to form governments only in the Provinces of the Punjab and Sindh, and needed to protect this base by not allowing discretionary power to the Governors, who were nominated by the President, to dissolve the Provincial Assemblies and these governments.

ZAB went to Peshawar for the PPP Central Committee meeting, but was urgently summoned back by the President on 30 November. At their two meetings, the President gave no indication of the military action to be launched from the West

Wing on 3 December, although after a banquet on 26 November he had informed two foreign journalists, as reported in the *Daily Telegraph,* London, 'In ten days I might not be here. I will be off fighting a war.' Discussions centred on a proposed coalition government. Nurul Amin,[59] a Bengali politician, attended one of the meetings as he was to be designated Prime Minister. ZAB said he would confirm his participation in this coalition after consulting his Central Committee.

Opinion in the Central Committee was divided on whether to join a coalition government. The majority considered that the grave situation required the Party's contribution, particularly as opponents would otherwise fill any gap, but some of us advised against it. ZAB felt that, once in the 'corridors of power', he could strengthen his position.

Following this meeting, ZAB told us the proposed composition of the Provincial Governments of Sindh and the Punjab. It had earlier been settled that Ghulam Mustafa Jatoi would be Chief Minister of Sindh, but now Mir Rasul Bux Talpur was selected, with Mumtaz Bhutto as Home Minister and Hafeez Pirzada as Law Minister in the Province. Jatoi was to be a Federal Minister. Talpur could hardly believe his good fortune, and we too were surprised. Jatoi was crestfallen. The Chief Ministership of a Province is to most politicians the top position after Prime Minister; he is king in his domain and has an opportunity to build a power base of his own. As expected, Mustafa Khar was nominated Chief Minister of the Punjab.

At the PPP rally on 3 December in Rawalpindi, ZAB called for the new government to be inducted the next day to meet Indian aggression. In his letter of 3 December, delivered the same day, he set out his position on various aspects of the coalition government, covering such points as the allocation of portfolios and non-interference by each Wing in matters which did not concern it 'essentially'. In particular, the letter stated:

According to our understanding, the Coalition Government will consist of seven Ministers from the East Wing and six from the West Wing. There will also be some Ministers of State. I gave cogent reasons to explain why it would not be possible to collaborate with a Minister or Minister of State from either Wing if he is a representative of the Jamaat-e-Islami. Mr Nurul Amin is to be the Prime Minister and I, as the leader of my Party, shall be the Vice Prime Minister; we will be included in the 7-6 ratio. Of the six Ministers from the West Wing, five will be from the provinces of Punjab and Sindh and belong to my Party, and one from the NWFP who will be an Independent, or from the Jamiat-ul-Ulema (Mufti Mahmood and Hazarvi Group) provided it does not join the United Coalition Party. Balochistan will be represented by a Minister of State who will be an Independent or, with the same proviso, from the Jamiat-ul-Ulema...Neither the Minister from the NWFP nor the Minister of State from Balochistan should belong to any of the Muslim Leagues...[60]

At the time, hardly anyone, including ZAB, anticipated the total collapse ahead. Certainly no one thought he would assume full power within seventeen days.

War and the Security Council

In the evening of 3 December, the Pakistan Air Force launched air raids from the West Wing which led to full-scale war. ZAB remained in Rawalpindi for the next two days expecting to be called and consulted by the President. That first night, Iqbal Shaban,[61] Burmah Shell's representative in Rawalpindi, invited some of us to dinner with Air Marshal Rahim Khan, who said the attacks on the Indian airfields had been so devastating that 'we will not see Indian planes overhead again'. This statement proved as wrong as everything else about the war.

Even the outbreak of war in the West Wing failed to bring Yahya Khan and ZAB together in a common endeavour. Their approaches continued to be motivated by suspicion and distrust. To ZAB's disappointment, Yahya Khan did not contact him. Despite this, ZAB sent his list of the PPP Federal Ministers: Sheikh Rashid, Mubashir Hasan and Mustafa Khar from the Punjab and Jatoi from Sindh. When I asked why Khar, who was to be the Punjab Chief Minister, was included and not Mahmud Ali Qasuri, he said that Qasuri had been given too much importance by the military and would be nominated later when Khar became Chief Minister. It was always politics with ZAB.

When the President addressed the nation following the outbreak of war on the western front, he indicated that a civilian government would not be immediately installed. On 5 December, ZAB learnt that no steps regarding the new government would be taken till Mahmud Ali of the Pakistan Democratic Party returned from the United Nations. Annoyed and insulted at not even being contacted by Yahya Khan, he left for Lahore with a few of us. Most PPP leaders returned to their Provinces by road as all commercial air services had been cancelled.

The controlled media churned out success stories about the war, though Indian troops made significant gains. In Lahore, we visited some of the affected areas; overhead, the Indian Air Force, with complete mastery of the skies, flew low with impunity. ZAB was enthusiastically cheered everywhere.

The following day ZAB was summoned to Rawalpindi and asked to go to the United Nations. During the drive from Lahore, we witnessed scenes of complete disorder. Many cars with 'Crush India' signs were driving away from the border. In Rawalpindi, before ZAB's meeting with the President, we reviewed the situation and decided it would be appropriate for him to represent Pakistan as an elected leader and not merely as an envoy of a military regime. Unlike the time he went to Beijing, he accepted our advice and conveyed it to the President. They agreed on announcing the formation of a civilian government, with Nurul Amin designated as Prime Minister and ZAB Vice Premier and Foreign Minister. There was to be no swearing-in. The same evening he told me to accompany him to New York. I asked to be excused,[62] but he and Mubashir Hasan insisted I put aside personal matters. While in the hotel we heard the Government announcement, which did not mention his becoming Foreign Minister; he refused to proceed unless this was corrected. Sidney Sobers of the US Embassy, who was meeting ZAB at the time, helped secure the clarification.

No one from the Foreign Office or any government agency briefed ZAB before our departure. On the morning of 8 December we left by road for Peshawar and continued to Kabul to catch a flight. The group accompanying him consisted of two East Pakistani 'nominated MNAs', an ISI colonel and myself. Two comparatively junior officials of the Foreign Office came to hand over our passports, and no escort was provided from Rawalpindi despite war conditions. The regime and its officials clearly did not take seriously the new civilian Vice Premier designate.

At Peshawar we stopped with Hayat Sherpao, the PPP Provincial President, and had lunch with the Governor, a serving general, who inquired about the war situation. We shared our ignorance. While at Peshawar there was a telephone call from Agha Shahi, Pakistan's Permanent Representative at the United Nations. ZAB asked me to speak to him. He suggested that, in view of the 'Uniting for Peace' Resolution calling for a cease-fire, which had been passed overwhelmingly by the General Assembly in favour of Pakistan, we need not now proceed to New York. He had consulted Islamabad. On learning this, ZAB burst into expletives: he was not a 'yo-yo', to be asked to go one day and not the next; he was on his way and refused to turn back.

From Kabul we proceeded to Tehran then Rome, where the Ambassador was not even informed about our arrival, and then Frankfurt for an overnight stay before flying to New York. Our itinerary has been spelt out in detail because ZAB has been unfairly criticized for lengthening the journey unduly in order to allow Pakistan's position at the UN to deteriorate. In fact, all travel arrangements were made by the Government.

Early on 10 December, before we left Frankfurt for New York, ZAB sent the President his preliminary assessment of the situation, indicating what steps were required. He predicted that the Soviet Union would mount increasing pressure on Iran to prevent effective assistance to Pakistan. He pointed out that the

'Uniting for Peace' Resolution of the General Assembly was only recommendatory; the Soviet veto would block any action in the Security Council and incapacitate the United Nations, unless Pakistan was able to strengthen its position on the ground against India. He urged Yahya Khan to hold East Pakistan till the bitter end and at the same time launch an initiative against India from West Pakistan. The US should be persuaded to pressure the Soviet Union to ensure that India kept the sea lanes open. Yahya Khan should also prevail upon China to intervene immediately and effectively.

When we landed at New York, it became clear that developments in East Pakistan had taken place far faster than we had anticipated or were given to understand. On the flight ZAB had prepared notes for a press conference, but, on arrival, we were greeted by Agha Shahi with a copy of a message from the East Pakistan Governor. This had been delivered at 9 a.m. on 10 December to Niaz Naik, a member of the Pakistan delegation, by Robert Guyer, Under Secretary-General for Political and Security Affairs at the United Nations. It contained the following startling proposals:

A. As the conflict arose as a result of political causes it must end with a political solution.
B. I therefore having been authorized by the President of Pakistan do hereby call upon the elected representatives of East Pakistan to arrange for the peaceful formation of the Government in Dhaka.
C. In making this offer I feel duty-bound to say the will of the people of East Pakistan would demand the immediate vacation of their land by the Indian Forces as well.
D. I therefore call upon the United Nations to arrange for a peaceful transfer of power and request:
   1. An immediate cease-fire.
   2. Repatriation with honour of the Armed Forces of Pakistan to West Pakistan.
   3. Repatriation of all West Pakistan personnel desirous of returning to West Pakistan.
   4. The safety of all persons settled in East Pakistan since 1947.

> 5. Guarantee of no reprisal against any person in East Pakistan.
> E. In making this offer I want to make it clear that this is a definite proposal for peaceful transfer of power.
> F. The question of surrender of the Armed Forces will not be considered and does not arise and if this proposal is not accepted the Armed Forces will continue to fight to the last man.[63]

Later, the President asserted that he had authorized the East Pakistan Governor to confine himself to a different text. Maj.-Gen. Farman Ali, who was at the time principal adviser to the East Pakistan Governor, on 13 December maintained that the message had the President's approval.[64] Whatever the truth, the delivery of this message to the UN Secretariat did great damage.

ZAB's immediate task was to control the damage. He met the heads of important missions including the Chinese Vice Foreign Minister, Chiao Kuan Hua, who told us that when the UN Secretary-General had conveyed this message, the Chinese representatives could not believe it. According to the Chinese, who knew the supplies position, the Armed Forces could have held out for three months in Dhaka. Chiao Kuan Hua pointed out that the Dhaka administration had taken the 'fantastic step' of calling upon the American, British, French and Soviet representatives to take over the city and East Pakistan, and had asked the UN representatives at Dhaka to inform the Chinese through the United Nations at New York that they should participate in a Five-Power take-over. The Chinese were offended that they had not been directly approached.[65] The incident demonstrated the confusion that reigned in Dhaka and Rawalpindi.

ZAB decided to send a tough message that night to the President. Pakistan would be friendless and ultimately finished unless we continued the war. He advised the President to press China to intervene; hopefully the Americans would also take some meaningful initiative. He threatened that he would return to Pakistan if the President did not rescind the 10 December message formally and immediately and direct Shahi to inform the United Nations' Secretary-General that this message should be disregarded.

The following day he met George Bush, then US Permanent Representative at the UN and, again, the Chinese Vice Foreign Minister in an attempt to evolve a common approach between the Chinese and the Americans. In a second message to Yahya Khan, ZAB referred to these talks and proposed that, if possible, the ground situation be improved for a week; in any event he urged waiting seventy-two hours before making any move in the United Nations. This delay, he felt, would diminish the impact of the earlier message from East Pakistan. Acceptance of a cease-fire resolution by itself, without the withdrawal of Indian troops, would be meaningless, merely enabling India to turn her full attention on the West Wing. He also maintained that the Chinese might hesitate to reverse their recent position calling for a cease-fire coupled with troop withdrawals which had been adopted by the General Assembly.

The President's reply on 12 December to ZAB's communication of 10 December claimed that the text of the message from East Pakistan did not have his approval. Although there was no question of surrender, East Pakistan could not withstand the Indian onslaught for long without supplies, reinforcement and air support. The President confirmed having asked the Chinese for immediate and meaningful support. He was also doing everything possible to secure political assistance and military equipment from the Americans. Although there was some movement in this direction it was not enough to meet the situation.

Later that day, ZAB received another communication from the President stating that the suggestion to delay the reference to the Security Council and to hold out for a week would be fatal and multiply difficulties. After the fall of East Pakistan, India would switch her forces to the West Wing and weaken our negotiating position. Yahya Khan sought fast action believing, erroneously, that the US, China and the Soviet Union were now thinking along similar lines on the question of cease-fire negotiations and withdrawal. As for ZAB's suggestion of a big push from West Pakistan, the President countered that it was for military commanders to deal with military matters and he could not blunder into such a situation. ZAB was asked to move a resolution urgently in the Security Council.

The exchange of messages resembled a cat and a mouse game. It seemed like a continuation of the negotiations on the constitution and the transfer of power, only this time involving national life and death.

The message from Dhaka to the UN had so dramatically changed the situation that, by the time ZAB addressed the Security Council on 12 December, he could do little but declare emotionally, 'We have not come here to beg for peace…So what if a city falls?…A battle lost is not a country lost'. He warned against the process of dismemberment and fragmentation which could occur elsewhere in Asia, in Africa, Latin America and Europe, and could affect even the Great Powers: 'There will not be a Bangladesh only in Pakistan.'

In a speech tracing the background of earlier wars between India and Pakistan, he pointed out that this time India was 'standing on the shoulders of a Big Power to look bigger': 'If the Soviet Union detaches itself from the Indo-Pakistan conflict, we are prepared to be pitted against India.' According to him, the 'fundamental' trouble started with the Indo-Soviet Treaty of Friendship. He thanked and praised China and the United States for acting in support of 'a principle'. He placated the United States by expressing sorrow for 'strained relations' in the past: 'I am prepared to do everything in my power to repair those relations in Asia for the United States and in my country where…I speak in my right as the authentic voice of the people…The time will come. We cannot forget it.' He appealed to France to take 'a positive moral position', and to Britain to 'play a role to at least preserve what they finally conceded to Pakistan'. The speech concluded on a jingoistic note: 'Listen Sardar Swaran Singh, the Golden Bengal belongs to Pakistan, not to India. Golden Bengal is part of Pakistan. You cannot take away Golden Bengal like that from Pakistan. We will fight to the bitter end. We will fight to the last man.'[66]

Two themes stood out in this speech: a declaration to the United States that ZAB could be their man; and a show of chauvinism intended both to force Yahya Khan into military action, and for public consumption in Pakistan.

The Pakistan delegation, including several senior Ambassadors, lobbied actively to secure a 'decent' Anglo-French resolution in the Security Council. However, these efforts were frustrated by the arrival of a second message from Dhaka, this time from Lt.-Gen. Niazi. It had been sent to the Indians and requested a cease-fire on the following terms:

(A) Regrouping of Pakistan armed forces in designated areas to be mutually agreed upon between the commanders of opposing forces;

(B) To guarantee the safety of all military and para-military forces;

(C) Safety of all those who had settled in East Pakistan since 1947;

(D) No reprisals against those who helped the administration since March 1971.

By some strange logic, the junta believed the surrender was best done as a local military decision. The authorities in Rawalpindi were either far removed from reality or desperately trying to distance themselves politically from final responsibility. The proposal was made by the 'Martial Law Administrator of Zone B (East Pakistan) and Commander Eastern Command'.

Now the time had come for ZAB to make an unenviable decision, whether personally to accept surrender or not. That evening he communicated to the President that he had come too late; as recently as 5 December he had been told to wait till Mahmud Ali returned, and only on 7 December had he been asked to go to the United Nations. ZAB pointed out that the messages from the Governor and Niazi left no bargaining position; the Indians had become instransigent and the Soviet Union had insultingly raised the price for a cease-fire; and, therefore, all Pakistan could hope to obtain was the least unfavourable cease-fire resolution. The confusion in Islamabad was typified by messages continuing to be routed through channels other than Pakistan's permanent mission, despite ZAB's repeated requests. It was a hopeless situation.

The evening before his final speech, he jokingly asked if any official, who had in the past claimed credit for his speeches as Foreign Minister, would now suggest points. Later, just the two of us discussed the situation. The military Government had lost the war and were conveniently using him as spokesman for the surrender. ZAB concluded that, as the representative of Pakistan at the Security Council while also being an elected leader, the only practical and honourable course was to make his speech and walk out, leaving the Government delegation to accept a resolution on surrender.

His speech before the Security Council on 15 December was short and deeply emotional. He castigated the Council:

> For four days the Security Council has procrastinated. Why? Because the object was for Dhaka to fall...So what if Dacca falls? So what if the whole of East Pakistan falls? So what if the whole of West Pakistan falls?...We will build a greater Pakistan...

He added:

> Even the Great Powers are pandering to India...India is intoxicated today with its military success...This is gunboat diplomacy in its worst form. In a sense, it makes the Hitlerite aggression pale into insignificance, because Hitlerite aggression was not accepted by the world.

He thanked the United States, China and the countries of the Third World, but attacked France and Britain for their lack of support. He concluded:

> I am leaving your Security Council. I find it disgraceful to my person and to my country to remain here a moment longer than is necessary. I am not boycotting. Impose any decision, have a treaty worse than the Treaty of Versailles, legalize aggression, legalize everything that has been illegal up to 15 December 1971. I will not be a party to it. We will fight, we will go back and fight. My

country beckons me. Why should I waste my time here in the Security Council. I will not be a party to the ignominious surrender of part of my country. You can take your Security Council. Here you are. I am going.[67]

He defiantly tore up the papers at hand. The walk-out was not spontaneous.

Just before, on 14 December, a draft Polish resolution had been circulated in the Security Council. There has been unwarranted speculation by Air Marshal Asghar Khan and other critics that ZAB tore up this resolution on walking out, resulting in Pakistan's surrender and the imprisonment of its Armed Forces. In the first place, ZAB did not tear up this resolution, but only some notes. However, the draft resolution had not been found acceptable at the time by anyone in the Pakistan delegation, including Shahi and other Ambassadors present. Later, in Pakistan, this resolution was blown up out of all proportion, without knowledge of its contents. In reality, it called for the immediate transfer of power to the Awami League headed by Mujibur Rahman, and only thereafter for an end to military action in all areas, with the cease-fire to commence for a period of seventy-two hours. Its acceptance would in effect have entailed recognizing the division of the country. In any case, there was nothing to prevent the Indians from delaying the resolution, or its acceptance, and capturing the whole of East Pakistan, the fall of which was imminent after Lt.-Gen. Niazi's message.

More moot is why, contrary to Yahya Khan's direction, he did not press for an immediate cease-fire resolution. The accusation was later made that ZAB deliberately delayed the cease-fire, resulting in the humiliating surrender of the Armed Forces. This was not evident to me. With India's dominant position, supported by the Soviet veto, any resolution to deprive India of the fruits of victory would have been frustrated in the circumstances prevailing at the Security Council. Also, ZAB could not, or did not, accept that the military had no capacity to make any thrust in the western sector or hold out any longer in

East Pakistan. In any case, a key factor seemed to be distrust of the military regime; as acting Vice Premier designate, he did not want to be left 'carrying the can' on their behalf. Despite his walk-out, ZAB directed that the Government delegation continue attending the proceedings, and therefore Pakistan's position remained unaffected and did not go by default.[68]

ZAB has been accused of failing to achieve an honourable resolution in the Security Council. However, from the time India invaded East Pakistan on 22 November, the Pakistan Government made no serious move to do so. The Government wanted to secure disengagement coupled with troop withdrawals. Indeed, even after the commencement of war on the western front, the Government was not pleased when the US, disregarding its views, called on the Security Council on 4 December for a cease-fire. The proposal was vetoed. A rapid deterioration in the military situation took place in the twenty-four hours preceding 7 December, which Yahya Khan communicated to President Nixon. ZAB was at the time in Pakistan, but was kept in the dark. On 7 December, the General Assembly passed the 'Uniting for Peace' Resolution by 104 votes to 11 with 10 abstentions. No initiative was taken by the Government on 7-9 December, the last opportune time, to bring the matter back before the Security Council. However, even then, a successful outcome would have been most unlikely in the face of the Soviet veto and Indian determination to dismember and defeat Pakistan. It is thus surprising that the Yahya regime told ZAB on 5 December he would have to await Mahmud Ali's return from the UN, then asked him on 7 December to proceed, and, while ZAB was on his way on 8 December, suggested that there was no need to continue. Indeed, once the Governor's message was received in the UN on 10 December, prior to ZAB's arrival, the earlier opportunities, if any, were lost, and ZAB could do little to avoid the humiliation the country suffered.

Following the surrender in East Pakistan to 'the Armed Forces of India and Bangladesh', President Yahya Khan made a strong statement about continuing the war in the West Wing till the bitter end, and yet, the very next day, 17 December, he accepted Prime Minister Indira Gandhi's offer of a cease-fire. Why the President took this course can possibly best be explained in the same way as ZAB's earlier walk-out decision. Yahya Khan wanted the responsibility for the cease-fire to fall on, or at least be shared by, ZAB. The power game had not ended. From the outset, the purpose of the war on the western front was to secure time for an honourable settlement in East Pakistan; with the fall of Dhaka and surrender of the Armed Forces, no purpose could be served in continuing the war. According to Henry Kissinger, the United States saved West Pakistan as the army could not have withstood the full Indian onslaught; the Pakistan Army generally felt otherwise. On this question, ZAB's views appeared to be similar to those of the Pakistan Army. He kept in tune with the military; his was a political answer to a military question.

Last Days of the Real War

This narrative is concerned mainly with political developments, especially as the outcome of the actual war was clear-cut; only its abrupt end was not expected. Several authors with military knowledge have written on the course of the war, and retired Maj.-Gen. Fazal Muqeem Khan was given access by President Bhutto to documents and information to write *Pakistan's Crisis in Leadership* [70] which sets out the military situation from Pakistan's viewpoint. Nevertheless, any analysis would be incomplete without at least a brief outline of the war's conclusion.

The outcome was really settled by 9 December in East Pakistan. Lt.-Gen. Niazi, the military commander, informed GHQ that the regrouping of troops and readjustment of battle positions had become impossible owing to enemy air activity and the extreme hostility of the local population. At the same time, the East Pakistan Governor sought the President's approval for peace proposals. Yahya Khan agreed in advance to 'whatever efforts you make in your decision to save senseless destruction of the kind of civilians that you have mentioned, in particular the safety of our Armed Forces'. He also said that Lt.-Gen. Niazi would be instructed to accept the Governor's decision and make necessary arrangements.[71] Maj.-Gen. Farman Ali, assuming the Governor had the President's approval, passed on the cease-fire proposal to the UN Representative, Paul Marc Henry.

A few days later, on 13 December, presumably following ZAB's plea to the President, GHQ urged Niazi to hold on. But the next day the Governor stated that further operations were futile, and resigned together with his Cabinet. They disassociated themselves from subsequent actions of the Yahya Government and sought the protection of the International Red Cross in the Inter-Continental Hotel. After hearing from the Governor, Yahya Khan the same day sent a signed *en clair* message to Niazi:

> You have now reached a stage where further resistance is no longer humanly possible nor will it serve any useful purpose. It will only lead to further loss of life and destruction. You should now take all necessary measures to stop the fighting and preserve lives of all armed forces personnel, all those from West Pakistan, and all loyal elements. Meanwhile, I have moved the UN to urge India to stop hostilities in East Pakistan forthwith.[72]

On the night of 15 December, the Army Chief of Staff confirmed that Niazi could accept the terms of surrender proposed by the Indians, which met Niazi's requirements.

As everyone realized, the surrender in East Pakistan could only be delayed by some success in the West Wing. This, however, was not forthcoming. The Pakistan Navy was incapacitated by India's Soviet-supplied Ossa vessels which could fire radar-guided missiles with deadly accuracy from twenty miles, while our destroyers had a limited range. Equally crippling was the fact that about 3000 out of 8000 officers and men in the Navy were from East Pakistan. The naval surface ships were withdrawn for their protection into Karachi harbour which the public felt was a shameful act as it increased attacks on an already vulnerable city. However, without air support, the Navy had no option; the argument that they should have put the vessels far out to sea was specious. The Air Force had no facilities for naval support operations, and maritime reconnaisance was carried out by Pakistan International Airlines and civil aviation aircraft.

The Pakistan Air Force (PAF) acquitted itself with distinction in East Pakistan but without effect in the West Wing. The Indian Air Force (IAF) was the fifth largest in the world, with nearly 2000 aircraft, of which 600 were transports, trainers, support and similar aircraft. They were deployed in 38 squadrons, 28 against the West Wing and 10 for East Pakistan. Against this, the PAF had about 250 first-line aircraft, many of them old— some F-104s, French Mirages, Chinese-built MIG 19s, B-57 Bombers, Sabres and T-33 jet trainers. Out of eleven squadrons, only one was in East Pakistan at the time. When, on 4 December, the IAF launched their main assault on the airfield in Dhaka, nine of their aircraft were shot down in air combat and a further seven Indian planes were destroyed by ground fire. The PAF lost three planes. The war correspondent of the *Daily Telegraph,* London, observed that 'spectacular daylight air battles over Dhaka airport and the military cantonment, in which the Indians lost sixteen aircraft against two Pakistan planes brought down, are not easy to explain'.[73] After this initial success, the PAF was grounded in East Pakistan, and they blew up their remaining eleven aircraft to prevent them falling into enemy hands.

In the West Wing, the first PAF attack was on several Indian airfields at dusk on 3 December. Within hours, the IAF hit back at six airfields including Mianwali, Sargodha, Rafiqui and Risalawala. Neither side inflicted significant damage. The next day the IAF continued their pressure and, as the Pakistani airfields were well-protected, changed their attack to communications and industrial targets to demoralize the urban population. They had the freedom of the skies. The PAF, handicapped by deserting East Pakistani airmen who informed the Indians about radar blind-spots, adopted a defensive strategy to preserve its planes, and was largely intact at the end of the war. It claimed that 141 Indian aircraft had been brought down, including almost half by ground fire: 104 confirmed destroyed, 15 unconfirmed and 22 damaged.[74] PAF losses were 10 in air combat and 5 by ground defences. These were mostly old Sabres and not modern French Mirages. In addition, four aircraft were destroyed in accidents.

Inevitably, in the subcontinent, the main battles had to be with ground forces. A West Pakistan counter-offensive was expected to be launched from 7 December onwards, which was the critical time both to release pressure on East Pakistan and to boost morale. It was also essential in the context of obtaining any cease-fire resolution in the UN Security Council. The Pakistan Army strategy was to 'suck' the Indian forces into the Shakargarh salient and then either cut them off or severely maul them. Not until the morning of 13 December did GHQ give approval for the counter-offensive to be launched in three days; shortly, that too was postponed till 17 December. On the evening of 16 December the signal 'Freeze Tikka' (the name for the operation) went out. It was never launched. Instead, a cease-fire in the West Wing was announced.

Some Thoughts on Post-War Events

The surrender in East Pakistan had come as a severe shock. In the West Wing, most of the forces had not even fired a shot, so neither they nor the general public could comprehend the cease-fire. It caused bitterness, indiscipline and insubordination against the army leadership. We learnt that Air Marshal Rahim[75] and Lt.-Gen. Gul Hassan were instrumental in prevailing on the Yahya junta to hand over power peacefully to ZAB. There were reports that one general, vehemently against ZAB, said that if the other generals could not cope he would undertake to sort out the PPP. Some brigadiers also threatened to march on Rawalpindi.

Months later, it was still unclear what had transpired within the junta. A Military Court of Inquiry subsequently found that six senior officers had been involved in a conspiracy two days before ZAB became President. Their retirement from service was announced by me on 9 August 1972. *Dawn* commented editorially on 'such extraordinarily mild action': 'A certain measure of mystery surrounds the events of the critical days of December.'[76] Several of the brigadiers maintained that their efforts were directed at getting rid of Yahya Khan and contributed to his agreeing to quit. What really happened remains unexplained.[77]

Speculation also centres on the last three days ZAB spent in the US prior to his becoming President. On 15 December he protested, 'Why should I waste my time here...My country beckons me'. Yet he did not leave till the late afternoon of 18 December, because he sought a meeting with President Nixon. After ZAB's message to Yahya Khan on 14 December, there was no important communication between them till the President's two messages of 18 December: the first in which ZAB was asked to return and not wait to meet President Nixon; and the second, following soon after, in which Yahya Khan said he had learned from the US Ambassador about the Nixon meeting, and ZAB should leave for Pakistan immediately thereafter.

Mustafa Khar telephoned from Pakistan to say the situation was getting out of hand following the cease-fire; ZAB's return was imperative; demonstrations were erupting which the PPP might not be able to control. He wanted to speak personally to ZAB to explain the gravity of the situation and to persuade him to return immediately to Pakistan. At the time, ZAB was hosting a lunch for George Bush and some other Permanent Representatives at the UN. ZAB told me to calm Khar. After lunch he pointed out that neither Khar nor I appreciated the importance of the opportunity of clarifying his position personally with President Nixon before seeing Yahya Khan. The meeting was finally arranged by pressing the request through Kissinger.

Immediately following it, we started on our return journey to Islamabad.

On the plane, he gave me the gist of his talk with President Nixon. He had explained that he was not anti-US; his pro-China policy, which had previously been misunderstood by President Johnson, had proved beneficial since Pakistan had been used as a bridge to Beijing by Nixon himself. He praised Nixon for the far-sighted new US approach to China, and expressed gratitude for the US preventing a full-scale attack on West Pakistan. In the new situation, with India and the Soviet Union working closely together, ZAB emphasised the vital role he could play. The military had failed and he now wanted to see a peaceful subcontinent despite his war-mongering reputation. He was clearly satisfied with the meeting.

At Rome we were informed that a plane was being sent to bring us back. No Air Marshal was sent to meet us as rumoured. ZAB was tense throughout. About one hour before landing at Islamabad, I was going to shave when he asked how I could think of such mundane matters at this time. We discussed the encounter-to-be with Yahya Khan. What if the President suggested the earlier arrangement with Nurul Amin as Prime Minister? I said that was not now possible, and he should categorically refuse. I thought this much might have been settled with Nixon, but he made no comment. He accepted that Yahya Khan could not make Nurul Amin the Prime Minister, but might want to continue as President with ZAB as Prime Minister. I said that would not work; the Yahya regime would collapse. Of course, nothing was certain with power and politics, but so far no military ruler had survived defeat, and Yahya Khan was not that power-hungry to risk major trouble in West Pakistan. ZAB said he knew the military mind better.

I have repeated this conversation to show that, although he had tried to clear his path to power with Nixon, his position was finally settled on his return to Pakistan. As we were landing, we saw a large crowd and an official reception on the tarmac. I remarked that the answer to his question was visible, especially as no one had seen us off twelve days earlier.

The international Press hailed the departure of Yahya Khan but gave faint praise to ZAB. One editorial commented, 'It is a measure of Pakistan's desperation that in its darkest hour it has to turn for leadership to the very man who helped bring disaster to the country'.[78] The domestic reaction was different. He was viewed as a saviour.

At the end of the previous Chapter, I concluded that ZAB did not fully take into account India's single-minded determination to benefit from the opportunities presented by the period following the 1970 elections and the civil war in East Pakistan. In a person who prided himself on his knowledge of foreign affairs, and was at the same time obsessed with India, this is hard to understand. He had roundly criticized the Government for failing to comprehend the significance of the Indo-Soviet Treaty of Friendship, but he too failed to understand the seriousness of its impact on China and the subcontinent. However, an objective assessment must bear in mind that not since the 1950s had any country, apart from Israel when it occupied Arab lands, been allowed to get away with the sort of aggression committed by India. Although the situation in 1971 was different from the 1965 War, he seemed caught in the mental trap of the earlier period. He had always regarded the 1965 War as having ended prematurely and did not want the same mistake to be repeated.

Critics who hold ZAB responsible for events of this period, resulting in the dismemberment of Pakistan, do so with little justification, as he was barely consulted by the Yahya regime during these nine months. An important part of his calculations concentrated on retaining the confidence of the military; it was not likely that he would risk deliberately bringing about their humiliating surrender, nor was it necessary for him to do so. Responsibility for the events of this period rests firmly with the Yahya regime.

NOTES

1. *The Pakistan Times,* Lahore, 27 March 1971.
2. Ibid.
3. Ibid., 28 March 1971.
4. *Dawn,* Karachi, 29 March 1971.
5. For allegations of atrocities committed in East Pakistan, *see, inter alia,* the following: Indian Council of World Affairs, *How Pakistan Violated Human Rights in Bangladesh: Some Testimonies,* New Delhi, 1972; D. R. Mankekar, *Pak Colonialism in East Bengal,* Somaiya Publishers, Bombay and New Delhi, 1971; B. N. Mehrish, *War Crimes and Genocide—The Trial of Pakistani War Criminals,* Oriental Publishers, Delhi, 1972, Chapter 3; and David Reed and John Frazer, *Nightmare in East Pakistan,* Readers' Digest, Volume 99, No. 596, December 1971, pp. 41-6.
6. *The Pakistan Times,* Lahore, 30 March 1971.
7. Ibid., 1 April 1971.
8. Ibid., 3 April 1971.
9. Ibid.
10. Ibid., 5 April 1971.
11. Ibid., 6 April 1971.
12. Ibid., 8 April 1971.
13. Ibid., 12 April 1971.
14. Ibid., 15 April 1971.
15. Major Zia-ur-Rahman subsequently became President of Bangladesh after a *coup.*
16. Most of the important documents, written statements and letters of ZAB and the PPP in 1971 were prepared by the author under ZAB's direction, and copies remain with the author.
17. *The Pakistan Times,* Lahore, 15 April 1971.
18. Ibid., 17 April 1971.
19. Subsequent writers exaggerated this rift; *see,* for example, Sisson and Rose, pp. 56-7.
20. *The Pakistan Times,* Lahore, 4 May 1971.
21. Ibid., 30 April 1971.
22. Ibid., and *Pakistan Observer,* Dhaka, 5 May 1971.
23. Ibid., 10 May 1971.
24. Ibid., 12 May 1971.
25. Ibid., 20 May 1971.
26. Ibid., 21 April 1971.
27. India maintained that the incident was engineered by a Pakistani intelligence agency, the ISI. The two 'commandos' were released following the December 1971 War and allowed to stay in Pakistan. *See also,* Sisson and Rose, p. 135-7.

28. *The Pakistan Times,* Lahore, 7 May 1971, carries the full text on p. 4.
29. Ibid., 16 May 1971.
30. *Hindustan Times,* New Delhi, 7 July 1971.
31. Copies of ZAB's letter of 16 June to the Secretary, Ministry of Information, Government of Pakistan, and telegrams of 21 June, are in the author's possession.
32. Copy of ZAB's letter of 21 June is in the possession of the author.
33. Stanley Wolpert, p. 158, incorrectly states that 'Yahya sent Zulfi to Tehran in July to seek more support from the Shah for the war in the east'. Similarly, there are other errors; for example, Wolpert says on p. 165 that ZAB left for New York on 8 December from Karachi after assuring 'his cheering countrymen who had come to Karachi's airport to see him off': *see,* p. 122 hereof for the correct position. Also, he refers to the author as ZAB's adviser principally on military matters (pp. 173 and 184), which is wrong.
34. J. A. Rahim, after his dismissal in July 1974, said this and denounced many actions of ZAB, even referring to him as a CIA agent, before friends in the presence of the author. But his prejudice at the time must be taken into account.
35. The interview in *Kayhan International* was partly reproduced by *Musawaat,* Lahore, on 20 July 1971.
36. Extracts from ZAB's letter of 15 July 1971; copy is in the author's possession.
37. *Dawn,* Karachi, 18 July 1971.
38. With President Yahya were A. R. Cornelius, M. M. Ahmed and Lt.-Gen. Peerzada; ZAB was accompanied by J. A. Rahim, Mahmud Ali Qasuri, Mubashir Hasan, Ghulam Mustafa Khar, Sheikh Rashid, Hayat Mohammad Khan Sherpao, Mir Rasul Bux Talpur, Hafeez Pirzada and the author.
39. The PPP team comprised J. A. Rahim, Mahmud Ali Qasuri, Hafeez Pirzada and the author. Justice Cornelius represented the Government along with G. W. Chaudhry and Col. Hasan.
40. Copies of these documents are with the author.
41. *The New Times,* Rawalpindi, 2 September 1971, carried an English version of the interview.
42. Extract from letter of 2 September from ZAB to Lt.-Gen. Peerzada; copy in the author's possession.
43. Copy of ZAB's letter of 8 September is with the author.
44. *The New Times,* Rawalpindi, 13 September 1971.
45. *New York Times,* New York, 30 September 1971, report by Malcolm W. Browne.
46. Letter of 4 October 1971 from ZAB to Lt.-Gen. Peerzada; copy is in the author's possession.

47. *Dawn,* Karachi, 9 October 1971.
48. Ibid. The 'non-Muslim' was Justice Cornelius.
49. Later, in December, Air Marshal Rahim Khan and Lt.-Gen. Gul Hassan were instrumental in asking Yahya Khan to step aside in favour of ZAB as President, which gave rise to speculation about a conspiracy during this visit. They were also old friends. In Pakistan, 'conspiracy' theories dominate much thinking at all times.
50. Mubashir Hasan's letter of 13 November 1971; copy is with the author.
51. Quotes are from ZAB's letter of 2 December 1971 to Lt.-Gen. S.G.G.M. Peerzada which was not delivered; *see,* p. 118 and Note (58) thereto.
52. *Dawn,* Karachi, 18 November 1971.
53. *Dawn,* Karachi, 25 November 1971.
54. Lt.-Gen. 'Tiger' Niazi, the military commander in East Pakistan, and others, subsequently said this was the strategy they had advocated, though it was rejected by Yahya Khan and GHQ. Niazi and the brigadier who was his PSO in 1971 explained this to the author with maps and actual plans in 1990.
55. *The Pakistan Times,* Rawalpindi, 26 November 1971. Subsequently, ZAB said the Government should have taken the matter to the Security Council earlier; he was sent too late—as indeed he was; *see* pp. 127 and 130.
56. These proposals, with some amendments, were contained in the Constitution that Yahya Khan attempted to publish on 20 December 1971.
57. Copy of ZAB's letter of 28 November 1971 is in the author's possession.
58. Stanley Wolpert, pp. 163-4, quotes the letter without knowing that it remained undelivered. The original signed letter remains in the author's possession.
59. Nurul Amin and Raja Tridev Roy, the Chakma Chief from the Chittagong Hill Tracts, were the only two non-Awami League politicians elected from East Pakistan in the 1970 polls. Nurul Amin was Chief Minister of East Pakistan in 1954.
60. Relevant extracts from ZAB's letter to Yahya Khan of 3 December 1971; copy is in the author's possession.
61. President Bhutto had Iqbal Shaban removed from his job: 'Who needs enemies with friends like Bhutto', remarked Shaban.
62. My wife was expecting a baby; ZAB broke the news to me later in New York that the new-born baby was lost.
63. Copy of the message as delivered to ZAB is in the author's possession.
64. Maj.-Gen. Farman Ali requested a court martial to clear himself; he maintained that Lt.-Gen. Niazi, Maj.-Gen. Jamshed and Admiral Sharif had agreed to the Governor's proposal. *See also,* pp. 131-2.
65. The author attended all meetings with ZAB in New York except for the talks with Henry Kissinger and President Nixon.
66. The full text of the statement made at the 1611th meeting of the Security

Council on 12 December 1971 was published in 1972 by the Department of Films and Publications, GOP, Karachi, *Zulfikar Ali Bhutto: Speeches in the United Nations Security Council in December, 1971.*

67. The statement made at the 1614th meeting of the S.C. on 15 December is published in ibid, Note 66.

68. Sisson and Rose, p. 220 and note 28 thereto, incorrectly state that Dhaka fell two days later, and that the walk out 'did halt all consideration of the Polish Resolution'.

69. At the 2002nd and 2003rd meetings of the UN General Assembly.

70. Fazal Muqeem Khan, *Pakistan's Crisis in Leadership,* Alpha & Alpha Publishers, New Delhi, 1973.

71. Ibid., p. 183.

72. Ibid., p. 187.

73. *Daily Telegraph,* London, 6 December 1971.

74. 44 SU-7, 43 Hunters, 19 Canberras, 8 Mig-21, 5 HF-24, 3 Gnat, 3 Mystere, 1 each of the Alize, MI-4 and ADP aircraft, and 13 jet fighters of unidentified type. For details, *see,* 'The PAF War Diary of Events', published in the PAF magazine *Shaheen,* Vol. XXII, No. 1, May-August 1972.

75. Rahim Khan confirmed this to me in 1972 after his removal, and then again in 1975 on a flight from Copenhagen to Paris when we both accompanied ZAB.

76. *Dawn,* Karachi, 12 August 1972.

77. In later versions there were opposing accounts: *see,* for example, Lt.-Gen. Gul Hassan Khan, *Memoirs,* pp. 341-3 and Brigadier F. B. Ali's critical article contradicting Gul Hasan in *The News International,* Karachi, 18 September 1993. Most surprisingly, although Chief of the General Staff at the time, Gul Hassan states, pp. 347-8, that he did not even know ZAB had become President till their meeting in the afternoon of 20 December 1971.

78. *The Ottawa Citizen,* Ottawa, 21 December 1971.

CHAPTER 5

CIVILIAN MARTIAL LAW
(20 December 1971 to 21 April 1972)

ZAB assumed the office of President with ease, and knew exactly how to function as Chief Martial Law Administrator. He had worked closely with Ayub Khan, and comprehended power and how to operate its levers. It was as if the Presidency was his by right, destiny fulfilled.

Several occurrences that first day showed how he would run the Government. Early arrivals were M. M. Ahmed and Mahmud Ali Qasuri. He got on well with M. M. Ahmed, a senior bureaucrat and Yahya Khan's Finance Adviser, who had helped him in the Ayub era and again recently, but he distrusted Qasuri. When ZAB asked him to be Law Minister, Qasuri replied that I should be given the position since he preferred to remain in Lahore. After they left, the President attacked Qasuri: 'See how he wanted to stay in Punjab to control the Province.' My comment that this was unlikely to have crossed Qasuri's mind so early, brought the response, 'I know how Punjabi politics work'. Suspicions about colleagues, which later were to cloud his administration, had appeared at the outset.

He then relaxed. 'We made it,' he said to me and recounted his meeting with Yahya Khan, who had first proposed remaining as President with ZAB as Prime Minister. When ZAB refused this, Yahya Khan had then suggested he should continue as CMLA and C.-in-C. of the Army, while ZAB became President. ZAB, however, sought full and effective control of the entire administration and insisted on assuming the office of President and CMLA, though it was the first time the latter position had been occupied by a civilian.

I suggested that Martial Law should be scrapped immediately. He rejected this as being inconsistent with his earlier stand that Yahya Khan could not withdraw Martial Law till replaced by a constitutional arrangement approved by the Assembly. He always thought quickly. He declined assistance for his address to the nation: 'This is my moment'.

I left before discussions took place about retiring a number of generals, on which the views of both Lt.-Gen. Gul Hassan and Lt.-Gen. Ghulam Jilani Khan, the Director-General, Inter-Services Intelligence, were sought. I met Jilani in the waiting room and was surprised to learn that, since returning from a mission to Dhaka about a month back, he had only met Yahya Khan once, while President Bhutto had called him immediately.[1] Yahya Khan's casual attitude to priorities was incomprehensible.

In a long, emotionally-charged broadcast that evening, ZAB said he had 'to pick up the pieces'. The attack against India continued, though with caution. He would rebuild the military machine, once 'the proudest and best in Asia', which had been ruined by ambitious generals interested in political power. Several generals, including Yahya Khan, Hamid Khan, Umar and Peerzada, were retired, and Lt.-Gen. Gul Hassan was appointed the 'acting C.-in-C. in the same rank'.[2] He promised a good, clean administration, a Government accountable to the people, and, above all, hope for the future. He spoke in English, from his heart, as he put it, calling for a 'new Pakistan'.

At the end of a very long day, Mustafa Khar and I were invited to dinner. Before discussing the appointment of Governors and Federal Ministers, ZAB recalled that I had earlier said I would not be interested in government office, so 'the position of Attorney-General should suit you'.[3] My refusal was not readily countenanced by one with complete power in the land. Immediately, he asked for other names. I tried to contact Yahya Bakhtiar but the telephone operator at President House had ceased work. The country was dismembered and in disarray, a new President had been sworn in twelve hours earlier, the enemy was confronting us, and yet there was no ready communication at midnight. With difficulty, a call was put through to Quetta and Bakhtiar became Attorney-General.

We then considered the question of Governors for the four Provinces. It was important to appoint the right man in each Province; the President himself could take care of the Centre. Khar was the obvious choice for the Punjab. I urged him to set aside personal differences and he agreed to appoint Mumtaz Bhutto as Governor of Sindh.[4] We discussed whether the Frontier and Balochistan should also have Partymen as Governors, particularly as the PPP was in a minority in the Frontier and virtually non-existent in Balochistan. Hayat Sherpao was selected for the Frontier and, finally, a few days later, ZAB appointed Sardar Ghous Bux Raisani as Governor of Balochistan. Participating in the highest level of administration was a fascinating experience. Many late-night sessions followed with the President, an indefatigable worker who could do with little sleep.

Over the next days, with much to accomplish, there was no time even to feel tired. The Cabinet swearing-in was held at two in the morning, to impress on all that hard work was the order of the day. The President mainly consulted Khar about Cabinet appointments, particularly those relating to the Punjab. When I phoned one Minister-to-be, he was so surprised, I could almost visualize him falling off his chair at the other end of the line. It was a small Cabinet with J. A. Rahim as Senior Minister, Sheikh Rashid, Mahmud Ali Qasuri, Mubashir Hasan, Ghulam Mustafa Jatoi, Meraj Khalid, Raja Tridev Roy, Rana Hanif, Justice Faizullah Kundi and Hafeez Pirzada. Nurul Amin had been sworn in earlier as Vice President. This post was not specified in the Constitution, such as it was under Martial Law, and had no practical function apart from allowing East Pakistan token representation in the Federation.

The same morning, Khar and Mubashir Hasan told me that the President wanted to see me and was not going to take no for an answer this time. Mubashir mentioned that even my designation as Special Assistant to the President, 'the same as

Thank you for all the trouble you took to lessen the tragedy in some ways than one. For this and other reasons, I have much pleasure in presenting the first copy of our effort to you

Zulfikar Ali Bhutto
September 28th 1971.

To Raj,

President Bhutto with President Hafez al Asad of Syria in Damascus in January 1972; the author is on the left.

President Bhutto and
author waiting in the
aeroplane at Ankara
airport in January 197
(see p. 161).

Prime Minister Zhou
Enlai and the author in
Beijing in January 1972,
(see p. 196).

President Bhutto and Prime Minister
Kosygin in Moscow on 16 March 1972.
Standing behind them, from the right,
are M. M. Ahmed, Aziz Ahmed and
the author.

alks between officials at Simla on 29 June
72. P. N. Haksar, Principal Secretary to
ime Minister Indira Gandhi, shaking
nds with the author. Behind them in the
cture are Indian Foreign Secretary T. N.
aul and Secretary-General Aziz Ahmed.

The Simla Accord being signed. Among those standing behind President Bhutto and Prime Minister Indira Gandhi are, from the right, P. N. Haksar, Jagjivan Ram, Ghulam Mustafa Jatoi, the author (behind whom is Director-General Aftab Ahmed Khan), Malik Meraj Khalid, Afzal Agha, Hayat Sherpao, Aziz Ahmed and Mian Anwar Ali.

Justice Hamoodur Rahman presenting his Commission's Report on the 1971 War to President Bhutto on 12 July 1972.

a cultural show
Larkana for the
of Iran: seated
e Shah's left is
AB and, on the
, Mustafa Khar,
1taz Bhutto and
the author.

h Mujibur Rahman, Prime Minister of Bangladesh, meeting Prime Minister
o at Lahore airport on 23 February 1974 during the Islamic Summit.
dent Fazal Ellahi Chaudhri is on the left.

Following the signing of the Constitutional Accord of 20 October 1972, photograph shows from the left, front row, Maulana Mufti Mahmood, Maulana Shah Ahmed Noorani, Sardar Shaukat Hayat, President Bhutto, Mir Ghous Baksh Bizenjo, and, at the back, Sardar Sherbaz Mazari, Maj.-Gen. (Retd.) Jamaldar, Abdul Qayyum Khan, Yahya Bakhtiar, the author (in front of whom is Ghulam

▲ Prime Minister Bhutto meeting Secretary of State Kissinger in Rawalpindi on 31 October 1974; the Pakistan team seated on the left are Lt.-Gen. (Retd.) Yakub Khan, the author, ZAB, Aziz Ahmed and Agha Shahi.

Prime Minister Zulfikar Ali Bhutto coming out of the Khana-e-Ka'aba Sharif on 2 September 1975.

Prime Minister Bhutto at the foundation stone laying ceremony of the Parliament building in Islamabad. With him is the author who was, at the time, Minister for Production, Industries and Town Planning.

Kissinger', had been settled. ZAB told me that the country was going through critical days and required my contribution. I pointed out that office might hinder my right to speak frankly, but he assured me this would not change. I accepted for one year. Maulana Kausar Niazi and Mairaj Mohammad Khan, both of whom had served the Party well, were appointed Advisers.

This political team was probably the best balanced and most competent ever in Pakistan. It included Party stalwarts, good orators and educated young talent, notwithstanding our inexperience and the inevitable weak links resulting from the need to represent various interests and regions. However, some of the senior bureaucratic appointments were less welcome, particularly within the Party. Aziz Ahmed, who had served ably in Ayub's administration as Foreign Secretary, was brought back from retirement, and Ayub Khan's Director of the Intelligence Bureau was also re-employed. Soon, poorly regarded, retired and even dismissed police officers were given important positions, including Haq Nawaz Tiwana, the first head of the newly-established Federal Security Force (FSF), Masood Mahmood, who later took over the FSF, and Said Ahmed Khan. The latter two testified against ZAB at his murder trial.

Several of these appointments demonstrated that, although contemptuous of Ayub Khan as a leader, ZAB remained greatly influenced by his period in office under him. The Ayub era also accounted for the fact that there were few people of middle age in the PPP Government. Rahim and Qasuri were in their sixties; several of us were in our mid to late thirties, including Mumtaz, Sherpao, Khar, Jatoi, Kausar Niazi and Mairaj Mohammad Khan. ZAB was one of the few in his mid-forties, and he had been Ayub Khan's Minister. The long thirteen-year period of direct or semi-military rule had deprived the country of normal politics from 1958 to 1971 and had taken its toll on a generation of politicians. Most of us had not been through the parliamentary mill nor held junior ministerial positions before being vaulted straight into high office.

The stage was now set for the most dynamic and stimulating period of the ZAB era. The team, although new to government, worked untiringly, and a sense of exhilaration was all-pervasive. The trauma of East Pakistan was forgotten surprisingly quickly, raising its head from time to time only in connection with our prisoners of war and occupied territory, and the question of the recognition of Bangladesh.

Although the PPP came to power earlier than anticipated, and there had been little consideration of the numerous problems ahead, ZAB had both an agenda and an astounding ability to think in advance. Armed with the authority of Martial Law, the President could do anything. During this period, economic and other reforms were introduced, an interim Constitution was settled, talks with India initiated, and a whole host of positive measures were undertaken. But ZAB's personal shortcomings also became visible, accentuated by the arbitrariness of Martial Law.

Reforms and Other Measures under Martial Law

The introduction and implementation of economic reforms promised during the elections were carried out with determination by Mubashir Hasan and J. A. Rahim. They enjoyed the full support of the President and the Party, so it is wrongly alleged that the 'mad doctor' and the 'loony left' were alone responsible for the assault on big business. Nationalization was a manifesto pledge, and at the time it was welcomed by the people. With defeat and dismemberment, the desirability, viability and even the basis of Pakistan were questioned. The nation needed shock therapy and the people a stake in the country. It was an economic and political imperative for the Government. Moreover, in the 1970s the prevalent mood was 'East is Red' and 'East is Best', following North Vietnam's firm stand against the might of the United States. Capitalism appeared in retreat. It was not till a decade later that such concepts as 'rolling back socialism', privatization and deregulation came on the scene even in the West.

The first major step was the Economic Reforms Order of 1 January 1972, which placed iron and steel, heavy engineering, the assembly and manufacture of automobiles, trucks and tractors, cement, public utilities like gas and electricity, oil refineries, petro-chemicals, and heavy and basic chemicals under state control. The acquisition of banks was discussed but deferred. Two novel and distinctive features were that management, and not shareholdings, were taken over; and, contrary to the practice in other countries, foreign investment was not touched. ZAB had only recently assured the US Administration he would follow a pragmatic and reasonable course if in government, and Pakistan also needed foreign investment.

The logic behind taking over management was that the Government exchequer could not presently bear the burden of compensation, while acquisition of management gave public control quickly at no cost. This 'nationalization' was a major blow to the 'twenty-two families' who had dominated the economic scene. The haste with which it was done appeared in the confused recital of aims in the preamble to the Order. It referred to Islam enjoining 'equitable distribution of wealth and economic power'; and to the duty of Government 'to ensure that the wealth and economic resources of the country are exploited to the maximum advantage of the common man' and to safeguard the 'interests of the small investor'.

Several of the units acquired were loss-making, having been financially stripped by their previous owners; in one case all that existed was a boundary wall; and another was only capable of making a single product so expensively that it could not be marketed at home or abroad. This unit had a foreign shareholding, unbeknownst to us at the time, and the offer to return it when I became Minister for Production in 1974 was hastily declined. Many problems could have been avoided, or at least reduced, with better preparation. At the time, however, speed was considered essential.

The burning question of unemployment compounded the difficulties of these industrial units. Provincial Governments and local politicians forced the absorption of an uneconomic number of employees in the state sector. The practice started in the Punjab and soon became commonplace, with even casual labour securing confirmed appointments. J. A. Rahim, a learned man in many ways, was not a practical Production Minister. He left management to their own devices, a good practice in principle, but one which gave little protection against improper demands. Some of the managers were neither adequately motivated nor sound businessmen. We shall leave to Chapter 8 a more detailed examination of the performance of these units.

A second onslaught was made on monopoly capitalism by stopping big business and industry from creaming off profits under a managing agency system. On 14 January, the Companies (Managing Agency and Election of Directors) Order abolished these arrangements and, for the first time, allowed small shareholders proportional representation on the boards of companies. Next, on 18 March, life insurance businesses were nationalized. This time foreign companies were included; Mubashir Hasan and Rahim maintained that no foreign investment was required in this field. ZAB agreed but informed the US Administration in advance because the American Life Insurance Company was being acquired, promising fair and prompt compensation. Three State Life Insurance Companies were established which successfully managed these businesses.

Some have argued that, representing feudal landlords who had previously dominated politics, ZAB sought to curtail the influence of the new industrial class which had grown under Ayub Khan. Industrialists who opposed him were indeed dealt with harshly, but the primary purpose of these measures was to implement the manifesto pledges and satisfy the people politically.

He pressed on relentlessly with his reform programme disregarding warnings from many quarters, including Chinese Premier Zhou Enlai. He was determined to accomplish as much as possible under the protection of Martial Law, in part to justify its existence and counter-balance arbitrary measures. Each Minister was told to expedite proposals, and each in turn tried to prove his 'revolutionary' qualities. As a result, several initiatives were ill-prepared.

One of my tasks as Special Assistant to the President was to examine the measures proposed and finalize any accompanying speech or announcement. Time for improvement or correction was always limited. For example, it was proposed that Pakistan should ratify all Conventions of the International Labour Organization, but this was hastily amended when I pointed out that even developed countries had adopted only a few, and such an undertaking would be impossible to implement. Instead, on 10 February, a 'New Deal' was announced for labour, promising dignity and fair wages; it reinforced security of employment and made arbitrary dismissal challengable before Labour Courts. These were much overdue measures, as were workers' rights to participate in management.

Most important, at least to ZAB, were the agrarian reforms announced on 3 March.[5] The 1970 Manifesto had stated that 'to destroy the power of the feudal landowners is a national necessity that will have to be carried through by practical measures of which a ceiling is only a part'. He reduced the ceiling on individual holdings from 1000 acres of unirrigated and 500 acres of irrigated land to 300 and 150 acres respectively. Various exemptions granted under Ayub Khan's earlier reforms were discontinued. Land was resumed without compensation and granted free to the tenants. Simultaneously, several feudal privileges and exemptions were abolished. Land revenue, water rates and the provision of seed were made the landowners' responsibility; the cost of fertilizers and pesticides was apportioned equally between the tenant and landlord; and all tenants were given the right of pre-emption of land under their tenancy.

The non-implementation of these agrarian reforms has been the subject of much criticism. Agriculture Minister Sheikh Rashid was assigned the task, which he pursued vigorously. Certain politically influential figures in the Punjab and Sindh managed to avoid these reforms, in some cases with help from Sadiq Hussain Qureshi and Ghulam Mustafa Jatoi when they became Chief Ministers, and even from ZAB himself. However, Sheikh Rashid's tenacity and determination yielded results. In all about 2.7 million acres were resumed by the Government, apart from half a million acres in the Pat Feeder area of Balochistan, and over 130,000 tenants received proprietary rights, in addition to the land given to 40,000 families in Swat, Chitral and Dir in the Frontier Province.

While it is not possible to describe all the reforms, two introduced just prior to the interim Constitution merit mention as typifying ZAB's actions. By President's Order No. 15 of 1972, the privy purses and privileges of rulers of Princely States which had acceded to Pakistan following Independence in 1947 were abolished and replaced by maintenance allowances which would not pass on to their successors. India had taken similar steps much earlier. In truth, these rulers had long ceased to govern; now their perquisites were dependent on the President's powers 'to grant to a ruler any privilege, facility or concession for such period and on such conditions as he may determine.'[6] This kept the princes 'on call'. Secondly, President's Order No. 16 of 1972 empowered the Central Government to revoke any industrial permission and related loan granted during the Yahya period.[7] This provided a weapon to deal with opponents. More damaging, however, it increased the insecurity felt among the business community, already greatly agitated by nationalization and arrests.

Martial Law was used to fulfil not only the Manifesto but also other 'promises' made by ZAB. He dismissed those who had been hostile in the past. Almost immediately, he removed, by Martial Law Order No. 28, retired Lt.-Gen. Habibullah Khan as Chairman of the National Press Trust and Z. A. Suleri as Senior Editor of *The Pakistan Times*. Both had been critical of ZAB. There followed, in quick succession, the removal of the Managing Director of Pakistan Progressive Papers Limited, the Chairman of the West Pakistan Industrial Development Corporation and the Managing Director of the National Shipping Corporation, the last two being close friends of Yahya Khan.[8] They were to 'cease to hold' office notwithstanding any law or contract to the contrary, and the orders were 'not to be called in question in or before any court including a High Court or the Supreme Court'. In the case of Habibullah and S.U. Durrani, Governor of the State Bank, arrest followed removal. They were treated shabbily and publicly displayed under arrest. Former supporters turned dissident were treated no better. Mukhtar Rana, a PPP MNA from Lyallpur (now Faisalabad), was arrested, tried under Martial Law and had his sentence confirmed, all within a matter of days.

Another example of arbitrary behaviour involved the National Press Trust (NPT). This had been established in April 1964, under Ayub Khan's direction, by several businessmen and industrialists. By 1971, it gave the Government control over several NPT-owned newspapers such as *Morning News* of Karachi, *The Pakistan Times* published from Lahore and Rawalpindi, *Imroze* of Lahore and Multan, and *Mashriq* published from Lahore, Karachi and Peshawar. While in opposition, the PPP had demanded the dissolution of the NPT, but, instead, on 13 February 1972, ZAB suspended the NPT board of trustees and the boards of directors of all three companies it controlled. The powers of over thirty trustees and directors, and four chief executives, were conferred on a new chairman. The pretext was that some of the original settlors and trustees were from East Pakistan. This gave the Government control of a sizeable part of the Press.

ZAB guarded these Martial Law powers which were exercised either by him personally or under his direction, as in the case of Mukhtar Rana. This was apparent from the State Arrangements Regulation, 1972—if the President and Chief Martial Law Administrator was absent abroad, the Vice President was to act as President, pointedly leaving out any reference to CMLA.[9] Similarly, when new Chiefs of the Air Force and Army were appointed on 3 March 1972, they were not made Deputy Martial Law Administrators, unlike their predecessors. Martial Law vested power absolutely in one individual.

Political Negotiations and the Interim Constitution

ZAB understood that his untrammelled powers as CMLA could not continue for long and had to be replaced by constitutional rule. He also realized that it would be difficult to secure early consensus on a constitution, so he settled first for an interim arrangement. For this some political *rapprochement* was required which he set about achieving. Earlier, when President Yahya Khan had informed us over lunch, on 26 November 1971, about banning the National Awami Party, ZAB had said it was the right thing to do as Abdul Ghaffar Khan, Wali Khan and his colleagues in the NAP were anti-Pakistan. However, on his first day as President, he lifted this ban to herald a new beginning. Within days, he met Wali Khan and other NAP leaders, but initial pleasantries were soon followed by acrimonious encounters with Wali Khan. For the same purpose, he also met opposition representatives from the Punjab. In those early days, they said that the country as originally conceived was finished; we should seek a suitable compromise with India, even suggesting confederation. These private conversations were quite contrary to some of their public, chauvinistic pronouncements.

He recognized that he could not use his absolute majority in the West Wing to force a constitution on the NWFP and Balochistan. Here, the majority were represented by the NAP-JUI alliance. Some from the PPP in the Frontier and the Muslim League (Qayyum) felt it was wrong to promote and give importance to the NAP-JUI, small in national terms, by negotiating with them as equals. They claimed that the Muslim League (Qayyum), along with the PPP and independents, could muster a majority in the NWFP Assembly. The personal animosity of Qayyum Khan and the NAP leadership dated back to 1946, and ZAB enjoyed playing each off against the other. On one occasion it was like watching a farce, with Qayyum Khan and Yusuf Khattak leaving the drawing room from one door while the NAP leaders entered through another. Talks with the Muslim League (Qayyum) about a political alliance also served to pressure the NAP-JUI in the negotiations on constitutional issues.

Preliminary meetings with the NAP-JUI were followed by detailed discussions on an interim Constitution at Peshawar in March. The PPP team consisted of the Chairman, Hayat Sherpao, Mustafa Jatoi, Hafeez Pirzada, Maulana Kausar Niazi and myself; the NAP were represented by Wali Khan, Arbab Sikander, Mir Ghous Bux Bizenjo and Sardar Khair Baksh Marri; and the JUI by Maulana Mufti Mahmood and Maulana Hazarvi. Sardar Marri was throughout abrasive. I do not recall anyone else speaking in this tone to ZAB, nor him ever being so restrained. These were early days—later, he would be less tolerant. At this point, he was seeking consensus on the interim Constitution.

On the third day of the talks, 6 March, we reached an Accord[10] which contained five main features. The interim Constitution was to be based on the Government of India Act, 1935, with consequential amendments, providing for a presidential form of government at the Centre and a parliamentary system in the Provinces. The National Assembly would meet briefly to pass the interim Constitution and for 'a vote of confidence in the Government and approval of continuation of Martial Law till August 14, 1972'. Thirdly, the Assembly was to appoint a Committee of the House to draft the permanent Constitution and, after 14 August 1972, it would act as both a constitution-making and a legislative body. Next, 'it was accepted that in the North West Frontier Province and the Province of Balochistan, the majority parties are NAP and JUI and they will be entitled to form the Governments'. Moreover, 'by way of compromise', during the interim period, Governors in these two Provinces would be appointed in consultation with the NAP-JUI. Finally, it was agreed that local bodies elections would be held soon after the convening of the Provincial Assemblies on 21 April 1972. In announcing the Accord, ZAB pointed out that in 1967 he had 'envisaged that we [the PPP] would be the Golden Bridge...Today we have passed through the dark tunnel, and I see the Golden Bridge'[11]

It was, however, not long before disputes arose over the Accord. Wali Khan maintained that the NAP-JUI were not required to vote for the continuation of Martial Law till 14 August. He was particularly agitated that the date for the appointment of the two Governors, 1 April, had not been honoured, resulting in what he termed an 'April Fool' joke on Arbab Sikander and Ghous Baksh Bizenjo, the two prospective Governors. Lengthy correspondence and public controversy followed, mainly involving Wali Khan and myself.

ZAB was deeply concerned about any curtailment of his Martial Law, but even more imperative was the need for consensus on a constitution. Accordingly, I persuaded him to put forward a new 'Basis for Negotiations'. On 8 April, he absolved the NAP-JUI 'from their commitment' to vote for the continuance of Martial Law. ZAB remained prepared to respect the NAP-JUI coalition as the majority in the two Provinces, but expressed unwillingness to have their nominees as Governors. The issue of the Governors was not settled until a fresh tripartite agreement was reached later, on 26 April, following the interim Constitution. On the other hand, he offered one Federal Cabinet position each to the NAP and JUI. However, the acrimony following the 6 March Accord was the first serious disagreement with these two parties and distrust increased subsequently.

The continuation of Martial Law till 14 August had by then become a personal issue with him. He was determined to secure it with the approval of the Assembly despite the NAP-JUI. The Governors of Sindh and Punjab were directed to obtain from all PPP MNAs a mandate expressing confidence in the President and sanctioning Martial Law. The refusal of the two MNAs from Sindh, Abdul Hameed Jatoi and Darya Khan Khoso, infuriated him. A. H. Jatoi continued as an outspoken critic. Hafeez Pirzada was of the view that Mustafa Jatoi, as Minister for Political Affairs, encouraged the recalcitrant MNAs. ZAB, like a true feudal, revelled in such differences and intrigues between his lieutenants. Although the public mood against Martial Law gained momentum, as became evident from the warm welcome given for the first time in Lahore to Wali Khan, ZAB still refused to give ground.

The text of the interim Constitution was finalized by Law Minister Qasuri, Law Secretary Justice Mohammad Gul, Hafeez Pirzada and myself after several difficult days. Much credit must go to the experience and painstaking contribution of Justice Gul. For the meeting of the Assembly, J. A. Rahim had been earlier assigned the task of preparing the Presidential Address which he, in turn, passed on to different individuals and groups. Only three days before the Assembly session, ZAB sent me a disjointed mass of papers with no theme, little substance and in poor English. He insisted I prepare it anew. There followed three of my most hard-working days, dictating the speech and putting together the various important elements. The President was available at all times, giving piecemeal approval and suggestions. Towards the end, I told the President I was exhausted and he would have to ad lib the remainder. In response, I received pep-up pills to keep me awake.

Two points of the speech deserve special mention. It stressed the concept, put forward by Allama Mohammad Iqbal in 1930, of Pakistan comprising the Provinces of Punjab, Sindh, Frontier and Balochistan. By this, we tried to emphasize that the loss of the East Wing had not destroyed the idea of Pakistan. Secondly, I again suggested that ZAB should try to achieve unanimity for the interim Constitution by lifting Martial Law. His first reaction was that this would make the MNAs who signed the resolution for continuing Martial Law look foolish. I was of the view that everyone would be pleased to end Martial Law, including these very MNAs. I finally convinced him by saying it could be conditional on the interim Constitution being passed unanimously, a formula which would have the benefit of making the Opposition responsible for any continuation of Martial Law if they did not agree.

His election as President of the Constituent Assembly had just taken place, though to his annoyance he was not unchallenged. Sardar Sherbaz Mazari contested in the 'spirit of democratic election', as he put it; but for ZAB this was a 'petty display of democracy'. He wanted to be elected unanimously. We inserted by hand the addition concerning the lifting of Martial Law as the Address had already been printed.[12] He relaxed at the prospect of cornering his opponents, recalling what *Time* magazine had written in 1947, as quoted in the Introduction.

The Presidential Address was well received and the reference to Martial Law being withdrawn was much applauded. The interim Constitution was passed without difficulty on 17 April and came into effect four days later, to great acclaim. This put an end to the question of the legitimacy of the Government, and the rump Assembly, and to the demand for early elections in the recently truncated country. Two other concerns, the Superior Courts and the Armed Forces, were also 'settled', at least for the present, leaving the President fully in control.

The Superior Courts

In Pakistan, the High Courts of the Provinces and the Supreme Court constitute the 'Superior Courts', and have what is commonly termed writ jurisdiction over legislative and executive authority. ZAB enjoyed being an all-powerful President, and the constraints of the Superior Courts led to friction throughout his period in office.

After years of autocratic rule under Ayub Khan and Yahya Khan, the Judiciary again asserted itself. Several cases had been instituted against the legality of Martial Law, and ZAB felt that these were really directed at the legitimacy of his position as Yahya Khan's successor. In the Asma Jilani Case, however, Chief Justice Hamoodur Rahman pronounced in April that 'in order to leave no room for doubt I wish to make it clear that this decision is confined to the question in issue before this Court, namely, the validity of the Presidential Order No. 3 of 1969 and Martial Law Regulation No. 78 of 1971, and has nothing whatsoever to do with the validity of the present regime'.[13]

Certain aspects of this decision disturbed ZAB, particularly the reversal by the Supreme Court of its earlier judgment which had validated Ayub Khan's 1958 Martial Law on the basis of Hans Kelsen's theory of the *grund-norm*. The Asma Jilani Case also witnessed a curious reversal of roles, with Attorney-General Bakhtiar defending Martial Law. However, Sharifuddin Pirzada's arguments largely prevailed, for which he was temporarily prevented from travelling abroad. The Chief Justice stated:

> In any event, if a *grund-norm* is necessary Pakistan need not have to look to the Western legal theorists to discover it. Pakistan's own *grund-norm* is enshrined in its own doctrine that the legal sovereignty over the entire universe belongs to Allah Almighty alone, and the authority exercisable by the people within the limits prescribed by Him is a sacred trust. This is an immutable and unalterable norm which was clearly accepted in the Objectives Resolution passed by the Constituent Assembly of Pakistan on the 7th day of March 1949.[14]

This view gave rise to concern that the Supreme Court might next give the Objectives Resolution 'supra-constitutional' effect; however, this proved unfounded.

The Supreme Court also firmly reiterated that, 'The Superior Courts are, as is now well settled, the judges of their own jurisdiction. This is a right which has consistently been claimed by the Supreme Court and other Courts of superior jurisdiction in all civilized countries'.[15] ZAB as CMLA had excluded the Courts' jurisdiction over his reforms and other measures. Article 281 of the interim Constitution also provided, *inter alia:*

> All Proclamations, President's Orders, Martial Law Regulations, Martial Law Orders, and all other laws made as from the twenty-fifth day of March 1969, are hereby declared, notwithstanding any judgment of any Court, to have been validly made by competent authority, and shall not be called in question in any Court.

Despite the double reference to 'any Court', designed to remove doubt in the matter, ZAB never felt secure.

The Supreme Court declared Yahya Khan a 'usurper', after nearly three years of accepting him as President and CMLA. ZAB sarcastically remarked, when greeting the Judges of the Supreme Court just prior to the interim Constitution, 'So now you can hold the British occupation of India as illegal'. Notwithstanding fine phrases concerning the independence of the Judiciary, he remained suspicious of the Superior Courts and spent his years in office trying to check them.[16] The Courts reciprocated by not following their strong words condemning Martial Law on the departure of Yahya Khan when the *coup* of General Ziaul Haq occurred five years later.[17]

The Armed Forces

The importance of the army in Pakistan, and for ZAB, has already been briefly touched upon. He played safe in selecting a friend and supporter as his first 'acting' C.-in-C. The appointment of Lt.-Gen. Gul Hassan was not welcomed because he had held the important position of Chief of the General Staff, and as such was commonly viewed little differently from the other senior generals responsible for the 1971 débâcle. ZAB set about pacifying the discontent in the Armed Forces by appointing, on 24 December, a Commission of Inquiry to examine the circumstances of the surrender in East Pakistan and the sudden cease-fire in the West Wing. The Commission consisted of the Chief Justice of the Supreme Court, Hamoodur Rahman, and the Chief Justices of the Lahore and Sindh High Courts. They submitted their Report on 8 July 1972[18] and, on the return of the prisoners of war and civil internees from India, were allowed eight weeks in May 1974 to record the evidence of these personnel. The establishment of this Commission marginalized the main issues concerning the dismemberment of the country but did not entirely contain the discontent in the Armed Forces. Subsequently, largely in deference to the wishes of the Service Chiefs, ZAB decided not to publish this Report.

The PPP Government's relationship with the Armed Forces required adjustments at the outset from both sides. Some leftists within the PPP still harboured hostility towards the Armed Forces, who in turn found difficulty in accepting the new civilian authority. Differences were inevitable and emerged early. For example, Mairaj Mohammad Khan and others decided to wear black armbands and observe the Quaid-i-Azam's birthday, 25 December, as a protest day against the 'past misdeeds' of the army. On learning that the army was disturbed by these reports, ZAB directed Mairaj to desist. At this time, Pakistan Television showed a documentary film of the Armed Forces surrendering in East Pakistan which ZAB wanted Information Minister Hafeez Pirzada to repeat. However, ZAB was taken aback when, at a dinner hosted by him for some senior army officers in Murree, they protested aggressively over the 'spectacle on TV'. The film was not shown again.

Two other incidents, one minor, which ZAB considered were 'challenges' to his authority occurred at this time and resulted in the removal of the Air Force and Army Chiefs, both friends of ZAB. The first happened when the Ambassadors appointed to Pakistan were invited to Larkana in January 1972, along with other guests including the three Service Chiefs and some Ministers. He had asked the Service Chiefs and myself for drinks before the reception, when Air Marshal Rahim Khan pressed for the release of retired General Habibullah Khan, who had been arrested earlier. ZAB did not concede, nor did he forget the incident, which he felt reflected an over-assertion of authority on Rahim Khan's part.

The second challenge was serious. A strike by the police in Peshawar was followed by a more threatening one in Lahore. ZAB had invited Sidney Sobers, the US Chargé d'Affaires, for dinner along with Mustafa Jatoi, Mumtaz Bhutto and myself. We discussed the Lahore strike and Sobers, seeing ZAB's discomfiture, offered his plane for use, if necessary, that night. After Sobers left, he asked his ADC to contact Gul Hassan, who was woken up, summoned, and asked to provide troops to prevent the strike situation deteriorating in the Punjab. Gul

Hassan insisted that none could be spared because of the situation on the Indian border.[19] ZAB concealed his anger but felt let down and did not forget the incident.

Gul Hassan's refusal resulted, incidentally, in establishing Khar's image as a strong man. Khar organized public support and groups to direct street traffic and maintain order; he then addressed a mass rally demanding that the police report back immediately or face dismissal. They meekly returned. Watching Khar's speech on television with me, ZAB said that people wrongly accused him of arrogating all powers to himself: 'Look how I have trained Mustafa and how outstandingly he has performed.' The incident also fixed in his mind the need for a Federal Security Force to deal with similar situations without recourse to the army.

As for Rahim Khan and Gul Hassan, their ouster was settled. The choice of successor chiefs and other arrangements were made in great secrecy and covered all contingencies. The appointed day, 3 March, was declared a public holiday to celebrate land reforms. The two officers were visibly surprised and shaken when asked to submit their resignations.[20] Subsequently, they were quietly taken from Rawalpindi to Lahore to ensure a smooth transition. In retrospect, this precaution amounted to overkill, but at the time ZAB was unsure of the army and could not afford to see his action fail.

After this successful 'coup', ZAB addressed the nation. He spelt out the Government's achievements, the negotiations both abroad and at home, but mainly he talked about the removal of the two Service Chiefs in 'the interest of the Armed Forces of Pakistan'. It was 'a sacred task' to have invincible Armed Forces: 'It must again become the finest fighting machine in Asia'. He was unstinting in his praise but equally 'determined to wipe out Bonapartic influences'. Clarifying the distinction between Bonapartic and Bonapartism, he said it was 'Bonapartic' because what had happened in Pakistan 'openly since 1958' was that 'some professional generals turned to politics not as a profession but as a plunder'. He delighted in such semantics and in displaying knowledge, which in this case he possessed as an admirer of Napoleon. He also emphasized two other changes: 'We will no longer have the anachronistic and obsolete posts of Commanders-in-Chief' but only Chiefs of Staff, as in other countries, and they would have 'a fixed tenure and under no circumstances will there be extensions of that tenure'.

ZAB's true feelings towards the Armed Forces surfaced on two occasions during our visit to Ankara in January 1972. Prior to our arrival, Martial Law had been imposed in several districts in Turkey, and at the Presidential banquet the presence of Ismet Onunu, whose party was in opposition, made ZAB cautious. However, he lauded the glorious traditions of the Turkish Army under the leadership of Kemal Ataturk in their struggle for national independence and integrity after the First World War. In private talks he pointed out that, in contrast, the leadership of the Pakistan Army played no part in the struggle for independence, and were 'more concerned with the distribution of regimental silver than the partition of the subcontinent'.

The second incident arose after we had boarded the plane to depart from Ankara. An urgent message was conveyed from Islamabad that, till further intimation, we should not proceed. It was the coldest winter in Ankara for several years, the Turkish President and others stood outside freezing as they waved goodbye while we sat in the plane, neither side knowing the real reason for the delay. ZAB said we should take off and circle near the airport to receive the message, but the pilot explained to me that the next destination, Morocco, was six hours away, and there might not be sufficient reserve fuel necessary for this VVIP flight; nor could the plane proceed far without losing radio contact. So, we waited. ZAB thought the reason for the cryptic message could only be a *coup d'état* or attempted *coup*. Kwame Nkrumah and other leaders had been toppled while abroad and similarly asked not to return. I argued that the military could not take over so soon, and finally convinced him that there was no chance of a *coup*. Those who suggest that ZAB disregarded the possibility of a military *coup* while in office are wrong.[21]

With this in mind, he was cautious in his appointment of Chiefs of Army Staff, in whom he looked for qualities of pliability and obedience, first in General Tikka Khan in March 1972, and then in General M. Ziaul Haq four years later. However, reports that he referred to Ziaul Haq as his 'monkey' or kept him waiting unnecessarily[22] are without foundation. On the contrary, he gave his Service Chiefs importance and at the same time sought reassurance from them, and also from us, about their loyalty. A typical example of this occurred prior to the appointment of Ziaul Haq as Chief of Army Staff. Earlier his choice had been Lt.-Gen. Abdul Majid Malik, to whom he had given a clear intimation when the three of us had dined together. However, in September 1975, when he specifically talked to me about the appointment of a successor to General Tikka Khan, he rejected Majid Malik for reasons of questionable loyalty to him. For similar, personal reasons, which he elaborated to me, he eliminated six other top generals, two of whom he considered were not suitably qualified for selection. When I remarked, 'Just be careful who you appoint—he may become the next President', he was initially taken aback, but he went on to say that I should not worry because he would choose carefully. He added that, if a small Bavarian could master the Prussian Army, what could prevent ZAB from doing the same with the Punjabi Army. At the time, I did not know he had Lt.-Gen. Ziaul Haq under consideration. In this respect, Hitler proved more successful and ZAB wrong—but we are moving ahead of our narrative.

NOTES

1. Lt.-Gen. Ghulam Jilani was accurate in his disclosures and reports, as the author learnt later on getting to know him. The PPP, wrongly in the author's view, accused him of working against ZAB; certainly, this was not the case to the author's knowledge till the end of April 1977.
2. Lt.-Gen. Gul Hassan subsequently claimed that this was at his instance: *Memoirs of Lt. General Gul Hassan Khan,* Oxford University Press, Karachi, 1993, pp. 348-50.

3. The author's resignation letter of 1 June 1973 referred to 'personal reasons' and his previous refusal; *see also,* Chapter 8 p. 270 and Note (8) thereto and pp. 304-5.

4. Several years later, in London, Ghulam Mustafa Jatoi said that ZAB had maintained that Jatoi was his choice, but the author had promoted Mumtaz Bhutto for Governor throughout the flight from New York. We did not discuss Mumtaz at all on the flight; but this was another example of ZAB pitting colleagues against each other. The talk on the night of 20 December 1971 was in the context of Rasul Bux Talpur and not Jatoi.

5. Land Reforms Regulation, 1972, Martial Law Regulation No. 115, published in the *Gazette of Pakistan, Extraordinary,* of 11 March 1972.

6. Article 7 of the Rulers of Acceding States (Abolition of Privy Purses and Privileges) Order, 1972, published in the *Gazette of Pakistan, Extraordinary,* of 18 April 1972.

7. Industrial Sanctions and Licences (Cancellation) Order, 1972, President's Order No. 16 of 1972, of 16 April, published in the *Gazette of Pakistan, Extraordinary,* of 20 April 1972.

8. A few of them were soon rehabilitated, one as Managing Director of Pakistan International Airlines. Altaf Gauhar, whom ZAB jailed, was later given government patronage.

9. Martial Law Regulation No. 110, published in the *Gazette of Pakistan, Extraordinary,* of 1 February 1972.

10. The text of the 6 March Accord, along with other documents and correspondence, was reproduced in *Constitution-Making in Pakistan,* published by the National Assembly Secretariat on 22 April 1975, pp. 73-5.

11. Ibid., p. 79.

12. The handwritten insertion in the original, with certain corrections and additions in ZAB's own writing, remains in the author's possession.

13. Asma Jilani v. The Government of the Punjab, PLD 1972 Supreme Court 139 at p. 208.

14. Ibid., p. 182.

15. Ibid., p. 197.

16. *See,* Chapter 6 pp. 185-7.

17. Compare the Asma Jilani Case above with the decision in the Begum Nusrat Bhutto Case, PLD 1977 Supreme Court 657.

18. Relevant documents were submitted to and considered by the Hamoodur Rahman Commission. The Commission's main Report of 8 July 1972 remains unpublished and confidential though, curiously, extracts and disclosures have appeared from time to time in the Press in Pakistan and in India, for example, in the *Illustrated Weekly of India* of 23 October 1988. The author submitted the PPP's position to the Commission, as directed by ZAB, to reduce the extent of his testimony before the

Commission; ZAB did not permit the author to show or discuss it with any other Party member. The author is probably the only person, apart from the members of the Commission, who has examined both the main Report of 8 July and the evidence on which it was based.

19. Lt.-Gen. Gul Hassan's version later was different and not entirely accurate; *Memoirs* p. 361-4.

20. Lt.-Gen. Gul Hassan's account in his *Memoirs,* p. 307-8, is inaccurate about what occurred in President House, according to the recollections of Mumtaz Bhutto, Mustafa Khar and the author.

21. In fact, the Foreign Office had delayed our take-off in order to convey to ZAB the Soviet Union's recognition of Bangladesh.

22. Compare Stanley Wolpert, p. 263, and Lt.-Gen. Gul Hassan, p. 396.

CHAPTER 6

CONSTITUTION-MAKING AND THE 1973 CONSTITUTION

The Constitution of a country, as Woodrow Wilson said, 'is not a mere lawyers' document but is in fact the vehicle of a nation's life'. Pakistan's constitutional controversies indeed mirrored her numerous problems, and led to dismemberment. As a lawyer and politician, and a key adviser of Ayub Khan in framing the 1962 Constitution, ZAB fully appreciated the difficulties.

Pakistan has been a veritable laboratory for experiments in tackling constitutional problems, either existing or created, from which students of constitutional law could learn much. Issues of concern include the federal structure, the quantum of provincial autonomy, the question of religion and the State, national language, whether the franchise should be universal or limited, whether minority religious communities should have electorates separate from the Muslim majority, direct or indirect elections, a presidential or parliamentary form of government, the division of powers between the executive and the legislature, the nature of fundamental rights, particularly the right to property, and the position of the Judiciary. Even Martial Law, abrogating the basic law of the land, has, uniquely, been upheld by the Supreme Court.

A whole treatise is needed to cover this vast field, which is not possible here. However, to understand the problems facing constitution-makers in framing the 1973 Constitution, some background is essential on at least three aspects: the federal structure, Islam and the State, and the executive authority. It places in proper perspective the problems faced by ZAB and

Mujibur Rahman in 1971 and, later, by the PPP Government. It helps those with the requisite background knowledge to focus on the issues, and to others it is an indispensible guide.

Federal Structure

In Pakistan, a geographically divided country, the ordinary difficulties of a federal system were compounded by the fact that the majority of the population were in the physically smaller East Pakistan, while the West Wing had the industrial wealth and the military-bureaucratic base of the Punjab. The search for an acceptable formula for provincial representation and autonomy dominated constitution-making. Failure to find a formula led to the break-up of the country.

Pakistan inherited a quasi-federal system under the Indian Independence Act, 1947, with Section 8 providing that, until a constitution was framed, the country would be governed by the Government of India Act, 1935. The 1935 Act was the longest ever passed by the British Parliament, consisting of 321 Sections and 10 Schedules, leaving a legacy of lengthy constitutional documents in the subcontinent. It was based on the 1930 Report of the Simon Commission, which stressed the 'unusual' significance of the 'change from a unitary to a federal system'. The Report stated that, 'it is only in a federal structure that sufficient elasticity can be obtained for the union of elements of diverse internal constitution and of communities at very different stages of development and culture...'[1]

The search for appropriate provincial representation began when the first Constituent Assembly of Pakistan, on 12 March 1949, appointed a Basic Principles Committee (BPC) with the Punjab and East Bengal as the main actors. In its Interim Report of 7 September 1950, the BPC stated that, 'there should be a Central Legislature consisting of two Houses: the House of Units, representing the legislatures of the Units; and the House of People elected by the people'. It did not determine the composition of each House and, in the BPC meetings that followed, leaders from the Punjab suggested that the two should be about the same size as they were to have equal powers under Clause 39 of the Interim Report. Finally, in late 1951, the Franchise Sub-Committee of the BPC proposed in Clause 30 of its Report that 'there should be parity as a whole between the two Wings of Pakistan'.

Later, in August 1952, the BPC proposed a formula for the composition of the two Houses under which the Punjab was the principal loser. Curiously, the representation of the Punjab was fixed in percentage terms of neither Pakistan nor of the West Wing. It provided for a form of parity reduction by which the Punjab, with a population of about 22 million compared to East Bengal's 50 million, was given 45% of East Bengal's representation, that is, 90 seats in a House of People consisting of 400 members compared to East Bengal's 200; and 27 seats compared to 60 of East Bengal in the House of Units. The other combined West Wing Provinces, with less than the Punjab population, would have received proportionately larger representation. This formula was set out in Clauses 38 and 43 of the BPC's Final Reports. It proved unacceptable to the Punjab, and no agreement was reached. Within months, on 17 April 1953, Prime Minister Khwaja Nazimuddin, a leading Bengali politician, was dismissed and replaced by Mohammad Ali Bogra, another Bengali, who was at the time serving as Ambassador to the USA. His appointment was the first occasion when overt American influence appeared to be exercised in the internal affairs of Pakistan.

Subsequently, in 1954, because East Bengal would have enjoyed a dominant role at the Centre, the Punjab argued that the powers of the Federal Government should be limited to foreign affairs, defence, communications between the two Wings, foreign trade, currency and foreign exchange. A 'Zonal Sub-Federation' was proposed for amalgamating the West Wing Provinces into a single unit. However, on 21 September 1954, the Assembly approved the Reports of the BPC with only minor amendments, and adjourned till 27 October for the Constitution Bill to be drawn up for final debate. Three days before that, on

24 October, the Governor-General declared a State of Emergency and dissolved the Constituent Assembly as it could 'no longer function'.

The very next month, the Prime Minister announced that the Government had decided to merge the West Wing Provinces and ten Princely States into a unified Province of West Pakistan, and 'One Unit' came into formal existence the following year on 14 October. As a result, the country now comprised only two Provinces—East Pakistan, formerly East Bengal, with a population of 50 million, and West Pakistan with 40 million. It suited both sides. In the West Wing, particularly the Punjab, One Unit was viewed as an administrative convenience to save expenditure and permit concentration on development; while East Pakistan hoped to gain greater devolution of powers. This unique two-unit federation, artificially balancing the two Wings, eased the task of framing a Constitution in a divided country, but sowed the seeds of eventual separation.

With the establishment of the second Constituent Assembly, again elected indirectly through the Provincial Assemblies, a new Constitution was presented on 9 January 1956. It was hurriedly passed and came into effect on 23 March. In all, the Constitution took over eight years in the making, the longest ever in any country. Essentially based on the Indian Constitution and the arrangements already existing in the country, there was little justification for such delay. More important, no Constitution meant no elections for all those years.

The 1956 Constitution provided for a Westminster-style parliament, with equal representation from each Wing. It did not work smoothly; there were numerous changes in government, without either no-confidence votes or elections. In the seven years after the assassination of the first Prime Minister in October 1951 until the Constitution was abrogated on 7 October 1958 there were in all six Prime Ministers and three Governors-General. The consequences of such instability were soon to become apparent. Under the 1956 Constitution, general elections were scheduled for February 1959, but, instead, Martial Law was declared in October 1958.

Initially, President Ayub Khan as CMLA ran the country virtually as a unitary state, and issues relating to provincial autonomy were relegated to the background. He then introduced a system of 'Basic Democracy', which was a four-tiered arrangement with 80,000 primary units, half each from East and West Pakistan, to be elected at grass roots level. His nominated Cabinet resolved that a vote of confidence in the President be held among these 80,000 members by way of a referendum and, if the majority were in the affirmative, 'he should also be deemed to have been elected President of Pakistan for the first term of office under the Constitution to be so made'. On 14 February 1960 Ayub Khan received 75,283 affirmative votes. As an 'elected' President he set up a Constitution Commission.

Making provision for provincial autonomy was not one of the special tasks assigned to this Commission. It was required to examine the 'progressive failure of parliamentary government in Pakistan' and to submit proposals for a democracy 'adaptable to changed circumstances and based on the Islamic principles of justice, equality and tolerance'. The Commission submitted its report in May 1961; Ayub Khan, however, had his own prescription. His 1962 Constitution provided for a presidential system and continued the two-unit arrangement, with each Wing having seventy-five representatives. The same 80,000 Basic Democrats, utilized as the electoral college for the President, were to elect the federal legislature. He described his proposals as a 'blending of democracy with discipline'. In reality, it was constitutional autocracy.

The 1962 Constitution appeared to give greater autonomy to each unit through a single list of federal subjects, with residual authority vesting in the provincial legislatures. However, Article 131(2) provided that if the national interest so required in relation to '(a) the security of Pakistan, including the economic and financial stability of Pakistan, (b) planning or coordination or, (c) the achievement of uniformity in respect of any matter in different parts of Pakistan', the central legislature would have power to make laws not enumerated in the federal list of subjects. Moreover, under Article 80, the executive authority of a Province was vested in the Governor, who was appointed by, and subject to, the directions of the President under Article 66, and could be removed at will by virtue of Article 118. Provincial autonomy was minimal. The President's powers were both wide and supreme.

This dispensation eventually led to an explosive situation because, under it, the West Wing prospered while East Pakistan suffered political and economic deprivation. Adding to this, the 1965 War with India witnessed the military separation of the two Wings. The earlier theory that the defence of East Pakistan lay in the strength of the West Wing was shattered. Clearly, only China prevented a possible invasion of East Pakistan, which India in any case did not then appear to seek. Sheikh Mujibur Rahman's Awami League made the most of the turmoil following the Tashkent Declaration to put forward, in March 1966, the Six Points proposal, which called for a complete change in Ayub Khan's semi-unitary constitutional system, and for economic emancipation.

The failure of the Round Table Conference to settle these issues resulted in Ayub Khan maintaining that he was 'left with no option but to step aside and to leave it to the Defence Forces of Pakistan, which today represents the only effective and legal instrument, to take over full control of the affairs of this country'. On 25 March 1969, General Yahya Khan, the C.-in-C. of the Pakistan Army, abrogated the 1962 Constitution, declared Martial Law, and took over as CMLA. Thus, both the 1956 and 1962 Constitutions ended with Martial Law.

Yahya Khan recognized the need for change to satisfy the feelings of deprivation and exploitation in East Pakistan. His Legal Framework Order, 1970, provided for greater provincial autonomy:[2] East Pakistan was given full representation by holding general elections, for the first time, on the basis of one-man-one-vote to a unicameral federal legislature. By then, however, it was possibly already too late. The 1970 elections produced yet another unique situation in our constitutional history: the unit with the largest population and the party with an overwhelming majority in the federal legislature sought complete provincial autonomy and the emasculation of the Federal Government and National Assembly; the less populous West Wing called for a Centre with greater authority.

Islam and the State

The creation of Pakistan was a practical expression of the assertion of the Muslim League that the Muslims constituted a separate nation and were entitled to a state of their own in the subcontinent. The demand for an Islamic Constitution followed naturally. How this was to be achieved, if indeed it was possible, was not thought through. An Islamic State was considered one with Islamic laws; but this described the laws rather than the nature of the State. Also, there was no consensus among the numerous religious schools of thought: for some believers, the *Shariat* (Code of Islam) existed exclusively in the Holy Quran; for others the basic laws of Islam were as enunciated in the early golden age, and only their application to twentieth century circumstances was required. To a third school, the *Shariat* remained a dynamic and not a static system; historical stages were available for study and guidance, and current and future developments were open to creative extrapolation.

As a result, we have throughout our history witnessed serious debate on religion-related issues such as the role of Islam in the State, how laws could be brought into conformity with Islamic injunctions, the form of government envisaged by Islam, and whether non-Muslims should vote jointly or separately with Muslims. The call of Islam which brought Pakistan into existence could not eventually prevent the separation of the East Wing.

The Pakistan Resolution of 1940 made no mention of Islam. After Independence, Quaid-i-Azam Mohammad Ali Jinnah was quite clear: 'Make no mistake: Pakistan is not a theocracy or anything like it.' He promoted the concept of a nation state, without regard to the religion of its citizens. After his death, we saw the first reference to Islam in a 'constitutional' document, the 1949 Objectives Resolution:

> Whereas sovereignty over the entire Universe belongs to God Almighty alone and the authority which He has delegated to the State of Pakistan through its people for being exercised within the limits prescribed by Him is a sacred trust...
>
> Wherein the principles of democracy, freedom, equality, tolerance and social justice, as enunciated by Islam, shall be fully observed;
>
> Wherein the Muslims shall be enabled to order their lives in the individual and collective spheres in accordance with the teachings and requirements of Islam as set out in the Holy Quran and the Sunnah...

However, speaking on this Resolution, Prime Minister Liaquat Ali Khan held out against theocracy:

> I just now said that the people are the real recipients of power. This naturally eliminates any danger of the establishment of a theocracy...I cannot over-emphasize the fact that such an idea is absolutely foreign to Islam...and, therefore, the question of a theocracy simply does not arise in Islam.[3]

Nevertheless, from 1949 onwards, the move towards an 'Islamic Constitution' gained ground. The 1956 Constitution incorporated the Objectives Resolution in the preamble and the country was described as 'The Islamic Republic of Pakistan'. Islamic provisions were, however, largely set out in what were unenforceable Directive Principles of State Policy. While the President was obliged under Article 197 to establish a body for Islamic research and instruction, nothing was done in its furtherance. Although Article 198 provided for bringing the laws into conformity with Islamic injunctions, no supervising authority was established.

After Ayub Khan became President, he attempted to reverse the trend by omitting the term 'Islamic Republic' in the 1962 Constitution. Soon, however, it was reintroduced under the First

Amendment. The remaining Islamic provisions of his Constitution were similar to those contained in the 1956 Constitution, and equally ineffective. Although they were not implemented, the subject itself remained contentious.

The Executive

The question of a presidential or parliamentary executive got entangled with the debate on the form of government Islam envisaged, and was further complicated by the wishes of military dictators.

In neighbouring India, the matter was settled immediately on Independence when Pandit Jawaharlal Nehru became Prime Minister and Lord Mountbatten continued as Governor-General. The parliamentary system, with executive authority vesting in the Prime Minister, has since been in force. In Pakistan, however, the Quaid-i-Azam became the first Governor-General, with popular support. As the Founder of the Nation, all power lay with him. On his death, executive authority devolved on Liaquat Ali Khan, who continued as Prime Minister. Until his assassination in 1951, the source of executive authority was clear. After this a diarchy set in.

Governor-General Ghulam Mohammad exercised certain 'inherent' powers in dismissing Prime Minister Nazimuddin and, later, dissolving the Constituent Assembly. Changes of government, either instigated or engineered, became a common phenomenon. The 1956 Constitution did not improve the situation as the President could, in his discretion under Article 37, appoint as Prime Minister the person who, in his opinion, was most likely to command the confidence of the majority in the National Assembly. The President retained an important role.

After October 1958, Ayub Khan ruled the country with dictatorial authority, to which the 1962 Constitution merely gave constitutional colour. The President's position was unassailable. His removal by impeachment under Article 13 required three-fourths of the total membership of the Assembly and, if it received less than half the votes, those who gave the initial notice automatically ceased to be members. Unlike, for example, the President of the United States, he did not share with the legislature the power of important appointments. He could also dissolve the National Assembly at any time under Article 23. He enjoyed extensive veto power over legislation and could return a Bill which then required a two-thirds majority of the National Assembly; and even then he could withhold assent and refer the Bill for a referendum to the Electoral College of Basic Democrats under Article 27. Moreover, the authority of the legislature over national finances was restricted, by Articles 41 and 42, to new and not recurring expenditure.

The President's powers to declare an Emergency were overwhelming. Such concentration of authority in the President inevitably resulted in abuse and upheaval. It discredited the presidential system and confirmed the view that a Westminster-style parliament was the proper solution, particularly with the country separated into two Wings. This was advocated in the PPP Foundation Documents and in its Manifesto. In spite of the changed circumstances involved in the loss of East Pakistan, ZAB was fixed in this position when it came to framing a new constitution.

Constitution-making under ZAB

When ZAB assumed power in the more homogeneous West Wing, or Pakistan as it became, the task of framing a constitution might have appeared less difficult. However, the earlier problems remained, even those concerning the federal structure. The Punjab now replaced East Pakistan as the Province with a population larger than the other Provinces combined.

Under the 1972 interim Constitution, discussed earlier in Chapter 5, there was a presidential system at the Centre, which suited ZAB, and a parliamentary system in the four Provinces.

The permanent solution required greater discussion, compromise and ingenuity.

We have earlier seen that the 6 March Accord provided for a Constitution Committee of the National Assembly, which then met for the first time on 22 April 1972 under Law Minister Qasuri. There was agreement on generalities; difficulties on particular issues were considered in the second session commencing on 18 May. Statements and discussions in the Committee, without any draft document, proved pointless, and the Committee asked Qasuri to submit a draft by the end of June. In the absence of specific direction or basic agreement, no draft could be prepared. On 13 August, the Committee was allowed till the end of the year for the submission of a report to the Assembly.

Not only was there no agreement on specifics between the various parties represented in the Constitution Committee, but there were also differences among the PPP members. Sheikh Rashid was determined on a 'socialist' constitution, while Qasuri called for greater freedom under fundamental rights. There were also the major problems of the Islamic provisions and how to incorporate East Pakistan in the new Constitution since Bangladesh had not been recognized.

From the outset there were differences between Qasuri and ZAB, which increased with passing months. He had directed his Military Secretary to have me present at all meetings with Qasuri; all files from the Law Ministry were routed through me, placing me in an invidious position. Qasuri and I shared a good relationship and believed that a parliamentary system was unavoidable, particularly in the light of the PPP's previous position. Over dinner with us, Daultana and Shaukat Hayat sought a *via media* by allowing ZAB to continue as President, while remaining answerable to the Assembly. We felt this was impractical. When Qasuri first resigned over not being adequately consulted, ZAB sent me to dissuade him; but on 5 October his second resignation was accepted.

By now ZAB felt irritated and worried by the lack of progress on a permanent constitution. He considered no one indispensible, least of all Qasuri. The new Law Minister, Hafeez Pirzada, and I discussed with the President the outstanding issues, pointing out that these could not be settled by the Committee. Accordingly, ZAB called a meeting of all parliamentary party leaders on 17 October. The next four days, in the month of Ramadhan, produced what is commonly known as the 20 October Constitutional Accord, which formed the basis of the 1973 Constitution. The PPP, ML(Q), NAP from both Frontier and Balochistan, JUI, CML, JUP, JI, Tribal Areas and Independents were represented.

ZAB conducted the meetings with distinction, agreeing at the second session to a parliamentary form of government with certain safeguards for himself as Prime Minister; he conceded similar protection for the four Chief Ministers. That apart, the principal decisions concerned a bicameral federal legislature, federal and concurrent legislative lists with residuary powers vesting in the Provinces, a Council of Common Interests, provisions regarding hydro-electric generation and gas to meet the NAP demands for NWFP and Balochistan, and Islamic provisions. Some of these will be considered later in the context of the 1973 Constitution itself. The Accord was hurriedly dictated by me, with Mubashir Hasan filling in certain financial aspects, in order to confirm points of agreement in writing before any of the participants had second thoughts.

ZAB was delighted with the outcome. We appeared in group photographs as if nothing could again go wrong. The Accord was reached more smoothly than we had anticipated. In a brief public statement he said:

> I am thankful to all my colleagues from all the parties who have participated in these discussions. They have all made a contribution and without their contribution, without their understanding, I don't think we could have brought about a satisfactory compromise. Each one of them has played a part and I am indeed thankful to them.[4]

This was probably the first, and certainly the last, time he expressed such thanks to the Opposition.

The task of the Constitution Committee was considerably simplified by this Accord. Still, finalizing the draft was a painstaking undertaking, performed mainly by Law Secretary Justice M. Gul, Parliamentary Affairs Secretary Justice Saad Saood Jan, Hafeez Pirzada and myself, with the Attorney-General also contributing. The Report of the Committee was rushed through to meet the deadline of 30 December. It praised Pirzada's impartial and successful conduct of the proceedings as chairman, and thanked me for the 'original and constructive proposals...which contributed in large measure to the evolution of the Draft Constitution'.[5] My work two years before in preparing constitutional proposals for discussion with the Awami League, which did not take place, had at last borne fruit.

Several Committee members appended notes of dissent to the Report, even on matters settled earlier by the 20 October Accord. Mainly, they were in respect of the Political Parties Act, fundamental rights in regard to property, the dissolution of Assemblies, separation of the Judiciary from the Executive, Islamic provisions, an independent Election Commission, greater say for the Senate, the removal of the Prime Minister, and provision for a caretaker government to hold elections. Despite this, we expected a reasonably smooth passage in the Assembly for the Constitution Bill, as it was based on the 20 October Accord. However, this was not to be.

In the Assembly, difficulties were exacerbated in February 1973 by the dismissal of the NAP-JUI Government in Balochistan, the change of NAP Governors in the two Provinces, and the resignation of the NAP-JUI Government in the Frontier. ZAB's attitude to the Opposition brought unity in their ranks and they jointly presented, on 17 March, a list of proposed amendments to the Constitution Bill. Relations deteriorated further following the bloody disruption of the Opposition public meeting at Liaquat Bagh, Rawalpindi, on 23 March. They had already boycotted the Assembly proceedings, and the following day published their eleven-point demand.[6] An agreed Constitution now appeared to be eluding ZAB and he held several high-level meetings of the PPP. By this time, the PPP

Central Committee seldom met, and the main participants in these strategy sessions with the President were Khar, Pirzada and myself, with Mustafa Jatoi and Yahya Bakhtiar also at times invited. Instead of replying formally to the Opposition's proposals, which might have resulted in a deadlock, it was decided to invite the leaders of all the parliamentary parties and groups to consider these eleven proposals at a meeting on 2 April. ZAB asked whether there were any additional points, and these were submitted two days later. There was some agreement on the first day, but major differences remained. Following the 2 April session, he asked me to prepare a suitable reply to the points raised by the Opposition. With the able assistance of Justice Gul and Justice Jan, a detailed *aide memoire* was completed and sent to the Opposition after the meeting of 4 April. It concluded with ZAB pointing out that 'concessions and important compromises' were made on the condition that the Opposition leaders fulfil their 'solemn obligation to the people' and return to the National Assembly on 7 April 'to stay there till the Constitution is framed'.[7]

The Opposition parties, mainly represented by the United Democratic Front, did not end their boycott. Instead, on 9 April, they issued a blistering and detailed rejoinder:

> We want a parliamentary form of Government and not the Prime Minister's dictatorship. We want a Constitution in which provincial autonomy cannot be destroyed under the guise of any emergency and a Constitution which guarantees to the citizens the enjoyment of their Fundamental Rights except during hostilities...

It was published in the newspapers over two days.[8] However, by the second day the text became redundant since we had unexpectedly reached a consensus on the Constitution. Mustafa Khar had used all his connections and persuasive powers in the Punjab to bring round some Opposition leaders, particularly from the Jamaat-i-Islami, while ZAB and his agencies were also hard at work. With a few minimal, almost cosmetic, amendments suggested by me, the boycott ended on 10 April, and the

Assembly immediately adopted the Constitution Bill without dissent, though not unanimously as ZAB claimed.

The speed of developments left even those close to events wondering how and why the Opposition, after such strong objections, should settle so far short of their demands. ZAB felt he had out-manoeuvred them. While not detracting from his role and that of others, it would be more fair to conclude that the Opposition leaders realized the importance of the Constitution to the recently dismembered country. The NAP-JUI members, who particularly had reason to oppose him after the recent events in Balochistan and the Frontier, showed magnanimity. The religious parties too, led by the Jamaat-i-Islami, displayed considerable accommodation in the national interest. The entire country wanted a permanent Constitution and no one was prepared to disappoint the people. Two days later, on 12 April, the members signed the document and it was presented to President Bhutto for authentication. His aim of an agreed Constitution had been achieved.

Permanent Constitution of 1973

The 1973 Constitution came into effect on 14 August 1973 and has proved to be ZAB's most significant achievement. Many of its provisions were contained in the 1956 or 1962 Constitutions, but numerous innovations were made covering the federal structure, the Islamic content, the Executive, and Martial Law and the Armed Forces, which deserve comment.

As far as the federal structure was concerned, one early complication was how to provide for East Pakistan, as Bangladesh had not yet been recognized. The Opposition insisted on its inclusion. Instead of having a Constitution requiring major changes subsequently—we wanted a permanent arrangement—I proposed we draw on the Constitution of the Federal Republic of Germany, and so Article 1, clause (3), provided:

> The Constitution shall be appropriately amended so as to enable the people of the Province of East Pakistan, as and when foreign aggression in that Province and its effects are eliminated, to be represented in the affairs of the Federation.

Several new features figured in the federal arrangements. For the first time, there was a bicameral legislature: the National Assembly was to be elected by direct vote; the Senate under Article 59 provided equal representation to the Provinces and was to be indirectly elected by the Provincial Assemblies. The Senate, however, was allowed no say in money matters. Moreover, Bills within the exclusive jurisdiction of the Centre could only originate in the National Assembly and then, if rejected by the Senate, required an absolute majority of the Assembly to become law. When the Federal Parliament considered matters over which the Centre and the Provinces enjoyed concurrent jurisdiction, in the event of disagreement between the two Houses, they were to go before a joint session. My suggestion that the Senate should have the right to initiate all Bills and be given some role in financial matters, at least as in Australia to debate them, was not accepted by ZAB on the ground that the Punjab would not agree.[9]

The division between the Federation and the Provinces over legislative subjects was settled by the 20 October Accord. The comparative table on the next page shows the extensive list of subjects over which the Centre continued to enjoy exclusive or concurrent jurisdiction.

The representatives of the smaller Provinces, particularly NAP, showed a surprising degree of accommodation in agreeing to such wide-ranging federal legislative powers in order to achieve consensus in October. ZAB indicated to them that the concurrent list of subjects could be reviewed in ten years time, but no one has attempted this in the past twenty-three years.

A major innovation was Part II of the Federal List, covering railways, oil and gas, the development of major industries, and water and power, which were subjects of inter-provincial concern. The Senate had equal powers and, on the administrative

List of Subjects	1935 Act	1956 Constitution	1962 Constitution	1973 Constitution
Federal	61	30	94 (Only 1 list)	Part I – 59 Part II – 8
Provincial	55	94		—
Concurrent		19		47
Residuary Powers		With Provinces	Provinces (but with overriding proviso regarding national interest)	With Provinces

side, Article 153 established a Council of Common Interests to formulate and regulate policies for these matters. Composed of equal numbers from the Provinces and the Centre, with the four Provincial Chief Ministers and four Federal Ministers, including the Prime Minister, it was subject only to Parliament in joint sitting. It also enjoyed exclusive jurisdiction over complaints and disputes between Provinces over water supplies, of vital importance in Pakistan, which were previously referable only to the Supreme Court.

Executive arrangements at the federal level were made applicable, *mutatis mutandis,* in the Provinces. The basic demands for provincial autonomy were met subject to Articles 232 and 234 relating to Emergency, whereby the Federal Government could assume powers in the Provinces. Such provisions are normal in a federation, and can be salutary if used properly. In Pakistan, they have throughout been weapons in the hands of the Federal Government to be used against the Provinces, especially those governed by opposition parties.

The Council of Common Interests was a new and potentially beneficial institution; unfortunately, only one formal meeting was held, at the end of 1976.[10] Instead the Ministry for Provincial Coordination, which was established to iron out routine differences between the Provinces, functioned as a mini-Council, and the constitutional issues of concern to the Council were left largely unattended. Despite all four Provincial Governments being PPP-dominated and supporting the Federal Government, power remained centralized in ZAB and was not exercised through the Council. In the rush to reach the Accord of 20 October, the Council had been conceded wide powers, and he did not want these implemented.

Although a federation in name, Pakistan has in reality always been run by a highly centralized Government. The delicate constitutional and federal balance achieved by ZAB through considerable effort was slowly dissipated. Interference in the Provinces became the norm. True to tradition, he centralized power and circumvented the carefully designed federal arrangements.

On the Islamic content of the Constitution, we have noted how earlier debates yielded little result. The 1973 Constitution settled some long-standing issues. As the leader of a socialist party in an Islamic State, ZAB was conscious of the need not to appear too secular; nevertheless, we projected the social gospel and dynamic spirit of Islam, which in essence is a practical, ethical religion, rather than a static set of prohibitions and injunctions.

Sheikh Rashid and a few other PPP members had earlier insisted in the Constitution Committee on using the phrase 'Islamic Socialism', which was opposed by the religious parties. To avoid deadlock, I proposed that Article 3 should require the State to ensure the elimination of all forms of exploitation and the gradual fulfilment of the fundamental principle, 'from each according to his ability, to each according to his work'. The religious parties accepted this, not realizing that it was a basic socialist doctrine. I had assumed this solution would meet with the President's approval, but, when I reported it to him, he was not pleased.

We also wanted the Constitution to facilitate social change and ensure that property rights could not be exploited to hinder reform. The 1973 Constitution accepted the right to property, but Article 24, clause (4), laid down that compensation for compulsory acquisition could not be questioned in any court. Parliament was given the final say in determining the quantum of compensation, and dilatory litigation was prohibited. Also, Article 8, clause 3(b), specifically provided immunity for all economic reforms introduced by the PPP Government prior to the Constitution. Moreover, Parliament was given the power, under Article 253, to legislate on the maximum limits of property holdings and to declare that any trade, business, industry or service could be nationalized by the Federal or Provincial Governments. These provisions regarding property rights were severely criticized from all sides, by Shaukat Hayat Khan of the Council Muslim League, by Ghafoor Ahmed, Maulana Shah Ahmad Noorani Siddiqui and Maulana Mufti Mahmood, representing the three main religious parties, as also by Amirzada Khan of the left-leaning NAP.[11]

The three main religious parties, the Jamaat-i-Islami, the JUP and JUI, had played a positive role in the Assembly to help bring about consensus. While insisting on a time-frame for the reports of the Council of Islamic Ideology, which was inserted after last-minute discussions, they mainly rested their case on their earlier notes of dissent. They described the draft Constitution as 'disappointing from the points of view of Islamic Provisions'.[12]

The 1973 Constitution concentrated executive authority in the Prime Minister, who was designated the 'Chief Executive'. ZAB directed me to ensure the complete end of the diarchy that had existed in the parliamentary system before 1958. The President was made a mere cypher who, according to one Opposition member, 'had less power than the Queen of England and none of her glory'. The most detailed provision concerning him related to impeachment, which gave rise to comments that the President had only one right, to be impeached.

The Prime Minister was to be elected by the National Assembly, and was to appoint the Federal Ministers. Previously this had been done by the President. Under Article 96, the Prime Minister could only be removed by a vote of no confidence which named his successor in the same motion. The Prime Minister's position was further fortified by discouraging floor-crossing in the Assembly. Earlier, the 20 October Accord had provided for the Prime Minister to be removed by a two-thirds majority. However, when this provision was subsequently strongly opposed, I suggested a compromise formula which, at the last minute, was accepted and incorporated in the Constitution. Under this, in the case of a vote of no confidence or on the budget, the vote of a member cast against the majority of his party would be disregarded. This provision would remain for a period of ten years or two general elections. But, although less objectionable than the two-thirds provision, it still ensured that in practice the Prime Minister was almost irremovable by the Assembly.

To forestall any possibility of the President acting independently, Article 48 required him to 'act on and in accordance with the advice of the Prime Minister and such advice shall be binding on him'. Moreover, the counter-signature of the Prime Minister was required for validating all the orders of the President. Some of these provisions followed the Constitution of the Federal Republic of Germany, but cumulatively the concentration of power admitted no flexibility. This lack of flexibility contributed to the return of Martial Law after the 1977 elections.

No analysis of constitutional issues in Pakistan is complete without some reference to Martial Law. Although extra-constitutional, the Supreme Court has bestowed legitimacy on it by recourse to the 'doctrine of necessity'. The Dosso Case[13] is particularly noteworthy in this connection. However, following the fall of Yahya Khan in 1971, the Supreme Court reversed this position and decided in the Asma Jilani Case that:

As soon as the first opportunity arises, when the coercive apparatus falls from the hands of the usurper, he should be tried for high treason and suitably punished. This alone will serve as a deterrent to would-be adventurers.[14]

The constitution-makers in 1972 were determined to ensure that the military did not again play a political role. Armed with this latest Supreme Court decision, it was provided for the first time, in Article 6, that any attempt to abrogate or subvert the Constitution by force would be punishable as high treason. Also, Article 12(2) made specific exception to the principle of non-retrospectivity of offences and punishments in the case of such high treason from the time of the 1956 Constitution.

Finally, in order to keep the Armed Forces in check, the Federal Government was conferred 'control and command' over them under Articles 243 to 245, and they were required to act in aid of civil power as directed by the Government. All members of the services had to take oath to 'uphold the Constitution' and not engage 'in any political activity whatsoever'. However, when the time came, these fine phrases did not deter the military *coup d état* of 5 July 1977. The Supreme Court subsequently modified its earlier decision in the Asma Jilani Case by holding the *coup* legal in the Begum Nusrat Bhutto Case.[15]

Constitutional Amendments and Consequences

Seven Amendments were made in the Constitution between May 1974 and May 1977. Mainly, they undermined the rights of individuals and the Judiciary. Only the Second Amendment, which defined a Muslim in the context of declaring Qadianis non-Muslims, was supported, if not encouraged, by the Opposition.

The First Amendment, apart from removing the reference to East Pakistan, allowed the Federal Government to ban political parties formed or 'operating in a manner prejudicial to the sovereignty or integrity of Pakistan', subject to a final decision by the Supreme Court.[16] Moreover, the writ jurisdiction of the High Courts, which could not previously be exercised in favour of a 'member of the Armed Forces', was extended to exclude any person 'who is for the time being subject to any law relating to any of those Forces'.[17]

The Third Amendment of February 1975[18] affected safeguards against arrest and detention under Article 10, the provision on which the Constitution Committee had laid stress. The Amendment also facilitated the continuation of a Proclamation of Emergency under Article 232. Nine months later, the Fourth Amendment[19] curtailed the writ powers of the High Courts under Article 199 in respect of preventive detention. Now, no order could be made prohibiting detention or granting bail to a person so detained.

Within ten months, the Fifth Amendment further restricted the Courts' powers under Article 199. More significantly, the term of office of the Chief Justices of the Superior Courts was to be determined not solely by age but, as an alternative, by a fixed period. ZAB's intention was to secure the extension in office of the Chief Justice of Pakistan and a premature end to the term of the Chief Justice at Lahore. Safdar Shah who was the Chief Justice of the Peshawar High Court became an incidental victim. The Executive was also empowered to transfer a Judge to any High Court for up to a year without his consent or even consultation with the Chief Justice concerned. Furthermore, in the past, a powerful Chief Justice of a High Court had at times refused to be appointed a puisne Judge of the Supreme Court; now, however, it was provided that, on failure to accept such appointment, he 'shall be deemed to have retired from his office'.[20] In addition, the Executive could appoint 'any one of the Judges', and not the most senior, to act as Chief Justice. In other respects, however, as in new appointments to the Superior Courts, ZAB accepted good advice in nominating qualified Judges on merit and not, except for one or two cases in the later period, on political considerations.

By the time the Seventh Amendment was passed on 16 May 1977, ZAB's Government was in the throes of post-election turmoil. It provided for a referendum to avoid re-elections. At the time the military had been called in to aid civil power in pursuance of Article 245, a form of 'mini Martial Law' in certain areas, and, when this was challenged, the jurisdiction of the High Courts was taken away in respect of ensuing actions.

It has been a common practice in Pakistan, both before and after the PPP Government, to seek in constitutional amendments and special laws a panacea for the problems facing the administration.[21] In doing so, the PPP sadly damaged its own creation, the 1973 Constitution. Such action was particularly unnecessary since the Government had sufficient authority under Emergency powers to cope with the aftermath of military rule and the débâcle in East Pakistan. These Amendments in the mid-1970s contributed considerably to the general view that ZAB could tolerate no opposition, not even from the Superior Courts, in his quest for dictatorial powers. The repeal of these Amendments became a principal issue in his last months in government. General Ziaul Haq later reversed a few of their provisions, which were inconvenient to him, but otherwise continued and even 'improved' upon them.

Despite some of these unnecessary and controversial changes, the PPP Government did indeed settle many problems that had long troubled constitution-makers in Pakistan. Nevertheless, the Constitution cannot tell the whole story, being only a legal form for a political system. In Pakistan, created from the diverse Muslim majority areas of India, constitution-making assumed overwhelming importance as a symbol of, almost a substitute for, national integration. A disproportionate amount of time and effort has been devoted to it, making it an end in itself rather than a means to an end—the right ordering of a nation's life. Pakistan's leaders and lawyers in government have throughout failed to look to the future, and the leaders have sought to constitutionalize their personal requirements.[22] However, once the over-protective provisions for the Executive had ceased after ten years, the 1973 Constitution would have provided a reasonable framework for building a democratic and parliamentary form of government. The tragedy of Pakistan is reflected in the fact that no system has been allowed to work, let alone succeed.

In the final analysis the 1973 Constitution can be termed a triumph for the PPP. General Zia as CMLA and President from 1977 to 1988 might have mangled and mauled the text, but even he, as a military dictator, could not abrogate the Constitution which had been so painstakingly achieved by the PPP Government.

NOTES

1. *See, Report of the Simon Commission,* 1930, Volume II, Parts 23 and 24, p. 14, and Part 27 p. 16, for details.
2. Legal Framework Order, 1970, President's Order No. 2 of 1970, *Gazette of Pakistan, Extraordinary,* of 30 March 1970, Section 20. *See,* Chapter 2, pp. 29-31 for further details about the LFO.
3. *Constituent Assembly of Pakistan Debates,* Official Report, 7 March 1949, p. 3.
4. *The Pakistan Times,* Rawalpindi, 21 October 1972.
5. *Constitution-Making in Pakistan,* 1973, published by the National Assembly Secretariat in April 1975, p. 118.
6. *Dawn,* Karachi, 25 March 1973.
7. *Constitution-Making in Pakistan, aide memoire* at pp. 147-59, quote at p. 159.
8. *Dawn,* Karachi, 10 and 11 April 1973.
9. General Zia subsequently permitted the initiation of all Bills in the Senate by amending the Constitution.
10. *See,* Chapter 9, pp. 316-7 for details.
11. *Constitution-Making in Pakistan,* Notes of Dissent of 30 December 1972 by Shaukat Hayat Khan at p. 122, by Ghafoor Ahmad, Maulana Shah Ahmad Noorani Siddiqui and Maulana Mufti Mahmood at pp. 130-1, and Amirzada Khan at p. 134. Article 23 referred to in the draft Constitution became Article 24 in the final text.
12. *Constitution-Making in Pakistan,* Notes of Dissent by Ghafoor Ahmad, Maulana Shah Ahmad Noorani Siddiqui and Maulana Mufti Mahmood at p. 129.
13. PLD 1958 Supreme Court 533. *See also,* Leslie Wolf-Phillips, 'Constitutional Legitimacy: a study of the doctrine of necessity', in *Third World Foundation,* 1979, London.
14. PLD 1972 Supreme Court 139 at p. 243.

15. Begum Nusrat Bhutto Case, PLD 1977 Supreme Court 657.
16. Constitution (First Amendment) Act, 1974, (Act XXXIII of 1974), *Gazette of Pakistan, Extraordinary,* Part I of 8 May 1974, Section 4.
17. Ibid., Section 9.
18. Constitution (Third Amendment) Act, 1975, (Act XXII of 1975) *Gazette of Pakistan, Extraordinary,* Part I of 25 November 1975.
19. Constitution (Fourth Amendment) Act, 1975, (Act LXXI of 1975), *Gazette of Pakistan, Extraordinary,* Part I of 25 November 1975.
20. Constitution (Fifth Amendment) Act, 1976, (Act LXII of 1976), *Gazette of Pakistan, Extraordinary,* Part I of 15 September 1976, Section 14, clause (2).
21. Of numerous examples, only three are here cited. First, Article 104 of the 1962 Constitution did not permit a Member of any Assembly to be a Minister, and President Ayub Khan, at the request of ZAB and some other Ministers, attempted to change this by President's Order No. 34 of 1962 published in the *Gazette of Pakistan, Extraordinary,* of 12 June 1962. The Supreme Court overruled this on the ground that it was beyond the scope of the 'Removal of Difficulties' provision under which Ayub Khan had acted. Second, Prime Minister Muhammad Khan Junejo under President Ziaul Haq enacted the Eighth Amendment to the 1973 Constitution, published in the *Gazette of Pakistan, Extraordinary,* of 11 November 1985, which, *inter alia,* considerably enhanced the powers of the President. Third, Prime Minister M. Nawaz Sharif rushed through Parliament the Twelfth Amendment to the 1973 Constitution published in the *Gazette of Pakistan, Extraordinary,* of 28 July 1991, which provided for the establishment of Special Courts 'to ensure speedy trial' of persons accused of 'heinous offences specified by law'.
22. Of several instances, two sufficiently highlight this point and demonstrate the uniqueness of Pakistan's Constitutions. In the 1962 Constitution, the preamble provides, 'Now, therefore, I, Field Marshal Muhammad Ayub Khan, Hilal-i-Pakistan, Hilal-i-Jurat, President of Pakistan, in exercise of the Mandate given to me...by the people of Pakistan...*do hereby enact this Constitution.'* (emphasis added). Article 226(1) further provided, 'In accordance with the result of the referendum...Field Marshal Muhammad Ayub Khan...shall, notwithstanding anything in this Constitution... become the first President of Pakistan...' (*see,* p. 169 concerning this 'Mandate' and 'referendum'). Similarly, President Ziaul Haq inserted clause (7) to Article 42 of the 1973 Constitution by President's Order No. 14 of 1985 with effect from 2 March 1985: 'Notwithstanding anything contained in this Article or Article 43, or any other Article of the Constitution or any other law, General Mohammad Ziaul Haq, in consequence of the result of the referendum...*shall become the President of Pakistan....'* (Emphasis added.)

CHAPTER 7

FOREIGN AFFAIRS

Foreign affairs had always held pride of place for ZAB. He read widely and kept abreast, if not ahead, of developments. Much of his book, *The Myth of Independence,* was devoted to this subject. In it, he had maintained that 'Pakistan must determine its foreign policy on the basis of its own enlightened national interest uninfluenced by the transient global requirements of the Great Powers'.[1] He envisaged an important role for Pakistan because of our strategic geo-political position but, in the first two years of his Government, his plans to place the country and himself centrally on the world stage were overwhelmed by considerations of the war's aftermath and 'picking up the pieces' with India.

Pre-Simla: 20 December 1971 to 27 June 1972

In his broadcast immediately after becoming President, though restrained by defeat, he attempted to maintain his anti-Indian image as the leader who in 1965 had called for a thousand-year war:

> Mr Jagjivan Ram should know this is not the end of war. This is the beginning of war, of a new state of affairs. He should not gloat over temporary victories. India should not get intoxicated with the military take-over of East Pakistan. We will continue to fight for the honour and integrity of Pakistan.[2]

While the tone of this was understandable given the need to boost the morale of his domestic audience, particularly the

Armed Forces, it did not send the right signals to India, where politicians and journalists dwelt on these words to question his subsequent actions and intentions.

Rhetoric apart, the situation confronting ZAB was no different from that which had forced Yahya Khan to accept the cease-fire in the West Wing. East Pakistan was lost, there were over 90,000 Pakistani civilian and military prisoners of war, and the Indian Army had occupied 5,000 square miles in Sindh and Shakargarh, a sensitive area of the Punjab. The armies of the two countries faced each other in the trenches and any incident could have triggered further conflict, for which the Pakistani troops were ill-equipped, without even adequate winter clothing or blankets for the January cold in the north. In contrast, the morale of the Indian troops was high after their victory, and their mountain divisions—established with US assistance after China stunningly defeated India in the 1962 Border War—were now fully in place. Unlike in 1965, China had applied no pressure on the northern borders to engage Indian troops, whose full might remained poised against West Pakistan.

ZAB had to come to terms with India but, before undertaking negotiations, he had not only to strengthen his domestic position but to repair international relations. The Pakistan he had inherited was isolated internationally as a result of the civil war, the errors of the military Government and immense Indian propaganda. Defeat does not make friends. He sought international sympathy and support by visiting numerous, mainly Muslim, countries, and consolidating his position on the domestic front. So far as the superpowers were concerned, he was determined to avoid what he regarded as the errors of Tashkent which had resulted in the fall of Ayub Khan. While he recognized the important role the Soviet Union could play in influencing India, he sought Soviet assistance to initiate direct talks but not mediation. This, as we later learnt, was also India's intention.

As far as the US was concerned, he could not gain the support he hoped for. Previously, the Nixon Administration, for its own geo-political considerations, had warned India against destroying Pakistan militarily in the West Wing. However, the US seemed both unable and unwilling to proceed further. They lacked leverage on Prime Minister Indira Gandhi; possibly, they also did not want to exercise pressure directly on her to get talks under way between our two countries. They tried to correct the so-called 'tilt' towards Pakistan during the war. Throughout this early period, except briefly, Washington did not have ambassador-level representation in Pakistan. Relations were maintained at the level of the Chargé d'Affaires, Sidney Sobers, who was on cordial terms with ZAB.

The Soviets had punished Pakistan for arranging Kissinger's secret mission to Beijing, considered an attempt to isolate the Soviet Union, and for rejecting Brezhnev's Asian Plan, but they did not want to see India dominating the region, with the consequent loss of their own influence. They sought to continue their key role in South Asia, and pressed for the early recognition of what they termed the reality of Bangladesh. Soon after ZAB assumed office, the Soviet Ambassador delivered a *démarche* from his leadership to this effect, calling for early talks with India.

Although ZAB had clear ideas about foreign policy, he lacked confidence in the ability of existing institutional arrangements to implement them. Despite bringing back an experienced retired official, Aziz Ahmed, to head the Foreign Office as Secretary-General, he nevertheless wanted to maintain another channel through me as his Special Assistant. He told Sidney Sobers that if there was anything urgent or important to convey and he was not available, it should be communicated directly through me. On my suggesting that this facility should not be confined to the Americans alone, the Soviet and Chinese Ambassadors were similarly informed. Moreover, to ensure a political approach and avoid a post-Tashkent scenario, he again placed me between the Foreign Office and himself in the talks with India.

In his conduct of foreign affairs he also indulged his penchant for dramatic effect. An example of this occurred in his decision to withdraw from the Commonwealth of Nations. He felt it was necessary to take some steps to dissuade states from according early recognition to Bangladesh; he adopted the 'Hallstein Doctrine' by which West Germany had tried to prevent countries from recognizing East Germany, and severed diplomatic ties with those recognizing Bangladesh. It soon became apparent to us that this could not be done indefinitely, and an early test came when the United Kingdom recognized Bangladesh. He was angered by the hasty acceptance of Bangladesh by Britain and other Commonwealth countries. In this case he did not apply the Doctrine, 'just as West Germany continued diplomatic relations with the Soviet Union'. Instead, he decided to leave the Commonwealth, and summoned me to tell me about 'the dramatic move' as the Foreign Office made the short announcement. In a long diatribe, he described the Commonwealth as anachronistic and serving no useful purpose; far from allowing a free exchange of views, its disparate membership hindered discussion and decisions on various issues; disputes with India could not be raised because they were bilateral. Third World matters were not dealt with satisfactorily as the Old Commonwealth countries did not share the views of developing countries, and even anti-apartheid policies could not be followed because of their links with South Africa. In particular, the emerging importance of the Muslim countries and South-East Asia left no scope for the Commonwealth; even Britain had turned to Europe. He had for some time seen no merit in the organization, and now, he pointed out, was a good opportunity to make a 'historic' decision by 'breaking away from the colonial legacy of the past'. He said it would go down well with the public.

He expressed surprise at my lack of response and asked if I were upset by the decision. I replied that I was not, though I failed to understand the undue significance he attached to this glorified English Speaking Union. There might be no great advantage in membership, but equally there was no particular disadvantage; we were friendless after the 1971 civil war, and it could provide a useful international forum. The President described the difficulties he had experienced as Foreign Minister, and the political importance of 'kicking away colonial remnants'. Our talk ended unsatisfactorily, certainly from his point of view, as I remained unconvinced. He arrested Altaf Gauhar, the Chief Editor of *Dawn,* a leading Karachi daily, partly for expressing views similar to my own.

His was a love-hate relationship with Britain, scornful yet admiring many qualities of her people. He prided himself on his Oxford past and derived immense pleasure from the fact that his two older children followed in his footsteps, valuing their Oxford careers much more than their earlier American education. He was consequently deeply hurt when, on 7 February 1975, the Oxford University Congregation denied him an Honorary Doctorate by a narrow vote, despite every effort on his part,[3] mainly because he had supported the military action in East Pakistan in 1971.

There was an unease in his relations with the British and their acceptance of him. Soon after the decision to withdraw from the Commonwealth, he was to play host in Larkana to all the Ambassadors accredited to Pakistan and thought the British Ambassador might refuse the invitation. I said it would be a breach of protocol and he would come unless recalled for instructions. Sir Laurence Pumphrey did attend. Surprisingly, in spite of his reticence about the British, he asked the same Ambassador, a few years later, for his opinion about the Punjab Chief Minister, Sadiq Hussain Qureshi.

Over a year after his decision to leave the Commonwealth, he still took pride in it: 'as a forward-looking nation, we reject any legacy of the past which has outgrown its usefulness. Hence Pakistan has recently left the Commonwealth, which had long since ceased to have any practical meaning'.[4] Seventeen years later, in 1989, his daughter Prime Minister Benazir Bhutto claimed Pakistan's acceptance back into the Commonwealth of Nations as a personal achievement, especially as General Ziaul Haq had earlier been refused this 'honour'. There must have been smiles in the British Foreign Office.

More constructively, ZAB began his tours abroad in an effort to stave off the early recognition of Bangladesh by the Muslim World. His visits to Iran, Turkey and the Maghreb countries proved successful and contributed significantly to strengthening

Pakistan's position in subsequent negotiations with India. He took the opportunity to explain to their friendly leaders the errors of the previous regime which had resulted in the civil war and defeat, and also about his plans for a 'new Pakistan'. Of all the leaders of the Muslim World whom I met, President Boumédienne of Algeria and President Asad of Syria impressed me the most. They were down-to-earth, lived simply without any sense of false grandeur, and at the same time showed a wide understanding of world developments.

Next, at the end of January 1972, he went to China, a country of unquestionable importance to Pakistan. After a long flight, with magnificent views of the Karakoram mountains, the highest range in the world, we went straight into the first session of talks. Air Marshal Rahim Khan, the Air Force C.-in-C., sat on the President's right and Lt.-Gen. Gul Hassan, the acting Army C.-in-C., on his left. The same two officers had accompanied him to Beijing in November 1971, and were reputed to have helped bring him to power. Our civilian government thus appeared 'propped up' by the military.

ZAB greatly admired Premier Zhou Enlai, who was indeed a most impressive man. Zhou Enlai inquired about various members of the Pakistani team and their ages. When he learnt my age, he joked that everyone on his side was twice as old. This was apparent as the meeting progressed. The hall was over-heated. Feeling tired after the long flight, I dropped off to sleep for a few seconds. Embarrassed on awakening, I noticed that the Chinese Deputy Defence Minister, a veteran of the Long March, and another person sitting opposite were sound asleep. They woke as if sleeping at meetings was routine, stretched, called for warm towels, and continued their participation after refreshing themselves. This was indicative not only of age but self-confidence.

The first meeting adjourned before dinner. ZAB recalled telling me earlier that, after the 1965 Indo-Pakistan War, there had been a major debate among the Chinese leadership over whether to continue their policy of close friendship with Pakistan against India. Senior leaders, including former President Liu

Shao Chi and the Mayor of Beijing, considered Pakistan was feudal, with a degenerate leadership, and not worth supporting as we made little effort to stand on our own feet. At that time Premier Zhou Enlai, Foreign Minister Chen Yi and others won the argument in favour of Pakistan. He thought the same debate might be taking place again following the 1971 surrender. He was disturbed that no mention had been made of a meeting with Chairman Mao Zedong, and that the customary ceremonial drive in an open car had not taken place on his arrival. I was asked to mention both matters to the acting Foreign Minister Chiao Kuan Hua, my counterpart host.

While proceeding to dinner I told Chiao Kuan Hua that we were delighted with the airport reception; I had seen pictures in the past but the reality was more impressive. However, on this occasion, the state drive seemed missing. I also inquired about Chairman Mao's health, saying that since my youth I had greatly admired Mao and Zhou. He smiled and took me towards Premier Zhou for a photograph. At the next session the Premier excused himself briefly. When he came back he said that Chairman Mao would see the President.

During Premier Zhou's absence, ZAB had asked me to raise the question of equipping our proposed mountain division, giving me a list to refer to the acting Foreign Minister. When I mentioned some items, including padded uniforms, the acting Foreign Minister quietly pointed out that the cotton used in them came from Pakistan. So much for the brief prepared for us. So much for Pakistani self-reliance, which probably confirmed the Chinese leadership's view of Pakistan.

After the President met Chairman Mao, discussions with Zhou Enlai proceeded smoothly. By then the military 'props' had also been removed and ZAB looked in control. The Chinese were impressed that he remembered to call on the widow of his old friend, the late Foreign Minister Marshal Chen Yi. They respected their heroes and, for quite a while, his successor was only designated acting Foreign Minister.

Premier Zhou Enlai advised us not to rush through reforms, not to move on too many fronts, to take a step at a time and consolidate at each stage. He stressed the importance to Pakistan of good relations with the United States, and the need to improve ties with the Soviet Union. He pointed out that, for the next ten years, China could not significantly counter any Soviet measures in the region. He was clearly referring to the Indo-Soviet Treaty of August 1971, and the Soviet support for India in the recent war; he was conscious of China's lack of backing for Pakistan. A ten-year period was probably considered necessary for China to make sufficient economic and military progress, although the subsequent disorder caused by the 'Gang of Four' delayed this. He encouraged our going to Moscow in search of solutions to the war's aftermath. He was a true pragmatist.

At the end of the visit we set out in a motorcade for the airport. After a while, our car stopped. On my inquiring about the reason, the acting Foreign Minister told me the two leaders were transferring to an open car; there had been no state drive on our arrival because of the severe winter and Premier Zhou's influenza, 'but, though he is still not fully recovered, he wants to pay full honour to the President of Pakistan at the time of departure.' He smiled.

We next had to brace ourselves for the important three-day visit to Moscow, for which careful preparations were made. The rhetoric against India was toned down. In an interview with Ian MacIntyre of the BBC on 18 February, ZAB talked of a *modus vivendi* with India: 'Even when my posture was different when I was preaching confrontation, there was a theme in that confrontation. That confrontation would be inevitable until the fundamental disputes are settled. Now after this war the fundamental disputes are still to be settled, if not by confrontation, by consultations and by negotiations.'[5]

On arrival in Moscow on 1 March, we were given the minimum reception. The President's eldest son, Mir Murtaza, joined the delegation and he was installed with me in a grand suite of rooms in the Kremlin. I got to know him quite well, and we toured Moscow together. He was personable and intelligent, and, I felt, underestimated by his father.

The Soviet leaders insisted we talk to India and accept the reality of Bangladesh. At both banquets, Premier Kosygin, rather than speaking of South Asia or the subcontinent, referred to 'the Hindoostan peninsula', the emphasized 'o' sounding even more ominous in Russian. The Soviet Union had no regrets about its recent role, he pointed out, and would do the same again if necessary. Whether this was intended as a threat of further dismemberment, or merely a manner of speaking, was difficult to tell.

The private talks between the two leaders, with limited numbers present, were more congenial. Premier Kosygin suggested an early meeting between Pakistan and India; their two leaders should appoint envoys, who enjoyed their full confidence, to hold preliminary talks. Premier Indira Gandhi had indicated to him her nominee, and President Bhutto should name his. ZAB wanted time to consider this as his close confidantes were young Party members who were inexperienced in diplomacy. Looking towards me, Kosygin said that the President's young aides had gained experience and could conduct negotiations properly, especially as he would arrange the talks in Pakistan. This was to assure ZAB's presence and guidance for the negotiations. The Soviet Ambassador had apparently relayed the earlier mention ZAB had made about me as a channel of communication. Equally likely, Kosygin wanted no delay in the response.

ZAB told me that D. P. Dhar, a senior Kashmiri diplomat and close confidante of Indira Gandhi, was her nominee, and I should represent him. I pointed out that he had given me a great many other tasks; the issues with India, particularly Kashmir, had been thrashed out over numerous meetings and had developed their own terminology; it would be difficult to prepare for such important talks at short notice without detracting from other work. Most important, however, someone from the Punjab should represent Pakistan. Aziz Ahmed was the obvious choice, but ZAB felt he lacked the flexibility to handle such a delicate situation, which was not the same as in 1965. He insisted that I would have to be present at all meetings with Dhar. This subsequently led to complications, some amusing.

During the talks in Moscow the question arose of relocating the planned Steel Mills. J. A. Rahim had felt that the proposed site outside Karachi was not environmentally acceptable; the beaches would be spoiled and the winds would carry pollution across the entire city. This was probably the first time that a Third World country had delayed a major project for environmental reasons. The feasibility study had been prepared at considerable cost. The Soviets believed the examination of a different site was an excuse to shelve the entire project, since they suspected that we did not welcome their involvement in a project affecting our entire economy so soon after the Soviet-backed military onslaught by India. They, however, were keen to proceed with the project because of prestige and trade. After some discussion, they agreed to undertake a feasibility study of a new site, which subsequently became the location of the Steel Mills. This meant a delay of over two years, which proved costly because massive inflation followed the oil price rise after the Ramadhan War in October 1973.

The Soviet leaders took us to a performance at the Bolshoi Theatre; unfortunately, a modern opera and not a classical ballet. We sat in what was formerly the royal box. Our Foreign Office team were negotiating the text of the joint communiqué with the Soviets; during the performance, a Director-General informed ZAB that the Soviet text went too far on the question of Bangladesh. ZAB asked me to go out and ensure that it was politically acceptable. The draft dealt in considerable detail with Bangladesh, and we were not in a position to make the requisite commitments at that stage. The Soviet side initially insisted that their text remain intact. When I said we would do without a joint communiqué, and only a short statement need be issued when the visit concluded, they became conciliatory, inquiring about what was unacceptable.

I pointed out the specific paragraphs. One official promptly crossed them out, asking whether the rest was all right. I went through the remaining text and, apart from a few minor changes, agreed. The Soviets had obviously anticipated the possibility of deleting those particular paragraphs, because the revised text read smoothly even though a large part had been omitted. They were heavy-handed but could also be charming.

Despite the very strained and difficult circumstances which preceded the talks, ZAB succeeded in neutralizing the Soviets. This was reflected in the arrangements for our departure, which were with somewhat more fanfare than on our arrival.

Next on the agenda came preparations for bilateral talks with India. ZAB spoke about these for the first time in an interview with Dilip Mukerjee of the *Times of India* and B. K. Tiwari of the *Indian Express* at Larkana on 14 March: 'I am allergic to third-party intervention. It is high time the nations of the subcontinent solved their disputes without having to turn to outside umpires for help.'[6]

To avoid public pressure and publicity, Murree, a hill station about twenty-five miles from Islamabad, was chosen for the talks at the end of April between officials of the two countries. The Indian team leader, D. P. Dhar, was accommodated in President House. Aziz Ahmed instructed the Foreign Office to arrange that only he and Dhar should be together on the drive to Murree, but the President insisted I be present at all times even though three was not a comfortable number in a car. Dhar was versatile, highly articulate, and proved to be a formidable adversary. His official position was only Chairman of the Planning Committee of the Indian Ministry of External Affairs, yet he spoke with great authority.

D. P. Dhar's main thrust was that in Kashmir we should accept a new Line of Control, as a result of the recent war, instead of the Cease-fire Line arranged earlier under United Nations' auspices. He proposed complete normalization of relations between our two countries, with free travel for relatives, businessmen and journalists, together with trade and cultural exchanges. He insisted that our prisoners of war in Indian camps had been moved from Bangladesh for their own safety, and that India alone could not negotiate their release as the surrender was to the joint India-Bangladesh military command. The participation of Bangladesh was required on the prisoners issue, but India was ready to talk about the return of West Pakistani territory.

We advocated a step by step approach to the peace process and did not accept the Line of Control, which would derogate from the United Nations cease-fire arrangements in Kashmir. The talks were deadlocked. After lunch the next day D. P. Dhar told me that no resolution was possible without agreement on the new Line of Control; at the next formal meeting he would press this issue and, if we did not accept, the Indian team would return home. He understood I was representing the President, to whom this position should be communicated. I suggested Dhar meet the President in the course of our discussions rather than formally at the end of the visit, as planned. I telephoned the President that evening about the main sticking point being the Line of Control. He agreed to see Dhar the next morning in Rawalpindi, and asked me to brief him fully over the telephone as we were to drive from Murree directly to the President. Aziz Ahmed and I were present with ZAB and Dhar for the formal briefing about our discussions, after which ZAB talked to Dhar alone for about half an hour. Before we returned to Murree, the President took me aside and told me to ensure a summit: 'Use whatever persuasion this may require.'

Dhar was encouraged by their private meeting. He considered ZAB most reasonable for accepting in principle the Line of Control, provided other issues were resolved. ZAB had maintained that only by first normalizing relations could there be increased exchanges between the two countries, and then within Kashmir, which would result in normal neighbourly relations; with the passage of time the Cease-fire Line or Line of Control would become irrelevant. Dhar conveyed the impression that all issues could be resolved. Talking of Kashmir in a different tone, he said that the people on both sides could start by crossing the line freely, separated relatives and friends could get together, and then greater accommodation could be achieved. Later, I learnt that this was generally ZAB's view. He told Kuldip Nayyar in an interview published in *The Statesman,*

Calcutta, on 27 March: 'We can make the Cease-fire Line as the basis of initial peace. Let the people of Kashmir move between the two countries freely. One thing can lead to another. Why should it be ordained on me and Mrs Gandhi that we resolve everything today. We should set things in motion in the right direction. Others can pick from it. We cannot clear the deck in one sweep.'[7]

ZAB wanted to avoid deadlock and to smooth the way for a summit. In Dhar's description of the meeting, his fluency with words may have got the better of him, but he was clearly not discouraged by ZAB. Dhar was also told that if there were any political issues to be resolved, he should speak to me; as Special Assistant, I knew the President's 'thinking on these matters'.

The remaining sessions between the two delegations went well. Dhar made a point of being friendly with me and seeking my comments, to my embarrassment and Aziz Ahmed's annoyance. We settled on a summit and, in principle, an agenda. ZAB's main aim was achieved.

After four days of talks, Dhar left on 30 April with a thumbs-up gesture at the airport: 'I am greatly satisfied with the conclusion of talks.' He said President Bhutto would be received in India with 'open arms and open heart, as a brother, as one of us'. He was 'impressed by Mr Bhutto's sincerity to establish peace', finding 'great consistency in Mr Bhutto's entire approach to problems in the subcontinent'. He quoted from Faiz Ahmed Faiz:

Aayie hath othain hum bhee,
Hum jinhen rasme dua yad nahin,
Hum jinhen soze muhabbat ke siwa,
Koi but, koi khuda yad nahin.
Aayie arz guzarain ke nigare hasti,
Zahre imroze main sheerinie farda bhardey.[8]

Before proceeding to the Summit in Simla, President Bhutto spent the next two months in efforts to gain further international support and to keep the public informed about developments on issues with India. He also undertook a twelve-day tour of fourteen countries, mainly in the Middle East,[9] from 29 May. During this tour, he held a meeting at Izmir in Turkey of some twenty-five Pakistani Ambassadors to discuss and coordinate diplomacy, information and trade through our embassies, and to streamline their efforts. Several Ambassadors maintained that it would reduce their status and effectiveness in the royal courts, to which some were accredited, if they became directly concerned with such matters as information and trade. Diplomacy has its own logic of convenience. Despite some tough talking by the President, this attitude never changed.

The tour of fourteen nations from Ethiopia to Nigeria came to be known as a 'Journey of Renaissance'. It helped to improve Pakistan's poor international image. There might be an independent Bangladesh, but the message went out loud and clear that Pakistan remained relevant and important.

Returning to Pakistan, the President held a series of meetings with various delegations preparatory to the Simla Summit. He invited a nationwide debate on the current situation in the subcontinent and on Pakistan's relations with India and 'Muslim Bengal', calling on 'intellectuals, writers, journalists, teachers, students and workers' to discuss the issues in the changed postwar circumstances. His aim was to inform and involve the public, and to avoid a situation similar to Ayub Khan's isolation following the 1965 War and Tashkent.

During this period, in several important interviews with foreign journalists, he explained Pakistan's position on the peace negotiations to the world, to India, and to the people at home through the reproduction of his views in the domestic Press. He covered four main issues: relations with India, the prisoners of war question, Kashmir, and our attitude to 'Muslim Bengal', as he insisted on calling the new state of Bangladesh. These interviews also sought to portray him to the outside world as a reasonable and pragmatic leader, ensuring that the blame for any subsequent summit failure would fall on India.

He told *Der Spiegel* of the Federal Republic of Germany that he wanted

> ...peace which is not imposed, peace which is not in violation of established principles...so the main thing is the intention to live in peace and come to the conclusion that war is not really an answer for the settlement of our outstanding differences.[10]

He informed Peter Grubbe, the Editor of the German weekly *Stern,* in an interview on 27 May:

> I would venture to welcome collaboration with India and our whole effort is going to be to have a new era between the two peoples on the basis of mutual self-respect...we can only go slowly, slowly, step by step.[11]

On meetings with India and Mujibur Rahman, he informed George Vergese of the *Hindustan Times* on 4 May:

> I told Mr Dhar that when we begin our negotiations, let us have a preliminary discussion between ourselves—between your Prime Minister and us. And then at any stage that we feel that it's appropriate to have some kind of discussions with Mr Mujibur Rahman, that can be done.[12]

Earlier, referring to India, he had said:

> Let us restore diplomatic relations...I am ready to send someone tomorrow. We have even got the person in mind...We are completely ready for that. And even for the other matters, the question of post and telegraph, travel, communications.[13]

He dealt with the prisoners of war on ABC's 'Issues and Answers' programme, also published in *The Pakistan Times* of 15 May 1972:

> It's a basic human problem involving about eighty thousand to ninety thousand people. But we went out to the families...to explain to them that they had to be patient. These were the consequences of war—of a lost war...I am very grateful to them that they have seen the point and they've exercised great restraint and discipline.[14]

Later, at Quetta on 22 May, he told Gerald Stone of the Australian Broadcasting Commission:

> Indians will tell you that Pakistan's problem is the prisoners of war. I would have told you that two months ago, but I salute our people's courage...Today that problem is not the most pressing one...In any case, to keep human beings as hostages has diminishing returns.[15]

He couched the Kashmir issue in cautious terms:

> This question reverts back to Nehru's time. Take the right of self-determination. This commitment was given by both countries...We cannot take it away from them. It is their own inherent right.[16]

Earlier, on 14 March, he had emphasized the same point in speaking to Dilip Mukerjee and B. K. Tiwari:

> If I say that there should be no self-determination for Kashmiris, it does not solve the problem. If the people of Kashmir believe in self-determination, no one can stop them.[17]

He expressed optimism about the future with 'Muslim Bengal' despite existing bad relations:

> I hope we can restore our links—and I do not define them...We can come closer. If we are making efforts to have good relations with India, why should we not try to have good relations with the people who were part of our country and have been separated by military conquest.[18]

His numerous meetings with Pakistani delegations served to inform the public about the position he would adopt in the talks with Indira Gandhi and to gauge opinion as to how far we could proceed with, or concede to, India. We soon learnt that sufficient self-confidence had been regained in the past six months for each delegation to emphasize firmness on the Kashmir question. This tied ZAB down, and affirmed his determination not to repeat Ayub Khan's mistake of ignoring domestic opinion.

The Simla Summit

At Simla itself several important issues remained to be settled, although considerable ground had been covered in the Murree talks, and privately between ZAB and D. P. Dhar. We set off on 28 June for Simla via Chandigarh. Despite all the preparations, ZAB was deeply concerned about the impact of the talks on the domestic situation, fearful that political hawks in the Punjab would seize on anything resembling 'surrender'. We even worked out signals about the message I would convey to the Punjab Governor, Mustafa Khar, by telephone about the appropriate arrangements to be made on our return to Lahore, depending on success or failure.

He still did not feel fully secure. The country only had an interim Constitution; the vexed issue of provincial autonomy remained. The Armed Forces, though subdued after the removal of the Army and Air Force Chiefs, remained restive, not fully accepting civilian authority. The prisoners of war loomed large, as did the sore question of the recognition of Bangladesh. The economy was still suffering from the effects of the civil war. The IMF, World Bank and Western countries were pressing for a resolution of the Bangladesh question; they refused to apportion the earlier accumulated debt, and to give further financial assistance, until this was settled.

The delegation was chosen carefully to cover a wide spectrum of public opinion, including politicians from both Government and the Opposition, Members of the National Assembly, senior bureaucrats, including Foreign Office and intelligence officials, and newsmen.[19] He also took along his young daughter, Benazir, as a personal friendly touch, recalling how Indira Gandhi accompanied her own father in her youth.

For the opening session, the Pakistani team was large, the only limitation being the size of the room. ZAB wanted to associate the maximum number of political leaders with the peace process. The Indian side included the Prime Minister's senior colleagues and officials. After introductions and exchanges of courtesies, Indira Gandhi's opening words of welcome were gracious and showed understanding: 'It was not an easy matter' for the Pakistani President to visit. ZAB replied, 'I want to say, believe me, we are interested in peace. That is our objective and we will strive for it. We want to turn the corner. We want to make a new beginning.' The large meeting clearly did not permit meaningful dialogue and it was concluded after about twenty minutes. The two leaders withdrew to another room, asking Foreign Minister Swaran Singh, P. N. Haksar, her Principal Secretary, Aziz Ahmed and myself to accompany them. We were surprised that D. P. Dhar was not included, but he was not well. We sat in a small room on armchairs. This was my first close encounter with Indira Gandhi. Initially, neither leader was relaxed. He was uncomfortable representing a defeated nation, while she was probably still upset by his earlier remarks in an interview with Oriana Fallaci, an Italian journalist. He had told Fallaci he did not think much of Indira Gandhi and how, in her days as Information Minister, she would sit at the back of meetings like a schoolgirl noting down every word that was said. These and other remarks were made when we met socially for dinner, after the formal interview, and should not have been reported. The publication of the remarks nearly caused the postponement of the Simla Summit; it was only saved by a firm contradiction.

I had expected to meet a hard lady. Indira Gandhi had in the past two years despatched all her rivals in the Congress Party, seasoned men who had brought her in as a compromise Prime Minister thinking they could manipulate her. She had also dealt devastatingly with the Pakistan Army and dismembered our country. Instead, she appeared shyly smiling, almost diffident, without any arrogance in victory.

The discussions dealt briefly with the ground we had covered at Murree. We talked informally. ZAB stressed the need for a step by step approach to improve relations between the two countries, how war had been detrimental to progress and development, which Pakistan needed. Indira Gandhi smiled, for these words came from the person who had wanted to wage a thousand-year war against India. She accepted that a total package settlement was not possible with so many outstanding issues, but was unwavering on re-designating the UN Cease-fire Line as the Line of Control. She did not want to go into details and suggested discussions between officials the following day. The meeting lasted about forty-five minutes.

On 29 June, our official-level talks produced no result. The Indian side had given our Foreign Office a draft text which was not acceptable to us, so we decided to meet again the next day. Basically, the Indians sought a conceptual framework for 'a durable peace' before dealing with specific issues like troop withdrawals, while we wanted a step by step approach prior to any general declarations such as a 'no-war pact'. That night Indira Gandhi invited the senior members of the Pakistan delegation to a small dinner at *The Retreat*, her official Simla residence. Although described as a 'working dinner' to the Press, there was no discussion on any specific issue. It was interesting to note the difference in the two sides—the Indian officials were voluble in the presence of their Prime Minister, while the Pakistanis were mainly silent, speaking only when required by the President.

D. P. Dhar was now in hospital following a mild heart attack. This was a setback as he was an extremely able diplomat with the ability to achieve agreement where others might have failed. ZAB, Aziz Ahmed and I visited Dhar, who appreciated the gesture. His inquiries about the talks showed he knew exactly what had transpired, and exercised influence over his Prime Minister. He was encouraging.

P. N. Haksar led the next round of talks, but there was no progress. Indian Foreign Secretary Kaul was particularly hawkish. The Indian position on the prisoners of war issue requiring the participation of Bangladesh had by then been accepted by our side, but their insistence on the new Line of Control, linked to the requirement that all issues between the two countries should be settled bilaterally, without qualification, proved a stumbling-block. Clearly, they sought to negate the United Nations resolutions which had brought into being the Cease-fire Line to be followed by a plebiscite in Jammu and Kashmir.

On 1 July, in the afternoon, the two leaders met, with Swaran Singh, Haksar, T. N. Kaul, Aziz Ahmed, our Foreign Secretary Iftikhar Ali and myself present. As the *Sunday Statesman* of 2 July reported, 'It became clear that the talks had failed'. No doubt, this was leaked by the Indian team. Discussions revolved around a future summit, to be mentioned in a joint communiqué which would also refer to general principles of good neighbour-liness and the desire to live in peace. ZAB admitted to the Press on 2 July that there was 'some kind of deadlock'. The official-level talks also produced no result.

On the last evening, he had an hour-long meeting with Indira Gandhi, their famous walk in the garden. It was their main one-to-one meeting. Reports later circulated that it was this which led to the successful outcome at Simla. However, despite all his assurances of improving relations and settling disputes between the two countries, she remained adamant on the text of the agreement prepared by her officials. Clearly she did not trust him; at least this was what he told me.

While the two leaders were walking, Aftab Ahmed, our Foreign Office Director-General, discussed with me the serious consequences if the talks failed. He suggested that, as a lawyer and draftsman with some expertise, who knew the President's approach, it should be possible for me to find some suitable phraseology to protect our legal and political position while acknowledging the change of nomenclature from Cease-fire Line to Line of Control, on which India insisted. To ensure that the reference to bilateral dealings should not be an exclusive arrangement, precluding the United Nations or any other international forum, I thought that the insertion of the words 'without prejudice to their internationally recognized position' would be a sufficient safeguard for Pakistan. An excited Aftab Ahmed said I should discuss this immediately with the President.

ZAB was very disturbed when he described his meeting with the Indian Prime Minister. I suggested to him that, if the talks failed, he could return saying he did not want another Tashkent, and disclose what really happened in Tashkent to Pakistan's detriment. For the first time he admitted that there was nothing of particular relevance to convey to the public about Tashkent that was not already known.

I mentioned the talk with Aftab Ahmed and explained how this phrase could be both a solution and to our advantage, as our position was recognized through UN resolutions, unlike India's claim that Kashmir was an integral part of that country. He liked the idea but did not think the Indians would accept it. He said the word 'internationally' should be omitted. Before we left for the official dinner, he reminded me to telephone Khar in Lahore, through our pre-arranged code, that the talks had failed, and we were returning the next day. Khar was surprised, but said everything would be under control.

Towards the end of the formal dinner, Indira Gandhi and ZAB left the banquet hall. Within minutes, Haksar and I were called to join them. I was taken to a small room where ZAB was sitting alone as, by then, Indira Gandhi had moved into a separate room where Haksar went. ZAB told me he had mentioned, at the dinner, the new phrase and she wanted to examine it with Haksar. Aziz Ahmed was called and the three of us sat waiting. The dinner was over but the guests did not leave as the two leaders were still in the building. There was also anxiety about the Summit's outcome.

After a while Haksar came to our room and underlined strongly how serious the set-back to relations between the two countries would be if no agreement were reached. He went so far as to say that he would place his *pugree* (turban-like headwear) at ZAB's feet to achieve a settlement. We conceded that an agreement was of the utmost importance, but maintained that a treaty which was not politically acceptable to the people of Pakistan could not be implemented and would eventually be rejected. ZAB asked Haksar to appreciate the political problems faced by our civilian government, which had only recently been inducted, and inquired about the latest proposal. Haksar said that the Indian side had settled the text and this new phrase would upset the entire arrangement. ZAB told him that I had explained how the formula did not require revision to the remainder of the text; it merely permitted an acceptable solution.

We briefly discussed the additional words. Haksar said he would explain this to his Prime Minister and her colleagues. We continued to wait. Silence prevailed as there was nothing to say to each other. After a few minutes, Haksar returned saying I should personally explain the addition to the Indian Prime Minister.

ZAB instructed me to go with Haksar. I had the impression that this was a sort of final gesture by India before admitting failure, to show that we were given a full hearing. Haksar led me into a billiard room where the Indian Prime Minister and her inner cabinet of five, including Swaran Singh and Jagijivan Ram, and a few officials were present. The draft of the text was on the billiard table. I was asked to explain, but I had barely started to speak when I was interrupted and told that there could be no change, even if inconsequential, which this was not. I said I should at least be heard. No one present was prepared to consider any change, or even to listen to me, until the Prime Minister took charge. With complete authority, Indira Gandhi said I should be allowed to explain the change properly, because, after all, that was why I had been called.

I outlined why we wanted these additional words and where they should be inserted. There was, at first, some discussion on the latter issue. I pointed out that there was no other logical place, otherwise the addition would upset the rest of the text and cause delay. Then Prime Minister Indira Gandhi pointedly remarked that she understood how the additional words, 'without prejudice to the recognized position of either side', helped Pakistan overcome difficulties caused by the new Line of Control, but asked how could I say that it was not one-sided and should be acceptable to both parties. I said that our recognized position on Kashmir differed from that of India, and the addition merely confirmed this; it did not say without prejudice only to Pakistan's position. The Indian Ministers present objected that these words would change the essence of the agreement that all issues be settled only bilaterally. Amidst interruption, I said that was a different matter; while we understood India's reasons for the new Line of Control, we could not abandon our position.

Indira Gandhi again calmed the heated discussion, and I was asked about the exact place where the words were to be inserted. I read out the whole sentence, including the addition, pointing out that other clauses were not affected. She thanked me. The discussion was obviously over so I returned to the room where ZAB and Aziz Ahmed were waiting. Tension was so great that they jumped up as I came in.

I narrated how the Indian side did not want the additional words, and only Indira Gandhi had given me a chance to speak. Really, now, it was her decision. We waited. Suddenly the President asked whether I had conveyed the message to Khar, although I had confirmed this earlier. We continued to wait like expectant fathers. Then, in came a smiling Haksar. His Prime Minister had agreed to the addition.[20]

Relief was writ large on everyone's face. A typewriter was brought to complete the Simla Agreement as finally settled. The signing ceremony took place in the early hours of the morning. The date of the agreement was wrongly typed as the preceding day; by now it was the early hours of 3 July. I later gave Khar the good news.

It was the wrong date but the right agreement. While we could not avoid the new Line of Control, and reference to the settlement of disputes bilaterally, Pakistan had protected her recognized position and could also proceed internationally.

Much has been said about a secret agreement at Simla. The only occasions when ZAB met the Indian Prime Minister alone were the hour-long talk and walk, and then at the banquet. Events that took place preceding and subsequent to the midnight meeting do not lend support to a secret agreement as such.[21]

We returned amidst jubilation to a big reception at Lahore arranged by Khar. ZAB sought to avoid any anti-Government campaign similar to that following the Tashkent Agreement, and assured this gathering on 3 July, 'I had told you that there will be no second Tashkent', and then emphasized, 'I tell you as a Muslim and I swear on oath—I swear in the presence of Almighty Allah—that there has been no secret Agreement'.[22] The next day, at Islamabad Airport, he repeated this: 'I told everybody that nothing was secret',[23] as if it were otherwise at Tashkent. He had become almost paranoid about Tashkent.

The foreign Press portrayed this as a great diplomatic triumph for ZAB, who had held 'no aces' and yet played a 'winning hand'. In fact, insufficient credit has been given to Indira Gandhi's magnanimity and decisiveness, which largely contributed to the Accord. When he came to learn that I had expressed these views to Hayat Sherpao, Mubashir Hasan and others, he remarked that it was strange that while everyone praised his success, I gave credit to the Indian Prime Minister. I pointed out that but for her there would have been no Accord, though this did not detract from his role.

ZAB has been wrongly accused of sacrificing the prisoners of war and wanting to keep the military personnel in India for political ends. He could not secure everything from India, the victors. They naturally dictated the course of the talks, and our efforts were directed towards salvaging whatever we could after the military defeat. An erroneous view also existed that India was mistaken in handing back territory, which she could have held on to longer as a bargaining lever, instead of the prisoners, who could not be detained indefinitely. But India consistently maintained that they had surrendered to the joint forces of India and Bangladesh and could only be repatriated if both agreed. Moreover, this served as a lever to secure the recognition of Bangladesh. India was also under pressure from the Soviet Union and the United States to vacate the territory acquired through war, while these countries supported the early recognition of Bangladesh. She received a suitable quid pro quo in the change of nomenclature to the Line of Control and the mention of bilateral settlement of disputes under the Simla Accord.

He remained rightly proud of his achievement at Simla, and stated at Peshawar on 10 September 1973:

We went to Tashkent with pride because we had thwarted the enemy's plan, taken a larger area of its territory and held a larger number of POWs. It was India which had pressed for a cease-fire then. Contrary to this, we went to Simla with the position that half the country was lost, 5,000 square miles of West Pakistan territory was captured, and 90,000 of our men were prisoners. In view of these objective realities, the Simla Agreement was signed and it was a positive gain to Pakistan.[24]

He gave little credit to Indira Gandhi. In fact, the Accord was really a victory for common sense.

No vital concession was made for the return of territory. ZAB asserted that our position on Kashmir had been secured by the words 'without prejudice to the recognized position', which reinforced our international rights and resurrected the Kashmir issue. Such talk upset the hawks in India, who felt we had misled them, though it pleased Pakistani ears. The so-called spirit of Simla died soon after the ink was dry on the Accord. In trying too hard to avoid a post-Tashkent scenario, ZAB lost the opportunity to open a new era in the subcontinent. For this, the Indians too were to blame.

Under the Simla Accord, as a first step, the military commanders had to demarcate the new Line of Control which existed on the day of the cease-fire, 17 December 1971. In the month of Ramadhan, GHQ in Rawalpindi was the venue for regular after-dinner reviews of the demarcation proceedings; the meetings were attended by Aziz Ahmed from the Foreign Office, myself representing the President, Chief of Army Staff General Tikka Khan, and other military officers, and we were briefed by the Lahore Corps Commander, Lt.-Gen. Abdul Hameed, who conducted the actual negotiations with his Indian counterpart.

The sticking point in the negotiations became an area of about one-and-a-half square miles known as Thako Chak, near what was described as 'the Chicken's Neck'. Our military commanders maintained that this was of strategic importance to the Karakoram Highway connecting Pakistan to China. However, it appeared from maps that, after the advances made by the Indian troops, the Highway was even otherwise vulnerable.

Meantime, great expectations had built up among the one million refugees displaced from Shakargarh. The Punjab Government found it difficult to cater for these large numbers, particularly as the refugees had prepared themselves to return immediately after the Summit. Also, the troops, who had been in trenches for eight months, were beginning to tire; they were ill-equipped for another winter, which added to the urgency of troop withdrawals. Some sections of the army were already restive; the families of the prisoners had organized themselves. Earlier, they had been told to be patient till the Summit, and had shown restraint, but now they had to wait with no near prospect of the prisoners' release. This situation was further aggravated by Bangladesh repeatedly asserting the right to hold war trials.

The stalemate on troop withdrawals had to be overcome. Official-level talks were proposed between the two countries to resolve the outstanding issues and to demarcate the Line of Control, but our internal debate concerning Thako Chak had first to be settled. I reported regularly to the President about the meetings at GHQ. We agreed that the new Line of Control would place Indian troops close to the Karakoram Highway at many strategic points apart from Thako Chak. The President thought that the military was placing the responsibility for difficult decisions on civilian shoulders, so he called a meeting at Lahore to inform the military commanders that, if the area really was of strategic importance, we would not concede it for military reasons. Faced with the alternative of no troop withdrawals, the commanders readily agreed to concede this small area.

We lacked the means to pressure India and Bangladesh to prevent war trials and secure the early return of prisoners. Withholding recognition of Bangladesh, our trump card, would soon be made irrelevant by the admission of Bangladesh to the United Nations. Agha Shahi, by then our Ambassador in Beijing, was in Pakistan, and, in a meeting with ZAB, suggested a Chinese veto in the Security Council to prevent this. We wondered whether China would exercise its veto so soon after its own admission to the UN, but we thought it would be worth trying to urge China to maintain its earlier stand 'on principle' at the UN concerning our position. After Shahi left, ZAB was irritated at not conceiving the idea himself. In the event, China did veto Bangladesh's admission, which greatly helped Pakistan—but more about that subsequently.[25]

Delhi Agreements and Relations with India

The stage was set for Aziz Ahmed to visit Delhi for talks with P. N. Haksar. ZAB insisted I also go despite my concern about the two-man leadership creating awkward situations. He said it was essential to reach a settlement, and he could trust me to talk to the Indian Prime Minister on the lines he wanted. I was to assure her that he meant what he told her at Simla, and he instructed me to give any undertaking to secure troop withdrawals and the return of Shakargarh. He wanted to create a new, lasting basis for good relations in the subcontinent, and would in due course proceed with the recognition of Bangladesh. She should appreciate his limitations, particularly on war trials, and not misunderstand his recent statements. I was to ensure there should be no opportunity for recording any discussion I might have with her.[26] He gave me a personal letter to the Indian Prime Minister stating that, although Aziz Ahmed was his Special Envoy, she should discuss political matters with me as his Special Assistant.

The Foreign Office announced that our visit was preparatory to the next summit envisaged under Article 6 of the Simla Agreement. We arrived in Delhi on 25 August and were housed in the Viceregal Lodge, presumably in return for our accommodating Dhar at President House in Murree.

At the time there was considerable publicity and speculation about the Simla Accord not working. The Indian Press referred to ZAB breaking an 'understanding' on the recognition of Bangladesh by blocking its admission to the United Nations. *The Hindu,* under a front-page headline, 'India to Ascertain Pak Sincerity on Simla Agreement', stated,

The Pakistan delegation...will be formally headed by Mr Aziz Ahmed, Secretary-General of the Pakistan Foreign Office, but the real spokesman for Bhutto at these talks will be Mr Rafi Raza, his special assistant who was able to establish some degree of personal rapport with the senior Indian officials at the Simla Conference...[27]

Apart from the Indians, I now had to play at diplomacy with a displeased Aziz Ahmed.

The talks with P. N. Haksar and his colleagues, Foreign Secretary T. N. Kaul and the Prime Minister's Secretary S. K. Banerji, started badly. Over dinner, Haksar aggressively asked why India should make any further concession after achieving nothing following Simla. He continued in this vein till, in the end, I pointed out that he had failed to appreciate the main gain, goodwill, which they had lost. The people of Pakistan were expecting the worst from the Indians after their defeat, but Simla showed that the Indian Prime Minister could be magnanimous in victory. Since then, however, one million refugees from Shakargarh, ready to return, remained homeless and, as a result, unfortunately, most Pakistanis had reverted to the view that no good could come from India.

On 28 August, we met Foreign Minister Swaran Singh and then D. P. Dhar, who had become Planning Minister. Once again he was helpful. We paid a formal call on the Prime Minister, and I gave her ZAB's letter. Aziz Ahmed did not know its contents. Realizing this, she looked up and smiled after reading it. Thereafter, the official talks proceeded smoothly. The next day we agreed on completing the delineation of the Line of Control by 4 September, and the date for troop withdrawals was extended to 15 September. Political leaders from Tharparkar would visit the areas of Sindh occupied by Indian forces to ensure that the Hindu inhabitants returning would be 'welcome to remain in or return to their homes in Pakistan in safety and dignity from camps in India'. We accepted an outline plan for their resettlement which, in ordinary times, would have been considered gross interference in our internal affairs.

We flew back to Rawalpindi on an Indian plane. Foreign Secretary Kaul, while walking with me to the plane, asked for serious consideration to be given to the question of expediting cultural and other exchanges between our two countries. I could not resist commenting, 'Yes, we should start with the exchange of 90,000 people.'

While we were in Delhi, ZAB visited Gilgit and Hunza to show Pakistanis there was no compromising on the Northern Areas. Radio Pakistan announced on 26 August that the next summit would be in a month's time, the date to be settled in Delhi, with ZAB adding, 'However, if the Indians want to come earlier, they are welcome'.[28] But the Indians at no time discussed any further summit. He also indirectly conveyed to India that he was moving towards recognition. He informed newsmen about his recent telephone conversation with Mujibur Rahman, who 'started talking about the recognition of Bangladesh', but was reminded by ZAB of their earlier understanding that they would discuss mutual problems 'some time later'.[29] In this way, Pakistanis too were being gradually introduced to the idea of talks with Mujib, as a first step towards recognition, which was still very unpopular.

ZAB was under foreign pressure to make progress in this direction and, in a subsequent interview[30] with the Editor of *The Blitz,* Bombay, suggested political-level talks between P. N. Haksar and myself, but there was no response. India and Bangladesh reiterated that recognition was a pre-requisite for peace in the subcontinent. Countering their arguments that a meeting with Mujibur Rahman could only be on the 'basis of equality', established through recognition, ZAB referred to the talks between the leaders of the US and China without recognition: 'The fact that we meet obviously means that we are meeting as equals.'[31] However, there was no second summit with Indira Gandhi, and the meeting with Mujibur Rahman came simultaneously with recognition.

After the Indian withdrawal from parts of Sindh and Shakargarh, and the demarcation of the Line of Control, domestic pressure mounted for the release of the prisoners, even though the public was not yet ready for the recognition of Bangladesh. At a rally at Liaquat Bagh, Rawalpindi, to commemorate the fifth anniversary of the founding of the PPP, ZAB for the first time specifically raised the question of recognition. The meeting was immediately disrupted. Ambassadors and other dignitaries who had been invited scurried away along with everyone else in total confusion. ZAB was taken from the ground under heavy escort. This major reversal delayed normalization with Bangladesh; more serious, it heralded a new dimension of violent confrontation in domestic politics. The public's refusal to accept his proposal for recognizing Bangladesh was read by ZAB as a personal rejection. It confirmed his view that the public were fickle and could not be relied upon in difficult situations—exactly what his security and intelligence agencies wanted him to believe, to increase his dependence on them and on the use of force.

By the time I got home after the rally, there had been several urgent calls from the President summoning me. Seldom had I seen ZAB so angry; he felt insulted, and vowed to settle scores with the Opposition at the first opportunity. This presented itself when the Opposition parties held a public meeting at the same venue, Liaquat Bagh, on Republic Day, 23 March 1973.

Despite domestic problems over recognition, ZAB stayed on course under international pressure. He secured clearance in the form of an advisory opinion from the Supreme Court, and a resolution from the National Assembly on 9 July 1973, authorizing the Government to accord formal recognition to Bangladesh at the appropriate time. This was followed by the first breakthrough on the prisoners' issue at the Delhi talks from 18-28 August 1973. By this time the Indians too were feeling the pressure of continuing to hold 90,000 prisoners, which was proving internationally embarrassing as well as financially costly after more than one and a half years. 'Trilateral' talks settled the repatriation from India of the main body of military prisoners and civilian internees, and of the Bengalis detained in Pakistan; there was also agreement that the 195 'war criminals' would not be tried during this repatriation period; and that the representatives of India, Pakistan and Bangladesh would meet again to settle this question. ZAB sought their release without recognition or war trials. Bangladesh, with India's support, insisted on participating in formal talks only on the basis of sovereign equality, which meant actual recognition.

The postponement of trials provided some relief in the domestic context, but no solution was in sight regarding these 195 prisoners. This was eventually achieved through the good offices and generosity of King Faisal of Saudi Arabia, who agreed to the holding of the Islamic Summit in Lahore in February 1974.[32] The Saudi monarch was the Chairman of the Islamic Conference, and ZAB became Co-Chairman as the venue was now Pakistan. Under the umbrella of the Summit and Islamic brotherhood, reconciliation between Pakistan and Bangladesh finally took place. It was skilfully orchestrated by ZAB.

At a meeting of Party legislators in Lahore, he announced his readiness to recognize Bangladesh. Mujibur Rahman agreed, in return, to participate in the Islamic Conference. President Gadhafi, who was considered a heroic figure, helpfully praised the wisdom of ZAB's decision. The Islamic Summit witnessed the recognition of Bangladesh on 22 February 1974. Most important, this was achieved with dignity and without India. Shortly thereafter, on 5 April 1974, Pakistan, Bangladesh and India agreed on the release of the remaining 195 prisoners. So, after much perseverance, the issues arising from the 1971 War with India were finally settled. It was a high-water mark, but was followed by rapid deterioration in Indo-Pakistan relations.

On 18 May 1974, India exploded a 'peaceful' nuclear device. There was no prior indication from any intelligence agency and it took not only Pakistan but the world by surprise. A major new dimension was added to the problems of the subcontinent. The following day, ZAB told a news conference at Lahore:

> Pakistan would never succumb to nuclear blackmail by India. The people of Pakistan would never accept Indian hegemony or domination in the subcontinent. Neither would Pakistan compromise

its position on the right of the people of Kashmir to decide their own future.[33]

Such sentiments were widely supported throughout the country. He also called off talks on the resumption of communication links and diplomatic ties, because India was 'effectively destroying the two years' relations between the two countries by nuclear blackmail'. India was accused of violating the 1963 Moscow Agreement prohibiting nuclear testing in the atmosphere, underwater and in space: 'It is very difficult to be assured of India's bona fides'.[34]

There was no doubt in Pakistan as to how to proceed following the Indian nuclear explosion. As early as January 1972, ZAB had held a convention on science, and had asked me to meet the person under consideration as chairman of the Pakistan Atomic Energy Commission. Over the next few years, emphasis was placed on Pakistani talent and expertise in this field. On the nuclear question itself, the arguments were not new and had been aired earlier in the European context. The British, after much debate in the Labour Party, pursued the nuclear path. President Charles de Gaulle had maintained that weapons of mass destruction had made alliances impossible and that each country should have its own deterrent. In view of Pakistan's past relations with India, and the failure of the US to assist Pakistan in 1965, or to prevent the dismemberment of the country in 1971, ZAB believed that there was no alternative but for Pakistan to pursue its own independent programme. His error, one that subsequently cost him and the country dear, was his unnecessary rhetoric on the subject. His personality refused to allow that he could be openly outdone by India.

The process of trying to normalize relations between the two countries could not, however, be postponed for long. The Foreign Secretaries met after a six-month delay, but without positive outcome. The next year, 1975, was also difficult with India. Sheikh Abdullah, the 'Lion of Kashmir', who had recently been released from jail, accepted the special status provided for Jammu and Kashmir under Article 370 of the Indian

Constitution. ZAB denounced this as a violation of the Simla Accord, and called for a strike on the last Friday of February throughout Kashmir and Pakistan. It was a total success.[35] There was complete unity on this issue.

The situation worsened when, in April 1975, India claimed the accession of the small Kingdom of Sikkim. ZAB described it as an illegal annexation, subsequently citing it while seeking US arms for Pakistan: 'Swallowing up the tiny Himalayan Kingdom of Sikkim, India has given new credibility to Pakistan's appeals for the United States to resume military assistance.'[36]

In May 1975, the Allahabad High Court found Indira Gandhi guilty of electoral malpractice, and disqualified her from holding public office for six years. She appealed to the Supreme Court and, shortly afterwards, declared a State of Emergency. While we were concerned that she might in desperation turn on Pakistan, at the same time India's discomfiture gave pleasure in Pakistan, enhanced by the fact, that, for the first time since 1948, the nation could glory in appearing more free than India.

The following May the Foreign Secretaries met again to review the process initiated by Simla. There had been some progress through the resumption of diplomatic relations, civilian over-flights, and rail and road traffic in passengers and goods. However, trade remained limited, and one example illustrated the difficulties. After nationalizing the export of raw cotton, we had unexportable stocks. India agreed to purchase two hundred thousand bales, but in turn wanted to sell goods whose import I, as Minister for Production and Industries, was not permitting. ZAB asked me to discuss this matter separately with both the Indian and American Ambassadors. I said there was no point in talking to Ambassador Byroade on an issue which was not his concern, but ZAB insisted because the US was eager to see Indo-Pakistan ties develop. I explained to them separately that, from the list given to us, all such imports would result in factory closures, particularly in the Punjab. Our industry, in both the public and the private sector, could not at this point compete, particularly with Indian engineering goods.

Ambassador Bajpai of India was intelligent and reputedly close to his Prime Minister. We discussed improving trade relations, and I stressed that India should take into account our state of industrial and political development, which required a fresh approach. We had, for example, considerable potential in fertilizer production, but we were short of cement in the north; our communications system got choked at certain times of the year, leading to problems of distribution. We could develop cross-border trade to ease this situation, thereby also complementing India's requirement for fertilizers and surplus in cement. Trade with India had to be seen as beneficial and not detrimental to Pakistan. But, despite such exhortations, there was no progress.

Apart from Simla itself and clearing the debris of the 1971 War, there were no significant or beneficial developments with India. ZAB remained in two minds: he did not want to lose his image as the warrior waging a thousand-year war, with its appeal to the Punjab electorate; at the same time he realized that economic progress depended on improved relations and cutting defence expenditure. The balance changed for him in 1974 with the Indian nuclear explosion, and because of the huge wealth generated by oil in the Middle East. Tapping this wealth, he felt, would enable Pakistan to develop militarily without compromising our economic programme. He was mistaken. His expectation of massive aid did not materialize, nor did the major role he envisaged for Pakistan after the Islamic Summit in February 1974.

As ZAB rebuilt our military strength, he again raised the Kashmir issue in Beijing in May 1976: 'Advance from normalization of relations to peaceful co-existence between India and Pakistan could be achieved only after a settlement of the Jammu and Kashmir dispute'.[37] Pakistan's official defence expenditure had, by 1975-6, increased to Rs 702 crores from Rs 444 crores in 1971-2 (1 crore = 10 million). He declared at Hyderabad on 25 January 1977 that the day was not far off when Pakistan would be a formidable nation, so strong that none would be able to cast an evil eye on her: 'Pakistan was bound to be the strongest nation in the subcontinent.'

Such talk was not conducive to improving relations with India, which shifted in inverse proportion to ZAB's desire to play a world role. The latter aim might have helped, in his view, to contain India, but it did not contribute to better conditions in the subcontinent. No solutions were attempted for the problems that existed prior to 1971.

Relations with Bangladesh

In his first speech as President, on 20 December 1971, he had called on the people of East Pakistan 'to forget and forgive' the wrongs committed by Pakistan's military regime. He had pledged to strive for a loose arrangement between 'the eastern and western parts of the country within the framework of one Pakistan', maintaining that East Pakistan was 'an inseparable and unseverable part of Pakistan'.[38] However, far from encouraging forgiveness, the references to a single nation further embittered Bengali feelings against ZAB. In any event, during this early period, Bangladesh was not in a position to act independently of India. Certainly, until recognition, and the subsequent release of the last 195 prisoners without trial in April 1974, Pakistan's relations with Bangladesh and India were totally intertwined. Thereafter, efforts were made to improve direct contacts with Bangladesh. Quite apart from the special significance of Bangladesh as a former constituent part of the country, it was essential to cultivate good relations with all the smaller nations around India.

This policy appeared to succeed when ZAB visited Bangladesh in June 1974. His enthusiastic reception surprised everyone, considering that only recently he was the most reviled Pakistani in 'Muslim Bengal'. An element of regret in the common man at breaking away from Pakistan, combined with animosity against the growing domination of India, seemed to find an outlet in his visit. However, the past was not altogether forgotten and, at the Martyrs' Monument outside Dhaka, he was greeted with black-flag demonstrations and slogans of 'Butcher Bhutto' and 'Bhutto go back'. This did not deter his attempt to mend fences:

> We lived as one nation, we are now separate and independent nations; but it does not mean we cannot join hands together to overcome poverty and improve conditions for our peoples for happiness and progress.[39]

He nominated Mujibur Rahman's friend, Mahmood Haroon, as Pakistan's first Ambassador to Dhaka, though he did not assume this post. Despite all the gestures made towards forgetting the bitter past, there was no progress on the two main issues between the countries, the distribution of assets and the repatriation of several hundred thousand Biharis stranded in Bangladesh.[40] ZAB was not in a position to satisfy either of these demands. Still, his visit marked the high point of relations with Bangladesh during his years in office.

He believed that Mujibur Rahman's dependence on, and inclination towards, India constituted a major stumbling-block to improved relations. He was convinced that Mujibur Rahman had organized the hostile demonstrations during the visit. He also felt that Mujib had let him down, breaking 'commitments' made during their two meetings of 27 December 1971 and 7 January 1972, before ZAB had set Mujib free. His opponents criticized him for this unilateral act of releasing Mujibur Rahman without any quid pro quo, saying that he should have sought our prisoners in exchange. In the circumstances, though, he had no option because of mounting pressure from the Soviet Union, the United States and other countries. Moreover, it was necessary to make a break from Yahya Khan's past actions, and Mujib's release was presented as a gesture for peace. Even then he could not resist securing some political advantage—he tried to tape his conversation offering Mujibur Rahman the Presidency or Prime Ministership of Pakistan, which was intended to show that his own earlier actions had not been for the sake of office.

Mujib, still a prisoner, was clearly cautious, committed himself little, and maintained that he had to consult his people before taking any decision. Unfortunately for ZAB, the faulty recording suffered from interruptions and gaps, and could not be used as he intended.[41]

For over a year following his visit to Dhaka, there was no significant development with Bangladesh. Then, suddenly, there appeared to be a new opportunity when Mujibur Rahman was assassinated on 15 August 1975. ZAB virtually hailed this as a victory for Pakistan, a vindication of the idea of one Pakistan. It was as if we had regained the ground lost in 1971. The first reports that emerged, headlined in Pakistan, indicated that the leaders of the *coup* would call Bangladesh an Islamic Republic, distancing themselves from secular India. Misled by this, ZAB ordered the state radio and television to play the nationalistic music and songs that had earlier dominated the airwaves during the 1965 War, and also in 1971. He immediately sent ten million yards of long-cloth, five million yards of bleached *mull*, popular among Bengalis, fifty thousand tons of rice and other assistance. He was the first to recognize the new regime, within twenty-four hours, and called on all Muslim and Third World countries to follow. Briefly, hopes were even harboured for some form of confederal arrangement with Bangladesh.

The euphoria soon died down—Bangladesh was to remain a 'People's Republic'. Thereafter, *coup* followed *coup*, which did not affect the close connections with India or prove beneficial to Pakistan, where jubilation had proved premature. Privately, ZAB regretted his hasty approval of the *coup*, conscious that it could provide an example to the military in Pakistan. During his remaining period in office, there was no serious dialogue or improvement of relations with Bangladesh.

Mujibur Rahman was the first of the three key political figures of the 1971 tragedy to meet a violent death. His brutal murder in August 1975 was followed less than four years later by ZAB's execution in 1979, while Indira Gandhi was assassinated in 1984. Only General Yahya Khan died peacefully. It is indicative of the politics of the subcontinent that no leader of a Third World country has ever achieved such a spectacular rise to prominence, and an equally rapid fall, as Sheikh Mujibur Rahman.

The Muslim World and the Third World

With the loss of East Pakistan, questions had arisen about whether Pakistan should remain rooted in the subcontinent, a fact of geography, or seek historical links with Muslim countries to the west. It was also clear that Pakistan could not count on either the Western countries or the communist bloc, apart from China, for support in its dealings with India. The foreign policy of every country is dictated by considerations of national security, and this is especially true for Pakistan, which has a hostile neighbour in India. The Muslim World was not only a source of immediate support, more moral and political than material, but Pakistanis also had emotional ties to it; so ZAB began his endeavours in that direction. A testament to the importance he attached to the Muslim World was the fact that his first major mission abroad was to several Muslim countries in January 1972, turning again to them when he was in trouble on his last tour as Prime Minister in June 1977. In between, the high point was reached during the Islamic Summit in Lahore in February 1974.

His first trip took in Iran, Turkey, Morocco, Algeria, Tunisia, Libya, Egypt and Syria, when he asserted the continued importance of Pakistan in the region and for the Muslim World. He exuded confidence and enjoyed his new role as Pakistan's principal spokesman, no longer under instructions from Ayub Khan or Yahya Khan. Pakistan gained support and, importantly, prevented the early recognition of Bangladesh. Again, in May 1972, he toured Muslim countries in the Middle East, but this time he widened his appeal, particularly in Africa, by including in his itinerary such Third World nations as Ethiopia and Nigeria.

In this relationship with the Muslim World, ZAB established full reciprocity whenever possible. A major opportunity presented itself in October 1973, when first Egypt and then Syria attacked Israel. During the Ramadhan War, Pakistan gave both material and political support to the Arabs, in sharp contrast to the position adopted previously during the 1956 and 1967 wars against Israel. At a press conference in Karachi on 20 October 1973, he declared that the 'whole of the Muslim World is on trial'.

Eventually, the Muslim World came to ZAB's rescue when he had to proceed with the difficult issues of the recognition of Bangladesh, and the trial of our prisoners, after the failure of earlier efforts. Here the contribution of King Faisal, in arranging for the next Summit of the OIC to be held in Lahore, was most significant. From 19 February 1974, officials of thirty-eight States prepared the ground and the resolutions for the Summit. Lahore witnessed the largest-ever gathering of leaders from the Muslim World. The Pakistan Army made arrangements with clockwork precision; everything went according to plan. The only awkward moment arose when Idi Amin, the President of Uganda, arrived uninvited, accompanied by one of his numerous sons.

Irrespective of whether a monarch, head of state, prime minister or foreign minister represented a country, both President Fazal Ellahi Chaudhry and Prime Minister Bhutto formally received them at the airport. This avoided embarrassment arising out of President Chaudhry receiving heads of state while ZAB met heads of government. Having permitted the President to enjoy the role accorded by protocol, ZAB held the field. Only the Shah of Iran did not attend; otherwise all the important Muslim leaders were present, including Sadaat of Egypt, Boumédienne of Algeria, Hafez al Asad of Syria, Gadhafi of Libya and Yasser Arafat leading the PLO delegation. The Arab states were united after the recent war with Israel, and most had good relations with Pakistan. In a stirring final address ZAB pledged: 'I declare here today that we, the people of Pakistan, shall give our blood for the cause of Islam…The people of Pakistan are soldiers of Islam and its armies are the armies of

Islam. Whenever any occasion arises the Islamic world will never find us wanting in any future conflict.'[42]

This was a major triumph for ZAB, but he was not permitted much time to bask in the glory of the Summit. In May 1974 the Qadiani issue at home,[43] and India's nuclear explosion, diverted the attention of Pakistanis. Moreover, the Arabs were not as forthcoming with their oil wealth as we had expected. Even earlier efforts to secure financial support from Libya— J. A. Rahim went in 1972 and I the following year to meet President Gadhafi—had not produced the desired results. There was talk of collaboration in the field of armaments, and I shared Gadhafi's concerns about the role of the US; but, in due course, ZAB distanced himself from Gadhafi to avoid unnecessarily annoying the US. While keeping his lines open to Arab neighbours, he also concentrated on Iran and Turkey.

Iran had proved an invaluable neighbour during the 1965 War with India, though fear of the Soviet Union prevented material assistance during the 1971 War, which, in any case, the Shah felt was a lost cause. ZAB made several overtures to the Shah of Iran but without any significant outcome. Iran was following the US in establishing contacts with Beijing, and the Shah's wife and sister visited Pakistan separately on their way to China. His twin sister, Princess Ashraf Pahlavi, also came twice to Pakistan. The Shah himself and Sheikh Zayed Bin al Nahyan of the UAE were the two leaders who visited Pakistan most frequently in this period; both gave generous financial assistance for various industrial projects. ZAB's only formal state visit as President was to Iran, in May 1973. We were throughout given red-carpet treatment and he successfully assuaged the Shah's concerns over Balochistan. He also secured fresh assistance, including 150 jeeps, which subsequently led to investigations under the Zia regime.[44]

Contrary to the general perception, he did not get on well with the Shah. Each envied the other: the Shah the public appeal of ZAB who, in turn, would have liked the Shah's financial resources and absolute authority. The hauteur of the Shah was evident in his dealings with ZAB, and his conduct contrasted markedly with that of Sheikh Zayed of the UAE. When we visited Qish Island as the Shah's guests in early 1974, ZAB was housed across the driveway from the Shah, who did not once walk over this narrow divide, even to say goodbye. How could the King of Kings call on a Prime Minister? On another occasion, the Shah declared he would only come on a private shooting visit to Larkana if the President of Pakistan was present, but Fazal Ellahi Chaudhry was at the time on his one and only official trip abroad, in Nepal. At ZAB's insistence, I explained the 'supreme constitutional position' of the Prime Minister to the Iranian Ambassador, but the Shah would have none of it. On the other hand, the following year, when we were Sheikh Zayed's guests at his palace near Rahim Yar Khan, the Sheikh arranged one banquet in his palace, followed by one in his own adjoining house where ZAB and a few of us were staying, so that ZAB could play at being host.

The arrogance of the Shah and his family led to a disregard for the feelings and opinions of others, and ultimately to the downfall of his regime. The family's insensitivity was displayed during the visit of Princess Ashraf Pahlavi and her daughter-in-law to Pakistan in 1975 during *Moharram*. Her programme provided for a banquet, with music and dancing, in Karachi at the Governor's House, on what was to be *Ashura*. At a dinner for them in Islamabad a few days earlier I told the Iranian Ambassador it would be most improper for them to have such a function on *Ashura*. He was perplexed and consulted Princess Ashraf's Court Minister who, in turn, asked the Princess to hear what I had to say. She could see nothing wrong. By this time, ZAB had joined us and told me in Urdu that, if they saw no objection, I should not press my point. I maintained that if the Iranian party were misguided we should not share their misjudgement. My view finally prevailed. Before leaving Islamabad an Iranian Foreign Office official assigned to her party said they had brought a gift of edibles which I would enjoy. I was looking forward to golden caviar when I returned home, but the gift turned out to be, of all things, a large ham. Presumably they thought that, being educated in the West, I would appreciate it. In fact, it was speedily despatched from my house. I told ZAB that, if this gift reflected the judgement of the Royal Family, they would not last long. At the time he dismissed my assessment.

ZAB continued to persevere with both Iran and Turkey. He could not leave CENTO because that would annoy the US so, instead, ZAB sought to instil fresh impetus in the Regional Cooperation for Development (RCD) with Iran and Turkey. As a prelude to a summit in Izmir, in a signed article, *RCD: Challenge and Response*, published on 19 April 1976, he stressed that Iran, Pakistan and Turkey 'constitute a single civilization'. Referring to the new situation created by detente between the two superpowers, he concluded, 'if we miss the opportunity to mobilize and integrate our resources in order to face contemporary challenges, the world will take no note of either our heritage or our aspirations.'[45] But the Shah's sights were set on becoming a major power in his own right, and President Koruturk looked to the integration of Turkey into Europe. ZAB's efforts to 'put teeth' into the RCD proved embarrassingly unrewarding; there was no response from the other two countries. In vain were such high-sounding sentiments as:

> The systematic consolidation and formalization of our joint will to defend our civilization against all challenges—economic, political, ideological or moral—is something different from adventitious arrangements which are apt to create suspicions in others...My perception of this association...is focused on the psyche of the contemporary age...the quest for ways to translate platonic levels of relationship into Aristotelian norms...[46]

Another blow to ZAB's ambitions had been the assassination of King Faisal on 25 March 1975. Pakistan lost an important supporter and the Muslim World a devoted leader. His successor, King Khalid, being pro-US, did not champion Muslim causes in the same way, though ZAB courted him assiduously, paying great importance to his state visit in October 1976. ZAB not only accompanied him personally throughout in Islamabad,

Lahore and Karachi, but, unprecedentedly, three Federal Ministers were attached as Ministers-in-Waiting, one for each of the senior Ministers of the Saudi delegation. The visit did not develop relations as planned, because the Saudis were dismayed by such events as the North Korean-style display at the stadium in Rawalpindi with banners reading 'Down with Feudalism', 'East is Red', and populist slogans eulogizing ZAB. The King did not oblige in respect of most of the vast sums of money we sought for various projects, including the nuclear programme.

With the oil-rich Arabs concentrating on their own economic programmes and on trying to secure the rights of the Palestinians, they had little incentive to look to Pakistan. Disappointed by this, and by his failure to provide a new dynamism for the RCD, he increasingly turned his attention to the Third World. This had long been the focus of his interest, even as Foreign Minister under Ayub Khan. He felt that the Third World had not fulfilled its potential as an effective force in international relations. Both President Nasser and President Soekarno were leaders he held in esteem. In fact, immediately after his interview with Oriana Fallaci,[47] when he had told her that the leader he most admired was Premier Zhou Enlai, he mentioned to me that he would liked to have said Soekarno but was hesitant lest unfortunate comparisons be made. At the time, the principal forum for the Third World was the Non-Aligned Movement (NAM); with Nasser and Soekarno no more, and President Tito of Yugoslavia now ineffective, India had a dominant role in NAM. India would not permit Pakistan to join because of our continued membership of CENTO, despite India's own Treaty of Friendship with the Soviet Union, nor could ZAB countenance playing a secondary role to India. So he adopted a different approach.

He looked instead to such leaders as Nicolae Ceausescu of Romania and Kim Il Sung of North Korea. Romania helped us to develop our oil-refining capacity and Pakistan enjoyed good economic relations with that country. In October 1975, he visited Romania; as he was Prime Minister, not Head of State, it was decided that his visit should also be as Chairman of the PPP so that Ceausescu could play full host to him as President and Head of the Romanian Communist Party. The tour was a success.

ZAB took up Third World issues in a more determined manner, concentrating on the reduction of foreign economic domination. He proposed a 'new economic world order...an equitable order so that world resources and technology, the common heritage of all mankind, were available for the benefit of humanity rather than be controlled by the few for their exclusive consumption'.[48] But it was on a visit to Pyongyang in May 1976 that he first publicly mooted the idea of a Third World Summit to dismantle the existing economic relations: 'The time has come for the Third World countries to take cognizance of their vital interests and to resolutely strive for fundamental remedial action to redress the grave injustice to the poorer nations of the world.'[49] He was deeply impressed by the complete control President Kim Il Sung exercised over his country and introduced some North Korean notions into Pakistan. He became increasingly conscious about his portrait being displayed, and commented that my office and house, unlike those of other Ministers, did not have a single photograph of him.[50]

In 1976, Colombo was to be the venue of a NAM Summit. ZAB chose this time to publish an article, *Third World—New Direction*, which proposed 'Third World Mobilization', and called for a Summit in Islamabad around March 1977. Pakistan had recently been elected Chairman of the Group of 77, and ZAB aspired to be the Chairman of this new Summit. The article had been prepared by Yusuf Buch and the Foreign Office, but ZAB sent it to me to be finalized. Several senior diplomats including Agha Shahi, who was at the time Secretary of the Foreign Office, and Iqbal Akhund, Pakistan's Permanent Representative at the UN, on learning that the article had arrived on my desk, pressed me to prevail on ZAB not to publish it. They presented cogent arguments. I told ZAB that we should discuss its publication, so he held a small meeting with only Yusuf

Buch, Agha Shahi and myself present. It proved interesting but not entirely surprising.

I put forward four main objections to the timing and the article itself. First, the Non-Aligned Conference was to be held shortly, and the article might be seen as an attempt by Pakistan to sabotage it, which would lead to misunderstandings with Prime Minister Srimavo Bandaranaike; our earlier efforts to develop close relations with our neighbour Sri Lanka might suffer. Second, ZAB was already Co-Chairman of the Islamic Conference and was nursing our nuclear programme; trying at the same time to assume the leadership of the Third World would undoubtedly make more nations jealous and inimical. We should operate without rhetoric and in low key, especially if we wanted to make progress in the nuclear field. We should avoid opening too many hostile fronts simultaneously. Third, we should first put our own house in order, because we could not seek to play such a major role while we went around with a begging bowl in hand. Lastly, and this argument also embraced the earlier three points, it was unlikely that Pakistan alone could muster sufficient support for such an initiative, and failure would make us appear foolish. It would be more appropriate, if he insisted on the move, to make it a collective effort with three or four other like-minded countries, each of which might also provide the support of additional nations.

ZAB reacted unfavourably; he said he consulted me like Ayub Khan did him as Foreign Minister, when he always adopted an aggressive, positive position. I remarked it was not a question of being aggressive or defensive but realistic. Once he had expressed his views, the others readily agreed with him and I suddenly found myself in a minority of one. I only succeeded in delaying the article, which was released on 5 September, soon after the NAM meeting in August 1976.

When he later realized that his proposal for a Third World Summit would not materialize, he scaled it down to a conference of Third World Education Ministers, to be held in 1977. This too gained little support, and the post-election agitation put an end to his idea.

Following the March 1977 general elections, ZAB was once again isolated. Domestically, the demand and agitation for *Nizam-e-Mustafa* (system of government of the Holy Prophet, [pbuh]) had gained momentum and weakened him. On the international front, the Western countries, particularly the United States, were not supportive, China could contribute little, and the Soviet Union failed to respond to his overtures. Once again he turned to the Muslim countries for help in reaching a settlement with the Opposition parties. In the midst of negotiations with them, on 17 June, ZAB suddenly left for Saudi Arabia, Libya, Abu Dhabi and Kuwait. President Gadhafi promised him 'full support', and Sheikh Zayed was, as usual, most friendly. ZAB regained his confidence, enjoying visions of another Islamic Summit in Pakistan and a proposed Treaty of Non-Aggression among all Muslim countries. Foreign excursions, even if not substantially successful, provide an antidote for domestic problems in the minds of most leaders, and ZAB particularly enjoyed the role of master-diplomat. After meeting Yasser Arafat in Kuwait, and President Sardar Daud at a brief stopover in Kabul, he returned to a still shaken government.[51] However, some Muslim countries, encouraged by those inimical to ZAB, disliked his trying to stride like a colossus through the Muslim World. They too strove for its leadership, and did not support his endeavours. His overthrow followed in a few weeks.

His main success in the Muslim World, apart from the 1974 Islamic Summit, was in improved relations with Afghanistan. Although slow in recognizing the regime of Sardar Daud after the overthrow of King Zahir Shah in August 1973, he subsequently cultivated President Daud, because Pakistan could not live with simultaneous danger from two hostile neighbours, India and Afghanistan. However, here too he missed an opportunity by allowing his domestic political time-table to dictate the release of pro-Afghan NAP leaders in exchange for Afghan acceptance of the Durand Line. Relations with Iran had earlier deteriorated when the Shah considered giving two billion dollars financial assistance to Afghanistan to support Sardar Daud in his effort to be less dependent on the Soviet Union. However, ZAB felt this was untimely because his own negotiations with Daud were not sufficiently advanced and such assistance would remove pressure on Daud to reach an early settlement over the Durand Line. He misunderstood the Shah's aim, judging it to be unfriendly. Iran's proposed assistance to India for the Rajasthan Canal Project also gave ZAB cause for concern. In the end, by July 1977, only President Asad, Sheikh Zayed, President Gadhafi and the PLO's Yasser Arafat remained steadfast to ZAB.

His efforts to achieve a unified Muslim World and a purposeful Third World were broadly correct. However, he lost sight of the undeniable fact that a leader can only become important abroad if he has a strong domestic base. Moreover, any objective observer would have correctly assessed that Pakistan was not yet prepared, so soon after being dismembered in December 1971, to play the grand role he envisaged, particularly on the Third World stage. In his dealings with the superpowers too a similar situation unfolded.

Relations with the Superpowers

United States of America

There has been a regrettable absence of balance and dignity in Pakistan's relations with the United States. Our leaders have invariably either sought the blessings of this superpower for their own actions, or blamed it for their failures. ZAB proved no different. Prior to his becoming President on 20 December 1971, and thereafter till April 1977,[52] he consistently showed great concern and consideration for, and went out of his way to favour, the US. Yet, on 28 April 1977, he virtually accused the US of conspiring against him and, two days later, he defiantly flaunted a letter from the US Secretary of State.[53] We need to look objectively at how his relations with the US developed.

He fully comprehended the importance of the United States in the domestic affairs of Pakistan. As the main supplier of assistance and military hardware, they had helped Ayub Khan, and, subsequently, through Yahya Khan, had maintained close contact with the Pakistan Army. In personal terms too ZAB understood their influence, claiming to have been ousted as Foreign Minister by Ayub Khan on the instructions of President Johnson.[54] As with Tashkent, he sought to avoid a repeat performance.

His relations with President Johnson had not been good. Philip L. Geyelin in *Lyndon Johnson and the World* described how, when the 'brash young' ZAB delivered Ayub Khan's condolence message on the assassination of John F. Kennedy, Johnson 'ordered' ZAB to sit down: 'For once, the vocal, volatile Mr Bhutto was nearly speechless; even US officials present were taken aback.' ZAB wrote at length to Geyelin on 4 June 1968, contradicting his version of the event.[55] Subsequently, when President Johnson's wife in her memoirs, *A White House Diary,* recorded the unfavourable impression she formed of ZAB after a dinner meeting in December 1963, he wanted to reply in the same vein as to Geyelin. At my suggestion, his wife Nusrat instead wrote on 25 October 1971 along the lines I had prepared:

> The policy of the United States towards Asia has in the past decade and a half, except for the intervening Kennedy years, been unrealistic. We bear no ill-will towards the West for the past, but equally we cannot countenance the continuation or repetition of the past.[56]

He thought that Johnson was high-handed with Pakistan. An excellent mimic, he illustrated this with an amusing story of how Johnson dismissed Ayub Khan when the Kashmir issue was raised: Johnson got up from the table remarking, 'Gentlemen, it is time to go for a pee', and the meeting concluded.

During 1970-1, he maintained regular contact with American Ambassador Farland and the Chargé d'Affaires, Sidney Sobers. We have seen how, after the fall of Dhaka, he spared no effort to meet President Nixon, even delaying his return to Pakistan.

To Nixon he conveyed friendship and the impression that only ZAB could 'deliver the goods' and overcome anti-US sentiments in Pakistan. This meeting undoubtedly served to reduce the American Administration's reservations about him.[57] Clearly he wanted to mend fences with the Americans. Earlier, soon after the 1970 general elections, he had been interviewed in Larkana by Peter Hazelhurst of *The Times,* London, who reported on 12 December 1970 how ZAB had stressed that 'the West, and particularly the Americans, misunderstood his internal and foreign policy', and quoted ZAB:

> The most angry people in Pakistan today are the communists for they know I have stopped the tide of communism…In fact I have done more to combat communism in Asia than the Americans in spite of all the resources and the money they have piled into this part of the world.

ZAB publicly expressed gratitude for the US having prevented an all-out assault on West Pakistan by India in December 1971. He told C. L. Sulzberger, a *New York Times* columnist, on 10 February 1972:

> I think that the world and my own people should know that the USA, in the interest of peace and civilized conduct among states, did put its foot down. If there had been no US intervention, India would have moved hard against Pakistan's occupied Kashmir and also the southern front in Sindh.[58]

As part of the effort to maintain good relations with the US, he regularly informed them about the Government's problems and programme. Before nationalizing the life insurance business, he apprised the US Government in advance and, because the American Life Insurance Company was being nationalized, they were promised fair compensation. Subsequently, he even suggested that Gwadar could be developed as a port or base for the US, but this did not materialize because Pakistan could not assure adequate economic development and support in the hinterland.[59]

The US sought early recognition of Bangladesh, and applied pressure by withholding aid and prevailing on the World Bank and IMF not to assist Pakistan. Massive devaluation of the rupee was undertaken at their insistence; later, they regularly monitored and attempted to control our economy.[60]

Pakistan withdrew from SEATO in October 1972; after the loss of East Pakistan, this Treaty was in any case not relevant, serving only as an unnecessary provocation to Beijing. But, contrary to the PPP Election Manifesto, we continued as members of CENTO, resuming participation in the Organization's military exercises and ministerial-level representation at its meetings.

The first high-level contact was when Under-Secretary of State Kenneth Rush and Assistant Secretary Joseph Sisco visited Pakistan in 1973. Rush, a businessman recently appointed following President Nixon's re-election, came on a fact-finding mission; Sisco was an experienced diplomat. ZAB emphasized Pakistan's 'geo-political relevance', and that the US should not place reliance only on Israel and Iran in the Middle East; Pakistan might not be as militarily important as those two countries, but a superpower needed to 'hedge its bets'. They were urged to maintain serious interest in our military and political position, but with little response. Over dinner, on ZAB's instructions, I too talked of this with Sisco, but equally without result.

He endeavoured to secure an early invitation from President Nixon, but was not lucky with his visits to the US. His first trip, arranged for July 1973, was cancelled at the last minute while he was en route to Washington, as Nixon was unwell—although two days later he received the Shah of Iran. When he subsequently visited the US on 18 September 1973, he regretted that it was as Prime Minister and not President, but was pleased to learn that protocol arrangements were similar for visiting Heads of State and Prime Ministers, the only difference being a 21-gun or 19-gun salute on arrival at the White House. He prepared assiduously, determined to impress his hosts both in Washington and New York. This he did, but there was no significant change in the attitude of the US Administration.

In his opening remarks in the White House East Room, where the ceremony took place because of rain, President Nixon stated that 'the independence and the integrity of Pakistan is *a* (emphasis added) cornerstone of American foreign policy'. The 'a' was trumpeted into 'the' by our controlled media even though the final joint Pakistan-United States statement issued on 20 September reiterated that 'President Nixon assured Prime Minister Bhutto of strong US support for Pakistan's independence and territorial integrity which he considered a guiding principle of American foreign policy'. The propaganda about 'the' cornerstone continued unabated.[61]

He praised the US President, his Administration and the American people. Replying to Nixon's welcome, he said:

> At one time it was said that in the recent past your Administration tilted towards Pakistan. That, Mr President, was a tilt for justice and a tilt for equity, which is characteristic of your distinguished career as statesman and a builder of peace.[62]

This was at a time when the Watergate scandal was building up against Nixon. The same evening, in reply to the President's toast, ZAB gave his own assessment of history's verdict on Nixon:

> As far as we are concerned, not only as a Pakistani, but as an Asian coming from Asia, I can tell you that at least my history books, the history books of my country will say that here was a great and lofty President who broke the barriers of prejudice and who chalked out a bold new policy according to the finest traditions of American history and brought peace to a tormented part of the world...So history will pay rich and glowing tributes to your statesmanship not only as an American President but as a world statesman.[63]

About Watergate, he said, 'the present trivialities will be set aside'.

He skilfully projected the 'new Pakistan' when addressing the National Press Club luncheon in Washington on 19 September and, again, two days later, before the Foreign Policy Association and the Asia Society in New York. This was a *tour de force*, covering various aspects of the global balance of power and emphasizing the importance of the 'new Pakistan'. Throughout, he presented a moderate image. In dealing with Islamic Socialism, he was particularly circumspect for his American audience:

> Our Socialism is in that sense restricted to the economic aspects of Marxism. That is further restricted by the conditions of the present time, which is further restricted by the amount of foreign assistance we can get, which is further restricted by the amount of foreign investment that can be made.[64]

His second visit was in February 1975, when President Gerald Ford removed the arms embargo which had been imposed at the time of the 1965 War with India. This represented a major improvement in relations with the US, our main arms supplier. While in New York, news came of the assassination in Peshawar of Hayat Sherpao. ZAB rushed back to Pakistan, deprived of enjoying to the full the success of his visit to the US.[65]

Secretary of State Kissinger twice visited Pakistan. On the first occasion, at the end of October 1974, I played host Minister as Aziz Ahmed was only a Minister of State in the Foreign Office; ZAB said Kissinger was very protocol-conscious. Attempts at conversation with Kissinger on the drive from the airport to the State Guest House only came alive when I mentioned his 'successful' shuttle diplomacy in the Middle East and he held forth on his great role. Kissinger was immensely pleased with himself, but his wife Nancy's charm and unassuming nature compensated for this. The official talks produced no tangible results. ZAB held private, wide-ranging discussions with Kissinger. No effort was spared to please and praise him; the large banquet and cultural show, and expensive gifts, befitted a head of state. But it was not only Third World leaders who feted Kissinger. When I represented Pakistan at the CENTO Foreign Ministers' Conference in 1976, I was surprised that Prime Minister James Callaghan was as deferential to Kissinger as ZAB had been.

Besides Kissinger, Treasury Secretary John Connally was also entertained lavishly in Pakistan. At the time he was considered a presidential hopeful, and it was thought that the investment in his visit might produce valuable results. But nothing substantial emerged from either his or Kissinger's visit.

The nuclear issue dominated the last phase of Pakistan-US relations. The agreement with France for a nuclear reprocessing plant was concluded on 17 March 1976, after approval in February by the Board of Governors of the International Atomic Energy Agency. Kissinger had earlier, on 10 March, maintained that the US was making strenuous efforts to prevent the spread of nuclear technology and equipment to what he described as 'sensitive' countries such as Pakistan. This was followed by a meeting sponsored by the US in June 1976 to discuss the strengthening of safeguards and control on the export of nuclear technology, about which there had earlier been informal agreement in November 1975 with the Soviet Union, France, the Federal Republic of Germany, Great Britain, Japan and Canada.

This issue was brought to the fore in the course of the election campaign of Jimmy Carter, who started by expressing the hope, on 13 May 1976 in New York, that a moratorium could be applied to 'recently concluded agreements'. Though directed at Brazil and Pakistan, it went largely unnoticed because he was not considered a serious candidate at the time. On the sale of the reprocessing plant to Pakistan, French President Giscard d'Estaing, in an interview on NBC's 'Meet the Press' in Washington later in May, after talks with President Ford and Kissinger, took the opportunity of pointing out that the transaction had 'all possible guarantees' against its use for the manufacture of nuclear weapons; the type of plant Pakistan was acquiring was normally used to produce fuel for power and not for armaments.[66]

Secretary of State Kissinger arrived in Lahore for a second brief visit on 8 August 1976 with his wife and young son, David. The talks were held directly between Kissinger and ZAB with no one else participating from the Pakistan side. It was later claimed that Kissinger threatened him that the US would make a 'horrible example' of him if he persisted with his nuclear policy. This is now part of the legend of ZAB and Pakistan, repeated and believed by all.[67] However, it is difficult to reconcile this with the content of the speeches made during Kissinger's visit or with the statements made at the final joint press conference at the airport. Nor does this claim tally with what ZAB conveyed to me at the time.

He hosted a grand dinner at the Governor's House, Lahore, and was more than fulsome in praise of Kissinger: 'The most erudite and articulate spokesman of the foreign policy of America.' He talked at length about Kissinger's statement in Tehran on the importance of Iran's role for the stability and equilibrium of the region, and said, '*Ipso facto* the US must come to the conclusion that the same applies to Pakistan. There can be no dichotomy in that approach.' He waxed almost lyrical about learning the art of diplomacy from the rivers of Sindh, praising his own knowledge and experience: 'It does not come naturally to people who do not have to tame rivers, fight them and make love to them...Please remember that we will be at our best tomorrow because our rivers are full of mischief.'[68]

Kissinger was not sparing with his compliments to ZAB: 'We know how he took over this country in a tragic and desperate period and how he has returned it to international respect, to self-confidence, and to a more important role than it has ever played before.' ZAB had already reformed his country and had since offered friendly advice to other countries: 'And those of us who have been exposed to this advice suffer from the fact that he almost invariably turns out to be right.' On touching upon 'fearsome weapons' in today's world, Kissinger emphasized the need to balance security against universal cataclysm.

At the airport the following afternoon, both ZAB and his wife accompanied Kissinger and his wife, in itself an unusual departure in protocol, and there was also an honour guard for the guest. Kissinger told the press conference that the reprocessing plant was a 'complicated issue'. Pakistan had been negotiating with France for several years and had concluded an agreement 'which had all the safeguards appropriate at that time'. However, 'our concern is not directed towards the intention of Pakistan,' but to the general problem of nuclear proliferation, which could have disastrous consequences for the future of mankind. He referred to the wide-ranging talks with ZAB: 'We have agreed to continue our discussions in the weeks and months ahead.' He also stressed that both sides would avoid confrontation. The US would try to elaborate general principles on reprocessing problems which would apply to all countries without discrimination. When an American correspondent pointed out that the US Congress seemed bent on confrontation over the reprocessing plant issue, Kissinger replied that he was 'hopeful' that some conclusions would be reached in future talks.[69]

ZAB expressed similar satisfaction with the discussions, reminding the Press that Kissinger had earlier underlined the fact that the US wanted to avoid confrontation, and that further talks would be held on the nuclear issue. He said the position on arms-supply was 'good'—all items except one had been cleared.

That day I was in Lahore reviewing the performance of the National Fertilizer Corporation, after which I met ZAB, who was completely relaxed; his meetings had apparently gone well. According to ZAB, Kissinger had pointed out that the Democratic candidate Carter was on record as strongly opposing Pakistan's nuclear programme. Kissinger suggested that this was a good time to deal with the Republican Administration, traditionally friendly to Pakistan, which was willing to supply modern aircraft and other military equipment. Carter would be heavy-handed and might 'run a freight train' over Pakistan. ZAB said he needed time to consider the matter, and it was left at that.[70]

ZAB felt that relations with the US were under control. I was less convinced of this, since he was increasingly criticized by the Western Press. In fact, on one earlier occasion, I had warned him that the US might be involved in an attempt to remove him. However, he confidently maintained that he was handling the Americans with meticulous care. Who, after all, could better understand the US following his experience as Foreign Minister in the mid-1960s? He also suggested I should curb what he termed my anti-American sentiments. I pointed out that I was not against Americans as individuals, and had several American friends, but I held the view that the US Government had invariably acted to the detriment of the Muslim and Third Worlds; they supported corrupt regimes when it suited them, and had not proved trustworthy allies.

Events took a dramatic turn immediately after Kissinger left Pakistan for France on a private visit. For some reason, he chose this occasion to refer to US pressure aimed at preventing the reprocessing plant deal. The French took this as interference in their affairs, angrily dismissing Kissinger's somewhat crude efforts to block the agreement. The American Chargé d'Affaires in Paris was summoned to the French Foreign Office, and Kissinger had to phone Foreign Minister Jean Sauvegnargues on 10 August to cool down the controversy.[71]

The next day, ZAB also contradicted Kissinger in a telephone interview with Radio Paris: 'We told him that we have signed an agreement with France and we will abide by this agreement. I do not consider this discussion as an ultimatum. We did not talk in terms of ultimatums. Dr Kissinger tried to convince me and I did the same.'[72]

While he was trying to calm matters, both the foreign and Pakistani Press were full of the subject. French Prime Minister Jacques Chirac reacted sharply: 'There is no question of accepting Dr Kissinger's proposal to settle with the United States an affair which concerns only France and Pakistan.' The French Foreign Ministry stated, 'Mr Gammon, the United States Chargé d'Affaires, expressed the Secretary of State's regrets over Press interpretations and commentaries which had given the impression that the US Government is seeking to exert pressure on the Pakistan Government.'[73]

Although on holiday, Kissinger told a press conference that the Symington Amendment 'imposes certain requirements on the executive', and explained, 'I made clear that it [the proposed reprocessing plant] is a matter we will discuss among friends in an atmosphere of non-controversial trust and cooperation'. Referring to France, he said: 'We deal with France as a friend. And we will not deal with France on the basis of pressure'; adding that the United States would look for a solution after the summer holidays when all the three parties would have an opportunity 'to exchange their opinions'.[74]

This sudden flurry of publicity placed ZAB in an awkward position. Kissinger made the issue public immediately after visiting Pakistan either as a tactical application of pressure or on the basis of some understanding with ZAB. In a later interview, he admitted suggesting that Kissinger should have direct talks with France where he was going from Pakistan.[75]

On 20 August, a French representative came to Peshawar to find out whether Pakistan wanted to proceed with the agreement—France would not back out first. When it was suggested that Pakistan obtain some critical parts for the reprocessing plant in advance as a precautionary measure, ZAB declined. Whether this was because alternative arrangements were in hand, to which Kausar Niazi later referred,[76] I did not inquire.

On 25 August, Jacques Chirac resigned as Prime Minister, and shortly afterwards, on 2 September, President Giscard established a Nuclear Export Policy Council to examine these matters. In early October, at the height of the American presidential campaign, Jimmy Carter again called on all nations to accept a 'voluntary moratorium' on the sale or purchase of enriched nuclear fuel or reprocessing plants which could be used to produce fuel for nuclear weapons. Specifically referring to the deals between France and Pakistan, and Germany and Brazil, he said, 'The contracts have been signed but deliveries need not be made'.[77]

Already angered by Kissinger going public, ZAB attacked Carter in his next interview with *De Telegraf* of Amsterdam on 5 October 1976:

> Today is the first time I am speaking to anyone openly like this, only because the day before yesterday, the Democratic Presidential candidate, Mr Jimmy Carter, said he is going to see to it that Pakistan does not get the reprocessing plant from France. Now this is contrary to our understanding with Dr Kissinger in Lahore that we will negotiate, we will discuss, we will talk, without confrontation and without going public. After all Pakistan and the United States are friends.

He added:

> In Lahore, the American Secretary of State promised that they will talk to us without confrontation, that they will talk to us as friends, that they will talk to us in private, that they will talk to us in confidence, then why do candidates, who are usually briefed on such matters, come out with such threats from public platforms.[78]

Aziz Ahmed was directed to meet Kissinger in New York, and they discussed the nuclear issue on 6 October. Newspaper reports continued to maintain that France had virtually dropped plans for the reprocessing plant sale, and would not object to Pakistan cancelling the deal. However, a French Foreign Ministry spokesman clarified on 12 November, 'There is no modification in the development of Franco-Pakistan relations nor in the implementation of agreements reached between the two countries.'[79]

There was no change in Pakistan's relations with the US, particularly concerning the nuclear issue, till the end of December 1976. ZAB then told Ambassador Byroade that the elections would take place on 7 March 1977, after which he would inform the US about his decision on the reprocessing plant.[80] Over the past months he had been considering how Carter, as a Democratic President, might change the US position, and how, in the light of Kissinger's advice, it might be advantageous to deal with the Republican President Ford. At one meeting he even suggested that he speak out in favour of Ford, in the belief that ZAB's international stature might make a difference in the American Presidential elections. It will, however, be more convenient to dwell further on relations with the US in Chapter 9 when considering developments over the elections, the subsequent agitation by the Opposition parties, and the military *coup* of 5 July 1977.

Soviet Union

The Soviet Union played an important role in securing the Simla Accord. ZAB's first visit to Moscow in March 1972[81] also saw progress in trade and economic relations. However, two factors prevented further major improvement: his concern about Soviet interference in our domestic affairs and his desire to please the US.

In the initial period the Soviets attached importance to Pakistan leaving CENTO, as much for political as for military reasons. They were disappointed by Pakistan's increased participation in CENTO activities, despite ZAB's election pledge to leave the Organization. He preferred military and economic ties with the US. It was, however, not long before the Soviets became preoccupied with their own *détente* with the US to counter improved Sino-American relations. For them, this new pincer pressure presented a more immediate threat than the previous 'encirclement' of which CENTO was a part.

The Soviet Union was throughout important to Pakistan, not only as a neighbouring superpower, but because of her close relations with our two hostile neighbours, India and Afghanistan, and with the NAP, which constituted a major part of the two non-PPP Governments in the Frontier and Balochistan. The removal of these two Provincial Governments in February 1973 met with Soviet disapproval. ZAB countered in an interview with *Kayhan International* of Tehran, published on 15 April 1973: 'We cannot be expected to take everything flat on our backs. We are more than ready to forge the best of relations with USSR provided they respect our dignity and integrity.'

By October 1974, when he next went to Moscow, his position both at home and abroad had improved. Following the successful Islamic Summit in February, Pakistan had emerged as an important country in the region. The position of India too had changed: her nuclear explosion of May 1974 appeared to have decreased her dependence on the Soviet Union. As a consequence, the Soviets indicated some readiness to support Pakistan's efforts to redress the balance in the subcontinent. Apart from the military potential of the Steel Mills, there was now talk of arms supplies. ZAB's reception at Moscow in October 1974, in stark contrast to his first visit, demonstrated our enhanced status. Throughout our visit, Prime Minister Kosygin emphasized the need for good relations between our two countries and how this could prove beneficial to Pakistan.

The main outcome of this mission was the agreement in principle for the Soviet-assisted Steel Mills; its financing was subsequently finalized when I visited Moscow in December 1974 as Minister for Production. ZAB's concerns about potential Soviet interference were reflected in our talk during the flight to Moscow in October 1974. I pointed out that the cost of the Steel Mills had soared with the sudden spurt in oil prices following the Middle East War, and would continue to escalate; moreover, one major spin-off benefit in the form of technically trained personnel would be lost as many were leaving Pakistan for the oil-rich Gulf countries. The Steel Mills should have been established earlier in better times; instead, the project would stoke massive inflation, because it would take several years to implement, and at the same time would divert resources from other quick-gestation projects. His immediate reaction was, 'the bureaucrats have got to you'. I replied that my comments were political as the benefits from the project would not materialize till after the next general elections. He wanted no further delay as this might confirm the Soviets' earlier suspicion that the re-siting proposal was just a subterfuge.[82] He wanted to avoid a situation where they might be provoked into interfering in our domestic politics again. In fact, once the project was under way, Pakistan had little cause for complaint as the Soviets were eager to develop further economic ties.

Throughout 1975 and 1976, his concentration on the United States, and on building his position in the Muslim and Third Worlds, led to the Soviet Union being given less importance. Apart from maintaining cordial relations, the Steel Mills remained the main link. Not until after the March 1977 elections did he approach the Soviets to counter a hostile US. By then, however, the handicap of his earlier ambivalence and present domestic weakness prevented him from playing one superpower off against the other.

When the PNA agitation was at its peak, ZAB conveyed to the Saudi Ambassador on 21 April 1977 his assessment that the United States was trying to destabilize Pakistan, while at the same time the Soviet Union and India were again attempting to dismember the country. The inclusion of the Soviet Union was to gain Saudi sympathy but, with the US the main source of his problems, he turned increasingly to the Soviet Union. On 10 May it was announced that Pakistan would be represented at the CENTO ministerial meeting to be held in Tehran by the Ambassador in Iran and not by our Foreign Minister. But, by then, such minimal gestures had lost their relevance.

Later, in May and June 1977, he spoke vehemently against the US and told Soviet Ambassador Azimov that he was considering leaving CENTO. His team negotiating with the PNA tried to arrange a joint decision to quit CENTO 'in the national interest'. Although this was also a manifesto pledge of the PNA, no agreement could be reached. The Soviets did not respond to ZAB's blandishments despite concessions, which it was rumoured included the offer of Gwadar port to the Soviet Union. Whatever the truth of the matter, the US too did not heed ZAB's warnings, given at this time, that the Soviets were persisting in their aim to reach the warm waters of the Arabian Sea. Thirty months later the Soviet Union invaded Afghanistan.

Neither superpower considered him reliable or capable of fulfilling his promises in 1977. The lack of importance attached by the Soviet Union to Pakistan under ZAB was evidenced by the fact that no significant Soviet dignitary visited Pakistan during his five-and-a-half years in government, despite his own two visits, not even in the last days when he turned to the Soviets. Unlike the United States, however, the Soviet Union did at least strongly advise General Ziaul Haq against ZAB's execution.

China

The vital importance of China, particularly to counterbalance India, has been recognized by all Pakistani governments. ZAB claimed to be the architect of Pak-China friendship, while his detractors point out that Pakistan was one of the first countries to recognize the People's Republic of China and, even subsequently, ZAB as Foreign Minister mainly carried out Ayub Khan's policy. The fact remains that relations between the two countries were never so close as when he was Foreign Minister from 1963 to 1966. Equally, the need of each country for the other was greatest in that period. China was still largely isolated, not a member of the United Nations, and seeking to make friends in the Middle East and Africa through Pakistan, which in turn had two armed conflicts with India, one a major war, for which Chinese assistance was required.

In his letter of 27 June 1971 to Lt.-Gen. Peerzada,[83] ZAB stressed the significance of China:

> Never before has a country become so dependent for its sovereignty and survival as Pakistan is on China. If the Consortium can dictate its terms, thinking it can get away with it, China will find it easier to do likewise if it chooses to assume such an attitude.

Fortunately these fears expressed about China did not materialize. Unlike during the 1965 War, however, China made no positive contribution when in 1971 India launched her 'war of liberation' in East Pakistan, not even by maintaining pressure on India's border in order to prevent the withdrawal of Indian troops facing China. Pakistan failed to comprehend that, following the Indo-Soviet Treaty of August 1971, China was not ready for any action that might involve confrontation with the Soviet Union. However, the following year, China exercised a vital veto over the admission of Bangladesh to the United Nations.[84]

The significance of this veto has not been fully appreciated in Pakistan. Almost all countries, apart largely from the Muslim World, had recognized Bangladesh. Encouraged by this, Bangladesh with India's support applied for membership of the United Nations in 1972. At the time, there was the real threat of Pakistani prisoners, or 195 of them, being tried for serious war crimes. While a small element in the PPP felt that it might be salutary to allow trials for actual atrocities committed, the general consensus supported ZAB's view that it would militate against national sovereignty and dignity to permit trials on foreign territory. Other elements suggested that we offer to hold the trials in Pakistan, and he hinted at this to India. However, fearing a reaction in the Armed Forces, he finally decided to resist all attempts to hold such trials.

China had only recently gained its seat in the United Nations, overcoming repeated vetoes by the United States. Bangladesh was a close neighbour. Nevertheless, at Pakistan's request, China did exercise a veto, stressing that Bangladesh should only be admitted after all Pakistani personnel had been repatriated. Previously, China had condemned the use of the veto, referring to it as superpower domination.

Had Bangladesh and India not received this setback, there was every likelihood that the trial of at least some prisoners would have taken place before the close of 1972. At the time, feelings in Bangladesh were still running very high against Pakistan, and India would no doubt have welcomed a further humiliation of the Pakistan Army. China's veto in 1972 and again in 1973 greatly strengthened Pakistan's negotiating position. With the passage of time and subsequent developments, the relevance of war trials diminished except as a bargaining point in the talks with India and Bangladesh.

Chinese assistance grew in the economic field following ZAB's January 1972 visit. China contributed substantially to establishing the first heavy-engineering projects in Pakistan, the Heavy Mechanical Complex and the Heavy Foundry and Forge near Taxila. Like India, they purchased raw cotton when, for a brief period, we carried large stocks after nationalization and could not readily secure foreign purchasers. When he next went to Beijing, on 11 May 1974, he covered such important matters as nuclear technology and power plants in meetings with Chairman Mao and Premier Zhou Enlai. In the course of a banquet speech, Vice-Premier Deng Xiao Ping assured his guests that, 'Come what may, the Chinese Government and people will, as always, firmly support Pakistan in the defence of national independence, state sovereignty and territorial integrity and against hegemonism and expansionism, and firmly support the people of Kashmir in their struggle for the right of self-determination'.[85] The Indian Ambassador walked out of the banquet.

Next year, on 20 April, Vice Premier Li Hsien-nien came on a five-day visit to Pakistan. He supplemented the Chinese attack on hegemonism and expansionism by stressing at Lahore that, in any future joint struggle, 'We will always stand by the Pakistani people'.[86] Pakistan had previously published a twenty-five year energy programme, including twelve nuclear power plants, and Chinese assistance for these projects received serious attention. However, by the time of ZAB's third visit to Beijing, on 26 May 1976, China had changed dramatically: Premier Zhou was dead and Chairman Mao critically ill, only capable of one brief formal meeting. Still, he secured military and economic assistance.

The repercussions of the change in China's domestic scene were not at first recognized in Pakistan. Previously, because of the Cultural Revolution and the later dominance of the 'Gang of Four', there had been no question of the Chinese Ambassador wearing a western-style suit or making any effort to speak

English. Within a month of Mao's death it was different. However, in Pakistan the Press and the Foreign Office did not for several months properly refer to the fall of the Gang of Four. ZAB felt it was too soon for any trial of Mao's widow and supporters, who could stage a come-back. This view changed following a meeting with the Chinese Ambassador which I reported to ZAB. China had throughout punctually supplied machinery for our projects, but recently there had been some delay, for which the Ambassador apologised to me. The delay was understandable because China had suffered some massive earthquakes. However, the Ambassador was at pains to attribute the delay not to the earthquakes but to the disruption caused by the Gang of Four. This was mentioned several times during our meeting. Thereafter we understood the full impact of the change in China.

Of the three Big Powers, it was only with China that relations remained good throughout his period in office. Very seldom have two countries, diametrically opposite in so many ways, sustained such a long and lasting friendship. However, because of internal turmoil in China, no major new initiatives could be undertaken during the ZAB era.

* * *

In the field of foreign affairs as a whole, ZAB brought Pakistan back to centre stage and regained its lost stature. He pursued a new direction for the Muslim World and the Third World, but instead of taking the country to the heights of which he was capable, his foreign policy plans suffered in part from the shortcomings of his own personality. As a result, following the Islamic Summit in February 1974, there was little positive progress.

NOTES

1. *The Myth of Independence,* p. 134.
2. *The Pakistan Times,* Lahore, 21 December 1971; Jagjivan Ram was the Indian Minister of War.
3. ZAB disclaimed the use of Government machinery in his efforts, but Q. U. Shahab, the Education Secretary, and several members of the Pakistan Embassy in London worked hard at it. On 25 February 1975, ZAB sent me a copy of Shahab's report on the subject with the comment, 'Might be needed by you to prepare our answers to critics on East Pakistan separation'.
4. 'Pakistan Builds Anew', in *Foreign Affairs,* Volume 51, No. 3, April 1973, p. 554.
5. *President Bhutto's Interviews to Foreign Correspondents,* published by the Department of Films and Publications, GOP, 1972, p. 6.
6. *Asian Recorder,* 22-8 April 1972; also reproduced in *President Bhutto on Issues in South Asian Sub-Continent: Extracts from the President's Interviews with Foreign Press Correspondents,* published by the Department of Films and Publications, GOP, 1972, p. 6.
7. *President Bhutto on Issues in South Asian Sub-Continent,* p. 25.
8. *The Sun,* Karachi, 1 May 1972. In English, this would read:
   Let us also raise our hands to pray;
   we who do not remember the ritual of prayer,
   we who do not recall any idol or god,
   but only the intensity of love.
   Let us solicit the deity of life
   to pour the sweetness of tomorrow
   in the poison of today.
9. The countries visited were Abu Dhabi, Kuwait, Iraq, Lebanon, Jordan, Saudi Arabia, Somalia, Ethiopia, Sudan, Nigeria, Guinea, Mauritania, Turkey and Iran.
10. *President Bhutto's Interviews to Foreign Correspondents,* 1972, pp. 53-4.
11. Ibid., p. 160.
12. Ibid., p. 80.
13. Ibid., pp. 78-9; interview published in the *Hindustan Times,* 8 May 1972, and reproduced in *The Pakistan Times,* Rawalpindi, 10 May 1972. Shahnawaz Khan, Pakistan's Ambassador in Tehran, had been selected as Ambassador to India.
14. Ibid., p. 111.
15. Ibid., pp. 142-3.
16. Ibid., p. 144.
17. Ibid., p. 13; *see also,* p. 200 and Note (6) thereto for this interview.
18. Ibid., p. 141.

19. Prominent among those included in the delegation from the Punjab were Chief Minister Malik Mairaj Khalid, Chief Secretary Afzal Agha, Mian Anwar Ali, the Director Intelligence Bureau, Sardar Shaukat Hayat Khan, a Muslim League Member of the National Assembly; from Sindh, Ghulam Mustafa Jatoi, Federal Minister for Communications and Political Affairs; from the Frontier Province, Hayat Sherpao, Federal Minister, and Arbab Sikander Khan Khalil, the NWFP Governor and a leading member of the opposition National Awami Party, Yusuf Khattak, a Muslim League Member of the National Assembly; and from Balochistan, Mir Ghous Bux Bizenjo, Governor and leader of the National Awami Party.

20. *Dawn,* Karachi, 3 July 1972, under the caption 'How it came about', accurately reported: 'The two leaders left the table at 10 pm and began their talks. They were joined within a few minutes by Mr Haksar, Mrs Gandhi's Principal Secretary, and Mr Rafi Raza, Mr Bhutto's Special Assistant. Soon after Mr Jagjivan Ram, Mr Y. B. Chavan and Mr Aziz Ahmed joined their respective leaders…The final draft of the agreement was settled between the billiard room and the drawing room, by Mr Haksar and Mr Rafi Raza in turn.'

21. Following ZAB's overthrow in July 1977, General Ziaul Haq tried to get some Foreign Office officials to 'confirm' a secret agreement at Simla, but without success. This narrative has set out precisely what happened to the author's knowledge. The revelation made some twenty-three years later by P. N. Dhar, who was at one time Secretary to the Indian Prime Minister, *see, The News International,* Karachi, 5 April 1995, and other dailies of that date, about a secret agreement is not correct; neither is his statement that the second and third sentences in Clause 4(ii) of the Simla Accord were added following agreement on the insertion of the words 'without prejudice to the recognized position of either side'. In fact, these second and third sentences were in the draft text from the outset. Possibly P. N. Dhar has played upon the inaccurate account ZAB gave of his meetings in an interview to Moti Ram, an Indian journalist, on 23 September 1976, reported in the *Business Recorder,* Karachi, on 15 November 1976.

22. *Zulfikar Ali Bhutto, Speeches and Statements, 1 July 1972 to 30 September 1972,* GOP, pp. 6-7.

23. Ibid., p. 12.

24. *Dawn,* Karachi, 11 September 1973.

25. *See,* pp. 252-3 concerning China's veto.

26. This private conversation cannot be repeated in full here.

27. *The Hindu,* Madras, 25 August 1972.

28. *The Hindustan Times,* New Delhi, 27 August 1972.

29. Ibid.

30. *The New Times,* Rawalpindi, 11 November, and *The Pakistan Times,* Rawalpindi, 10 November 1972.

31. Interview with *The Blitz,* Bombay, as reported in *The Pakistan Times,* Rawalpindi, 10 November 1972.

32. *See,* pp. 228-9 for details concerning the Islamic Summit.

33. *Dawn,* Karachi, 20 May 1974.

34. *The Pakistan Times,* Rawalpindi, 18 May 1974.

35. Stanley Wolpert, p. 252, incorrectly states that the call for *hartal* was in Jammu and Kashmir, and attributes its success to it being on a Friday; in fact, Friday became a public holiday in Pakistan only after 1 July 1977.

36. *The Pakistan Times,* Rawalpindi, 18 September 1975.

37. *Dawn,* Karachi, 30 May 1976.

38. *The Pakistan Times,* Lahore, 21 December 1971.

39. *Bangladesh Observer,* Dhaka, 29 June 1974.

40. Biharis had settled in East Pakistan both before and after 1947. The problem of Biharis has continued to disturb relations between the two countries to this day and is also a source of trouble in the Province of Sindh in the context of dealings between old and new Sindhis, or *mohajirs.*

41. The 'transcript' referred to by Stanley Wolpert at pp. 173-6 was neither a full nor proper record.

42. *The Pakistan Times,* Rawalpindi, 23 February 1974.

43. *See,* Chapter 8, pp. 291-6 for further details on the Qadiani issue.

44. When *The Guardian,* London, 16 January 1979, commented on the first of the four volumes of a White Paper detailing allegations against ZAB, it incorrectly stated that I supported the one concerning 150 jeeps from Iran. *The Guardian* on 19 January 1979 published my reply while I was in London: 'I have not seen the White Paper...However, I should like to make it clear that I have never supported any such allegation...I resent the attempt of the military Government of Pakistan to try to associate my name with their present campaign against Mr Bhutto and his family.' In fact, when the 150 jeeps were distributed, I was no longer Special Assistant to the President. Subsequently, I was asked by Benazir Bhutto's lawyer to provide a statement to the effect that the jeeps were a gift to ZAB. Benazir Bhutto in 1989, apparently without seeing his statement in the White Paper, rewarded with an ambassadorship a person who really supported this allegation against ZAB.

45. Z. A. Bhutto, 'RCD: Challenge and Response', in *Pakistan Horizon,* 29 (Second Quarter 1976), pp. 3-6.

46. Ibid., pp. 11-2.

47. *See also,* p. 207 concerning this interview.

48. *Dawn,* Karachi, 20 February 1976.

49. *Dawn,* Karachi, 26 May 1976.

50. The author told ZAB that he would display ZAB's photographs when he ceased to hold office; they are now displayed.

51. *See,* Chapter 9, p. 354.

52. *See,* Chapter 4, pp. 127 and 135-6; Chapter 5, pp. 147-8 and Chapter 9, p. 318.

53. *See also,* Chapter 9, pp. 351 and 357.

54. ZAB several times mentioned this to the author; *see also,* p. 135-6.

55. Copy of ZAB's letter of 4 June 1968 is with the author.

56. Copy of the draft of this letter is with the author.

57. *See,* Chapter 4, pp. 135-6.

58. *Asian Recorder,* 18-24 March 1972.

59. Both these matters were verified by the author in 1993 with V. A. Jafarey, who was the relevant Secretary to the GOP under ZAB.

60. *See,* Chapter 8, pp. 289-90.

61. Stanley Wolpert at p. 220 also uses 'the'.

62. Prime Minister Zulfikar Ali Bhutto, *Statements and Speeches, Visit to the United States, September 18-23, 1973,* Department of Films and Publications, GOP, Rawalpindi, printed at Ferozsons Printers, Karachi, 1973, pp. 8-9.

63. Ibid., p. 22.

64. Ibid., p. 37.

65. *See also,* Chapter 8, p. 276.

66. Extracts from the French President's NBC interview were reported in the *The Pakistan Times,* Lahore, 25 May 1976.

67. Stanley Wolpert, p. 229, Benazir Bhutto, *Daughter of the East,* Hamish Hamilton, London, 1988, p. 76, and others have cited this. NB: ZAB himself made no mention of it in *If I am Assassinated,* p. 138.

68. *Pakistan Times,* Lahore, 9 August 1976, and other Pakistani papers, carried extensive reports of the speeches of ZAB and Kissinger.

69. A detailed report of the press conference is in *The Pakistan Times,* Lahore, from which the quotations are taken, and other Pakistani papers, of 10 August 1976.

70. Mubashir Hasan also recalled to the author the term 'express train'.

71. *The Pakistan Times,* Lahore, 11 August 1976.

72. *The Pakistan Times,* Lahore, 12 August 1976, and other newspaper reports of the same day.

73. Ibid.

74. Ibid.

75. Interview with *Al-Jazira,* a Saudi Arabian newspaper, during King Khalid's visit, reported in *The Pakistan Times,* Lahore, 12 October 1976, and also *Dawn,* Karachi, 12 October 1976.

76. Kausar Niazi, *Zulfiqar Ali Bhutto of Pakistan — The Last Days,* Vikas, New Delhi, 1992, at pp. 113-15 describes, how ZAB 'duped' the US over

the French reprocessing plant while the real work was going on at Kahuta.

77. *The Pakistan Times,* Lahore, 4 October 1976.

78. The transcript of the interview with J. G. Heitink, Assistant Editor-in-Chief of the Amsterdam daily *De Telegraf* of 5 October 1976, was published in *Dawn,* Karachi, on 13 November 1976.

79. *The Pakistan Times,* Lahore, 13 November 1976.

80. *See also*, Chapter 9, pp. 318.

81. *See,* pp. 197-200.

82. *See also*, p. 199.

83. *See,* Chapter 4, p. 103 and Note (32) thereto.

84. *See also,* p. 216.

85. *The Pakistan Times,* Rawalpindi, 13 May 1974.

86. *The Pakistan Times,* Lahore, 24 April 1975.

CHAPTER 8

DOMESTIC AFFAIRS

The domestic scene witnessed much turbulence beneath the apparent calm of a PPP Government firmly in control. Having set the new direction in his early months as President, ZAB sought to leave his imprint on almost every aspect of national life. In his pursuit of national integration, he came to dominate the Provinces. He introduced change, much of it overdue and salutary, to a degree hitherto unknown in Pakistan, ranging from the transformation of the economy to revitalizing cultural expression and encouraging the rights of women. At the same time, he tried to establish his religious credentials. But in attempting to out-perform the left within his own Party, as well as the rightist, religious parties, he damaged his main objectives. His determination to give the country an unmatched lead was accompanied by a growing intolerance of dissent. As authority, and its concomitant arbitrariness, increasingly vested in him, so did discontent and opposition focus on him personally. While it is not possible to cover all aspects of domestic affairs, four in particular merit consideration: (1) the Provinces and the Opposition, (2) the economy, (3) religion and sectarianism, and (4) the Pakistan People's Party and preparations for the 1977 elections.

The Provinces and the Opposition

Provincial and separatist problems did not start or end with East Pakistan. The table below[1] shows the dominant position of the

Punjab and the continuing disparities in the 'new Pakistan' of 1972:

Province/ Territory	Area in sq. miles	1961 population	1972 population	Density per sq. mile
The Punjab	70,284	25,483,643	37,507,855	473
Sindh	54,407	8,367,065	14,007,722	257
NWFP	28,773	5,730,991	8,337,385	290
Balochistan	134,050	1,353,484	2,405,154	18
Federally Administered Tribal Areas	10,510	1,847,195	2,485,867	237
Islamabad Federal Capital Territory	350	94,000	235,749	674
Total	307,374	42,880,378	64,979,732	211

Basic issues concerning the Provinces had not been resolved since the creation of Pakistan. The Provinces had opted for Pakistan largely out of a commitment to Islam, combined with fear of Hindu political and economic domination, but they had not thereby abandoned their socio-cultural and historical antecedents. Provincialism was not superceded by a sense of national identity and remained a serious problem to which no solution was found—witness the separation of East Pakistan.

Although Pakistan was described as an ideological state, no concrete content was given to this 'Islamic Ideology'. Such contradictions were increased by the fact that Islam does not call for the establishment of a nation state, but underlines the need for all believers to live in a spirit of universal brotherhood. It could thus be argued that an individual can identify equally with his Province, with his country, or with the *ummah*. However, intellectual dishonesty, fear and cowardice ensured that anyone who did not conform to the thinking of the establishment was dubbed a traitor to Pakistan or anti-Islam.

ZAB fully comprehended the dimensions of the problems of provincialism, and, indeed, took pride in being the protector of the rights of Sindhis. However, as with other West Wing leaders who preceded and followed him, the demand while in opposition for genuine provincial autonomy was soon overtaken when in office by a desire for centralization. As a representative of a minority Province who had worked at the Centre for eight years, he might have been expected to behave differently, particularly in relation to the three smaller Provinces.

Sindh

Let us start with Sindh, ZAB's own Province, known as the land of 'Peace, *Pirs* and Poets', which had an old history of separatism. The expression 'Sindh and Hind' had long been used to denote the separate identity of Sindh in relation to former India. Sindh became a Province in our present-day terms when it was severed from Bombay by the British after the Government of India Act, 1935. Its capital, Karachi, was the only major port for the West Wing and witnessed rapid growth and industrialization after Independence. It was also the federal capital until Ayub Khan established Islamabad in the early 1960s.

In the December 1970 elections the PPP was not as successful in Sindh as in the Punjab. Out of 60 Provincial Assembly seats, it won only 28, even though it secured 18 out of 27 in the National Assembly. Several independents joined subsequently and, with the emergence of the PPP, Sindhi nationalism was relegated to the background as Sindhis sought opportunities beyond their provincial borders.

Mumtaz Bhutto's appointment first as Governor and then as Chief Minister of Sindh proved successful. An astute administrator with a traditional background, he dealt effectively with extremists among Sindhi nationalists led by G. M. Syed, who had disturbed previous regimes. Sindhis witnessed for the first time, under the direct supervision of their Chief Minister, a new rural development programme creating schools, dispensaries, roads and agricultural facilities. Mumtaz Bhutto improved the bad law and order situation, and was tough with labour clamouring *'gherao, jalao'* (besiege, burn).

But a far more difficult and divisive problem lay ahead. As in East Bengal, language was a sensitive issue in Sindh, which also enjoyed an old culture and traditions. Apart from the fact that the national language was Urdu, the massive influx of mainly Urdu-speaking *mohajirs* following Partition helped to entrench Urdu to the detriment of Sindhi. Sindh was the main Province which allowed *mohajirs* to settle in her cities, towns and even rural areas. While Sindhis learnt Urdu, and although settlers in the rural areas learnt Sindhi, the main body of *mohajirs* who came to Karachi and Hyderabad maintained their linguistic, cultural and ethnic separateness. Gradually, Sindhi began to lose importance since it lacked the status of an official language in the Province, which it had virtually been accorded in pre-Independence days. Not without justification was there concern that Sindhi was being eliminated, a fear which had caused trouble in the past. Mumtaz Bhutto prepared the Sindhi Language Bill, giving Sindhi official status in the Province, but waited for ZAB's green signal. ZAB felt that the most opportune time to introduce it was after the success at Simla, despite warnings of serious trouble.

The Bill was presented and passed by the Provincial Assembly on 7 July 1972. Karachi erupted. A leading Urdu daily, *Jang,* banner-headlined *'Urdu ka janaza'* (the funeral of Urdu). The Armed Forces were called in to aid civil power, but they were at first uncooperative. The President said Mumtaz should handle the problem, and directed me to visit Karachi to assess the situation and assist him. When I arrived in Karachi there was a curfew; it was like a ghost city.

ZAB addressed the nation late on 8 July: 'Urdu is the national language and its importance is undeniable. Sindhi is a centuries' old language, fully developed, but it has been denied its rightful place in recent times. The two languages must learn to live together in the Province of Sindh.'[2] He invited representatives of both groups to meet him in Rawalpindi on 10 July, appointing a small team to negotiate on his behalf. We held several days of detailed discussions; agreement was only reached when I suggested a Simla-like compromise to allay apprehensions through the promulgation of an Ordinance to the effect that 'the application of the [Sindhi Language] Act shall be in a manner that shall not prejudice the use of Urdu'.[3] When ZAB announced this solution on 15 July, he blamed the capitalists for stirring up trouble in the first place, though there was in fact no evidence of this. He had predicted trouble in the form of 'three Ls', labour, language and lawlessness.

The language riots, unlike the earlier police strike in the NWFP and the Punjab, directly challenged his position. A flood of posters with Ayub Khan's portrait appeared not only in the urban areas of Sindh but in the Punjab, affecting the Armed Forces. Just a few years earlier, Ayub Khan had been much reviled in Karachi, especially following the 1964 elections when a victory parade led by his son, Gohar Ayub Khan, resulted in the loss of several innocent lives. Karachi was in the forefront of the revolt against Ayub Khan in 1968-9. ZAB was incensed not only at this revival of Ayub Khan's name, but at being labelled a 'Sindhi President'. From this time, Karachi was the object of his antipathy and, indeed, the feeling became mutual.

Mumtaz Bhutto prepared a White Paper on the Language Issue to explain the historical background, but withheld it because of the delay in its completion. ZAB picked up this point: 'It should not have taken three months to produce this document...It has put the Sindh Government on the defensive.'

It has subsequently been argued that these riots were the forerunner of the ethnic troubles a decade later between *mohajirs* and Sindhis. However, the latter were caused largely by economic problems, the lack of employment opportunities resulting from the quota system, which limited the number of government jobs available to *mohajirs,* and the neglect and exclusion of Karachi from the corridors of power. There was no further trouble over language. Mumtaz balanced firmness with development. His schemes for Karachi began to materialize over the coming years, though some were later dropped or replaced by less desirable ones, resulting in the dreadful state of Karachi today. However, Mumtaz's efforts were neither generally known nor acknowledged.

Mumtaz was removed as Chief Minister at the end of 1973 because ZAB could not countenance his independence combined with his popularity. He was replaced by Ghulam Mustafa Jatoi, a courtly gentleman and major landlord, under whom the feudals, reined in by Mumtaz, regained their ground. ZAB interfered throughout. True, it was his Party's Provincial Government and had to carry out his mandate, but it would certainly have been more beneficial if he had allowed autonomy under the Constitution to take root. Instead, he intervened by giving direct orders to Provincial Ministers and officials. In his home District of Larkana, which he made a Division, senior officials took orders from no other person. Administration suffered and the law and order situation consequently deteriorated, particularly when each feudal had a major say in his own fiefdom. The *mohajirs* were pleased with the departure of Mumtaz and placated by Jatoi; but they were not won over, as events in 1977 showed.

At the same time, the Pir of Pagaro gradually became the pivot of Sindhi opposition. It was over a small but revealing incident that ZAB and he fell out. He expected the Pir to seek an interview and call on him as President. I said it would be a good gesture if the President extended an invitation: 'Keep your own backyard clear while dealing with other problems.' Mumtaz Bhutto felt the same, but ZAB refused. At issue was the question of who was the bigger feudal. There was an altercation between them on the telephone; ZAB was half-way through a sentence about 'fixing' Pagaro when the phone was put down at the other end. The Pir of Pagaro eventually became a leading figure in the ouster of ZAB in 1977.

Although Sindhis overwhelmingly continued their support for the PPP through later years, some were fierce opponents. G. M. Syed explained why the nationalists supported the PNA agitation in 1977: 'To wipe out Bhuttoism' from the country. After the announcement of ZAB's death sentence he said:

Even a Sindhi Prime Minister failed to recognize our genuine demands…Bhutto is a criminal and a symbol of crimes. He has not only plundered the country but also ruthlessly suppressed the Sindhis. He killed and tortured them and turned Sindh into a big jail.[4]

But neither G. M. Syed nor eleven years of General Zia's rule could wipe out 'Bhuttoism' from Sindh, where ZAB successfully arrested separatist tendencies.

Balochistan

Throughout the period of British rule the Baloch tribes had been difficult to control, and they remained restive from the very advent of Pakistan. The Quaid-i-Azam detained the Khan of Kalat for 'independence' efforts. Trouble continued, and subsequently both Ayub Khan and Governor Kalabagh had their leaders arrested. But no one succeeded in 'taming' the Baloch, not even ZAB with the use of the military.

In the new Pakistan of 1971, Balochistan was clearly important; it now comprised forty-three per cent of our total area and bordered on two of our four neighbours, Iran and Afghanistan. But its population was less than four per cent of Pakistan, with a little over half of Baloch origin. The Baloch and Brohi tribes were concentrated in Sibi, Chaghi, Kalat, Makran and Kharan, while Pathans overwhelmingly inhabited the Quetta-Pishin and Loralai areas. Quite apart from tribal rivalries and conflicts among the Baloch themselves, ethnic differences added a further dimension to the problems.

Baloch leaders were mainly from the NAP. Ghous Baksh Bizenjo, Sardar Khair Baksh Marri and Sardar Ataullah Mengal were three outstanding personalities. Bizenjo was in the mould of the Indian Congress leaders who had challenged the British. However, not being a big sardar in a tribal society, his role was regrettably limited, to the cost of everyone. Marri and Mengal, both proud and independent chieftains, blended these attributes, surprisingly, with enlightened leftist views.

Bizenjo was appointed Governor on 29 April and Mengal became the Chief Minister of the NAP-JUI Provincial Government on 1 May 1972. This was the first time these NAP leaders were part of any government. Formerly, they had opposed all governments. Apart from the early days of Mengal's nine-month Government, Centre-Province confrontation became a regular feature. This was partly due to the Baloch leaders' inexperience of formal government, though they were natural administrators. More important, however, was ZAB's failure to accept this NAP-dominated 'government of sardars' with their proud traditions, who rejected his assumption of superior wisdom and leadership. He only got on with Bizenjo who, although pragmatically accepting the concept of Pakistan, was no less a believer in the rights of the Baloch. On one occasion when he came upstairs to talk to me after meeting President Bhutto, I pointed behind my desk to a blown-up map of Pakistan, which also covered parts of neighbouring Iran, and jokingly inquired what was encompassed in his idea of 'Greater Balochistan'. He replied, 'Don't worry now, but it will happen one day'.

Chief Minister Mengal lived up to his people's expectations. He promised land reforms and the abolition of the *sardari* system, and sought to strengthen the cultural and ethnic heritage of the Baloch. He allowed freedom in the Province by withdrawing both press restrictions and Section 144. If he had received proper assistance in administration, many mistakes could have been avoided. His stubbornness, unnecessary at times, did not help, nor did reports of threats to throw out Punjabi government officials from Balochistan.

In September 1972, after the Simla Accord and the first Delhi Agreement, ZAB approved a publicity campaign about the 'London Plan' which, with NAP involvement, had allegedly led to the dismemberment of Pakistan in 1971, and envisaged a further break-up of what remained of the country. He felt this would serve to discredit the NAP while applying pressure on them to accept a constitutional arrangement, which he regarded as essential in the national interest. At the same time, he played on tribal rivalries and ambitions, both supporting and being encouraged by opposing factions. He placed great reliance on the advice of Said Ahmed Khan, his Chief Security Officer, who sent him a note from London, on 26 September 1972, suggesting that Akbar Bugti should be utilized to break the Bizenjo-Mengal Government with the assistance of Doda Khan Zarakzai and Ali Mohammad Mengal, and financial support from Nabi Baksh Zehri.

Trouble started in Lasbela, and tension mounted. ZAB invited Wali Khan and Bizenjo to Lahore on 28 January 1973 for discussions on the Balochistan situation and the Constitution. Next morning, in Lahore, Wali Khan said he was 'optimistic' concerning the Constitution and called for 'undiluted democracy'. On arriving in Islamabad he told the Press, 'We have set the ball rolling. The talks will continue',[5] and, without naming them, added that 'the Kissingers' would, hopefully, sort out matters. That day Governor Bizenjo and I met. The Press reported that he had 'requested' forces to restore law and order in Lasbela, though in fact Bizenjo had maintained, 'In the present trouble some vested interests wanted that President's rule should be enforced in Balochistan'.[6] The scene was set for the dismissal of the Provincial Government and, as the situation deteriorated, troops were in position.

On 10 February the Iraqi Embassy was raided for arms by federal forces. Subsequently there were rumours that the incident had been concocted by the Federal Government—such was the distrust with which the Opposition viewed ZAB. To raid any foreign mission constituted a major hazard, and we were concerned about the fallout among Muslim countries. A few days previously, ZAB had called Army Chief Tikka Khan, Aziz Ahmed and myself to a meeting at which Lt.-Gen. Jilani, DG, ISI, had presented photographic proof of crates of arms being landed, with the Iraqi Embassy as their destination. On the day of the operation ZAB directed me to be present in Jilani's office, which was near the Embassy, to settle any last-minute hitch that might arise if the President himself could not be contacted.

Later, Jilani light-heartedly recalled my repeating, 'I hope you are right about the arms being in the Embassy'.

The arms discovered were displayed in various parts of the country like booty from a conquest. On 15 February 1973, the Mengal Government was dismissed and President's Rule proclaimed. In an attempt to link the arms to Balochistan, at least in the public mind, ZAB told the National Assembly that 'some people say that these armaments were not to be used against Pakistan but were meant for a third country…[but]…the fact is that the arms were meant to be used in Pakistan and against us.' He justified the Proclamation against the Provincial Government on the grounds that it had 'embarked on a tribal vendetta' and had gone 'to the extent of not only killing but also taking calculated measures to starve Balochis in Lasbela'.[7] The Presidential Proclamation itself accused the Provincial Government of failing 'to take effective measures to check the large-scale disturbances in different parts of the Province resulting in loss of life and property on a massive scale and causing a growing feeling of insecurity among the inhabitants and grave menace to the peace and tranquility of the Province'.

Even without the Iraqi arms incident, which was never directly linked to the NAP in either Balochistan or the NWFP, ZAB fully intended to remove the Mengal Government. A month later, when I led the Pakistani delegation to the Islamic Foreign Ministers' Conference at Tripoli, I met the Iraqi representative, who attended for this purpose although Iraq was not a member. He maintained that the arms were intended for Iran, but did not explain why they should have chosen such a circuitous route through Islamabad.

Akbar Bugti replaced Bizenjo as Governor, and the Government of Chief Minister Jam Ghulam Qadir Khan of Lasbela was installed on 23 April 1973. It commenced as, and remained, a minority government. Within months, several army divisions were brought into the Province. Only days prior to the induction of the Jam of Lasbela's Government, the permanent Constitution had been approved by Members of the National Assembly, including NAP representatives from Balochistan and the Frontier. Shaukat Hayat had given a lunch for all MNAs, at which both he and Bizenjo had impressed on me the necessity of making some reciprocal gesture. I repeated their arguments to the President, pointing out it would be an opportune moment to restore the NAP-JUI Government and peace in Balochistan. But he was not to be convinced.

Two months later, in June 1973, the President invited Bizenjo and Mengal for talks in Murree. They occupied a cottage near where I was staying. Bizenjo came over and we had a long talk about the need for reconciliation between the PPP and NAP, and a political solution for Balochistan. We both agreed it was time for a settlement; otherwise, only harm would follow for all. Bizenjo was politically very wise. We not only became friendly, but I came increasingly to admire him. He took me to Mengal, who also agreed that the PPP and NAP could and should be able to work together. Mengal felt that the real problem lay with ZAB, who could not accept any power which did not emanate from himself, or which he could not control.

I undertook to speak to ZAB even though my position at the time was awkward; I had resigned on 1 June 1973 as Special Assistant to the President, though this was officially notified later in a different form.[8] I told ZAB that the Jam of Lasbela could not secure a proper majority in the Provincial Assembly, and that continued military intervention in the Province would prove disastrous. The new Constitution was coming into effect on 14 August and provided an opportunity for a fresh start. He did not disagree but asked what would happen if a future NAP-JUI government posed difficulties. I assured him that the new Constitution had sufficient safeguards, and he would thereby have established his good faith.

At the time, ZAB was preparing for his July meeting with President Nixon, which was later postponed, and met the NAP leaders after our talk. He agreed in principle to a settlement but wanted to make the announcement on returning from Washington. The NAP felt that this was a ruse, and that this meeting was merely to impress upon the US that he was trying to resolve issues through political discussions in a democratic way. Mengal insisted that any settlement should be announced before ZAB's departure since there was no reason for delay. As discussions got heated, Mengal thumped the table and implied that ZAB would not keep his word. ZAB retorted that he would not be threatened into an early announcement. Far from being productive, the meeting turned sour. Following this, there were no serious discussions with the NAP leaders, nor efforts to promote a political settlement. A further important factor was that ZAB had given undertakings to those at the helm in the Province, and also to the new PPP entrants; if he let them down now, he felt that there would be no hope of establishing a PPP government in Balochistan in the future.

On 20 July 1973, the Opposition UDF observed 'Balochistan Day' by way of protest. Trouble increased, especially in the Marri area, and criticism mounted. In response, ZAB had the Baloch leaders arrested under Emergency Powers immediately after the introduction of the 1973 Constitution on 14 August. By virtue of Article 280 of the Constitution itself, the earlier Proclamation of Emergency promulgated by Yahya Khan was deemed to continue. This was justified because the problems created by the 1971 War, such as the return of the POWs and the recognition of Bangladesh, were still outstanding. However, I myself was very doubtful about the wisdom of arresting the Baloch politicians and expressed my misgivings to ZAB, though I was not in government at the time.

By 14 February 1974, Prime Minister Bhutto felt the need to explain to the National Assembly that Balochistan was 'an old problem which existed before the creation of Pakistan. The down-trodden masses are being exploited by a handful of feudal lords and *sardars*. We took the help of the army for constructing roads, providing electricity and water to poor Balochis'.[9] Clearly such a massive deployment of the army in Balochistan could not be so simply justified.

My efforts in favour of the Baloch leaders continued when I again became a Federal Minister in July 1974. The Prime Minister sent me a letter of 18 August 1974 from the Balochistan Governor on which I commented two days later: 'On the subject of the arrest of the three in August 1973 I need not repeat the view I had previously expressed to the Prime Minister. Even as recently as May I thought their release would help alleviate the situation...' Now, despite the extreme bitterness that existed, I suggested they be released fifteen days before the expiry date of the extension for the laying down of arms, which was 15 October.

They were not set free. Instead, the Federal Government, on 19 October 1974, issued a *White Paper on Balochistan* claiming that the situation was normal and that the army contingents would soon withdraw. It gave the total number of 'rebels' who had surrendered as 5,501, and 385 as killed and wounded. The latter figure was an underestimate. The Baloch in reply said that 80,000 to 100,000 army personnel had been deployed, of whom about 3,000 had been killed and an equal number wounded. These figures were again inaccurate. However, many Baloch and army personnel were needlessly killed.

On 10 February 1975, the NAP was banned following the assassination of Hayat Sherpao. The Jam of Lasbela's Cabinet was shortly thereafter removed for failure to control tribal insurgency. There followed a further period of Federal Government rule, after which Mohammad Khan Barozai became Chief Minister on 6 December 1976, in time for the general elections, which were boycotted in Balochistan by the Opposition. These later developments, including the ban on the NAP, are narrated in connection with the NWFP as the fate of the two Provinces was intertwined.

Events in Balochistan affected me personally and contributed to my decision to resign on 1 June 1973. As we shall see, they certainly also contributed to the fall of ZAB. Neither he nor the Army Chief contemplated the withdrawal of forces till the 'mission' was completed. By the time he realized the error of reintroducing the army directly into the political arena, it was too late.

North-West Frontier Province

Prior to Independence, 'nationalist' and anti-Muslim League feelings had been strongest in the North-West Frontier Province; it had required a referendum in July 1947 to decide whether the NWFP would form part of Pakistan. Khan Abdul Ghaffar Khan and his 'Red Shirts' had boycotted the referendum as the choice was restricted to either joining Pakistan or staying with India. Despite the slogan that a vote for Pakistan was a vote for the Holy Quran, it was a tribute to the endeavours of Ghaffar Khan that the deeply religious Pathans boycotted the referendum in substantial numbers. He was a towering personality, known as 'the Frontier Gandhi', and took his message to all parts of the Province, where he was deeply respected. From Independence until 1972, most of his years were spent either in jail, under detention or in self-exile. At different times he had put forward varied proposals for the Pathans, including the removal of artificial boundaries and the integration of Pushto-speaking areas of British India in a state to be known as Pakhtunistan, the abolition of the exploitative feudal system, and equal representation of the Pathans in the socio-economic and political fields in Pakistan.

We have already recounted ZAB's early meetings with the NAP, the 6 March Accord, the appointment of NAP Governors in the NWFP and Balochistan, and the Constitution Accord of 20 October 1972. He sought to avoid problems in Balochistan and the NWFP by accommodating the NAP-JUI alliance in order to achieve three main objectives: agreement with the IMF and World Bank, who required massive devaluation, in order to secure rescheduling of debts and to get the economy moving forward; settlement with India; and, most importantly, agreement on a constitution. The NAP-JUI were involved in all three decisions. Their two Governors were appointed in time to participate in formal meetings on devaluation. Governor Arbab Sikander and others were included in the delegation to Simla, and the NAP-JUI were party to both Accords preceding the interim and permanent Constitutions. However, once he had achieved these objectives, ZAB's attitute towards the NAP-JUI changed.

By the time Ghaffar Khan returned to Pakistan on 24 December 1972, after eight years in Kabul, his talk of dialogue and cooperation with the Government, and proposals to launch a peace movement to promote harmony among the various regions of Pakistan, met with a negative response. The Government restricted him to his village. Except for an initial brief period there was constant tension between ZAB and Wali Khan, whose personalities clashed disastrously. Although Arbab Sikander, the new NAP Governor, had many commendable personal qualities, and Mufti Mahmood of the JUI as Chief Minister was an astute public leader, both were inexperienced in the ways of government and administration. They could not achieve much and the Federal Government gave them little help.

On 15 February, along with the removal of Governor Bizenjo and the Mengal Government, Governor Arbab Sikander was replaced by Muhammad Aslam Khattak. Earlier, Aslam Khattak had formed a United Front mainly from among independents in the Provincial Assembly. The Provincial Government of Mufti Mahmood resigned on principle to protest against the dismissal of the Mengal Government and the change of Governors. However, inexperience also played a part; the resignation was probably an error as it obviated the need for the new Governor to advise the Government's removal. It cleared the way for ZAB to promulgate President's Rule immediately in the Frontier.

With no agreement in sight on the Constitution, despite the 20 October Accord of 1972, ZAB once again went on the offensive against the NAP. Addressing the People's Student Federation at Peshawar on 3 March, he traced the history of the 'Red Shirts', referring to them as the 'militant hardcore' of the Indian Congress who continued to oppose Pakistan. He condemned Wali Khan: 'One who is the enemy of the land of the pure cannot be the upholder of undiluted democracy.'[10] He asserted that Wali Khan had taken his inspiration from East Pakistan and would follow in the steps of Mujibur Rahman.

On 15 April 1973, exactly two months after the introduction of President's Rule in the NWFP, a coalition government was formed with Inayatullah Khan Gandapur as Chief Minister. The coalition had the support of twenty-two out of forty-two members of the Provincial Assembly, including the United Front of Aslam Khattak, four PPP members, and representatives of the Muslim League (Qayyum). This Government was formed within days of the permanent Constitution being signed, and the NAP-JUI alliance considered it a poor return for their cooperation on the Constitution. Unlike in Balochistan, however, this ill-assorted Government at least enjoyed majority support in the Provincial Assembly. Aslam Khattak had delivered on his promise to ZAB, but the performance of the new Provincial Government was even worse than that of its predecessor.

The NAP stepped up its activities against the Federal and NWFP Governments. In July 1974 Wali Khan told a public meeting near Charsadda that the time of appeals had passed and they would now 'meet force with force' to realize their rights.[11] Within seven months, in the afternoon of 8 February 1975, Hayat Sherpao was assassinated.

His death came as a great personal shock, for he was a close and deeply valued friend and colleague. I had spent the previous two days in his house in Peshawar, and he was to have dined with me in Islamabad on the night of his assassination. The Prime Minister had sent me to Peshawar to look into the Provincial Government, which was not functioning properly. He knew of my friendship with Sherpao, who was then Senior Minister in the NWFP, and had given me a confidential note indicating that Sherpao himself was arranging bomb blasts in the Province. When I discussed the matter with him, Sherpao told me he knew about the reports and that was why he had not mentioned a recent blast behind the guest bedroom where I was staying. His death provoked my first and only angry altercation with ZAB a few days later: 'How do you think I feel when you told me to accuse him of being involved in bomb blasts just the day before he was blown apart himself.'

On being informed about Sherpao's assassination, ZAB talked to me from New York, directing me to take Interior Minister Qayyum Khan to Peshawar immediately; he ordered the arrest of all NAP representatives, and planned to ban the NAP, which he did two days later, on 10 February. The following week, Governor Aslam Khattak and the Gandapur Government were removed and the Federal Government imposed its rule in the Frontier, with retired Major-General Syed Ghawas as Governor.

When the time came to re-establish parliamentary government in the Province, a majority was available in the Assembly for a pro-PPP Government, but there were no suitable candidates for the two top positions. Qayyum Khan, Yusuf Khattak and Mir Afzal Khan declined the Governorship, and Mir Afzal did not want to be Chief Minister. Syed Ghawas continued as Governor until being replaced after more than a year by another Major-General who had been retired specifically to enable him to assume the post. Nasrullah Khattak was brought back from being an Ambassador to become Chief Minister of the PPP-QML-Independents Coalition Government. This was the worst Government the NWFP experienced in those years.

The Attorney-General presented the case relating to the ban on the NAP before the Supreme Court, which finally decided on 30 October 1975 to uphold the decision of the Federal Government. A week later the National Democratic Party came into existence in Islamabad under the leadership of Sardar Sherbaz Mazari. Former NAP workers and those not in jail comprised its members. Mazari and Wali Khan's wife, Nasim, took up the struggle against the PPP.

ZAB established a Special Tribunal at Hyderabad to try the NAP leadership. Said Ahmed Khan was placed in charge of the case. In the agitation that followed the March 1977 elections, this Tribunal became the second main point of contention in the abortive negotiations between the PPP and the Opposition.

The extent of the bitterness against ZAB is best illustrated by Wali Khan's statement on 10 December 1977, four days after he was released and the Hyderabad Conspiracy Case was dropped by General Ziaul Haq. He praised the General for 'rescuing the country from the clutches of a ruthless dictator... Unless traces of Bhuttoism are removed from the body politic of the country, no positive achievement could be made in any direction'.[12] This came from a person who had throughout his political career opposed military rule.

The Punjab

ZAB achieved his aim of securing PPP Governments in all four Provinces, but, in the process, by 1976 his popularity had declined rapidly except in rural Sindh. We have seen how ZAB dealt with the three smaller Provinces, and mention must also be made of his bastion of power, the Punjab. The uncertainty created by Mustafa Khar's being frequently installed in office then removed was the main cause for the absence of proper government in the Punjab from the end of 1973 onwards.

His relationship with Khar was indeed complex. In most ways they were different, the main common bond being their feudal background and politics. From 1967 up to the autumn of 1972 they had been the closest of friends, with ZAB entrusting Khar with several important and delicate missions. But then he began to distrust Khar's political ambitions, and rivals of Khar around ZAB encouraged their differences, which were further fuelled by various intelligence agencies. Khar ceased to be Chief Minister in March 1974 and was replaced by Haneef Ramay, only to return as Governor and then be removed again. Ramay's successor as Chief Minister was Sadiq Hussain Qureshi, a big landlord. Each change of Chief Minister was accompanied by major reshuffles, not only in government but also in the provincial PPP office-holders. The problem was further compounded by the fact that, unlike Khar and Ramay, Sadiq Hussain Qureshi did not have his roots in the Party, and allowed the feudals and well-to-do to hold sway. The result was the alienation of labour, a great source of support which ZAB had previously enjoyed in the urban areas. The intelligentsia and students had already become hostile. After Mustafa Khar, the Government in the Punjab remained weak, which was how ZAB wanted it.

Earlier, under Khar, and supported by ZAB, a culture of violence had pervaded the political scene; this continued till the end, with several opponents, including recalcitrant PPP MNAs, receiving rough treatment, and murders with political overtones in which the Government was accused of having a hand. Even in those instances when Khar sought to deal with the Opposition politically, ZAB resorted to coercion. For example, in the case of the Liaquat Bagh meeting on 23 March 1973,[13] I was surprised to learn on my return from Libya that ZAB had insisted on introducing force by testing out his newly-formed FSF.

In October 1975, one month after Khar was expelled from the PPP, a *Punjab Bachao* (Save Punjab) movement got under way. Khar, Ramay and others joined the Opposition in denouncing ZAB as a Sindhi Prime Minister who had usurped the rights of the Punjab.

One of the worst examples of ZAB's high-handedness was connected with the Lahore by-election in October 1975, when several of Khar's supporters were picked up and secretly incarcerated in the Dalai Camp in Azad Kashmir. This was done in complete disregard of all legal processes, and only a few of his henchmen, such as Masood Mahmood, knew of the detainees' whereabouts. In fact, following the July 1977 military *coup*, ZAB was more concerned that the disclosures concerning the Dalai Camp would give rise to a serious case against him than he was about the murder in November 1974 of Ahmed Raza Kasuri's father, for which he was tried and executed.[14]

Political violence in the Punjab had not hitherto been uncommon, but under ZAB it acquired a new, personal dimension, which in turn made him the target of the Opposition's pent-up fury in 1977. The Punjab eventually played a major part in his downfall.

For his last four years in office, his views on the Opposition were as he had expressed to the National Assembly on 10 July 1973. He described them as 'political charlatans' and 'a conglomeration of individuals who have got together to block the progress of the country…They have a common denominator which is hatred not only against me…[but] the concept of Pakistan, the struggle of Pakistan'.[15] The Opposition was as much to blame for this politics of abuse, though it was mainly the Government's responsibility to avoid the misuse of the machinery of state, encourage tolerance and create consensus where possible. It was not to be; otherwise, the course of Pakistan's history would have run differently, not only in 1977, but both before and subsequently.

The Economy

ZAB's years in office were throughout difficult on the economic front. Numerous problems were inherited, several created, and mother nature and international factors did the rest. The civil war and defeat in December 1971 reduced Pakistan's economy to a shambles. In fact, it had never really recovered from the September 1965 War, which was followed by the civil disturbances against Ayub Khan, and then electioneering throughout 1970.

The first formidable task for the PPP Government was to find new markets to replace those lost in East Pakistan. Over the previous two decades considerable inter-Wing trade had developed, mainly in textiles and other supplies from West Pakistan to the East Wing, which in turn earned a major portion of the nation's foreign exchange through jute exports. The abrupt disappearance of this trade required restructuring, especially as it further compounded the existing foreign exchange shortage caused by the cut-off in aid following the Yahya regime's declaration of a unilateral moratorium on international debt repayment on 30 April 1971. Moreover, the Government was compelled by the IMF-World Bank to undertake a long-delayed, massive devaluation of the currency on 11 May 1972. As if these were not difficulties enough, severe floods occurred in the summer of 1973. Soon there followed a truly major setback, when oil prices rocketed upwards as a result of the October

1973 Ramadhan War between the Arabs and Israel. The developed world could barely cope with these spiralling prices by supplying the oil-producing Arab countries with arms, engineering goods and luxury products. A new expression, 'stag-flation', was coined to describe the difficult circumstances. Third World countries like Pakistan, with no oil or such exports, suffered enormously. Pakistan was further hit in 1974 by drought, followed by fresh floods the following year and a major earthquake in the north.

Despite these adverse circumstances, economic development under the PPP Government was commendable, with the main impetus coming from agriculture and the public sector. The GDP rose from Rs 36,747 million in 1971-2 to Rs 46,223 million in 1976-7, an average increase of 5.5% per annum in those difficult years.[16] Significantly, all four Provinces and the Northern Areas enjoyed equitable distribution of this development. Roads and electrification in Balochistan and the NWFP, industries in rural Sindh, major projects such as the Steel Mills and Port Qasim near Karachi, and a range of industries including heavy engineering, fertilizers and cement plants in the Punjab were the main features. The electrification programme in Balochistan alone changed the outlook of the Province. ZAB sought to consolidate the Federation through development.

An important factor in the increase in agricultural output was the higher support prices for crops. In the past, industrialists and urbanites had benefited from the low-priced agricultural produce, which was a disincentive to maximizing crop production. From 1958 to 1971 the price of wheat had scarcely risen, from about Rs. 13 to Rs 17 per maund (40 kg) whereas between 1972 and 1977 it was increased to Rs 47, that is by almost 280%. As a result, several billions were pumped into the agricultural sector, mainly benefiting bigger landlords in the Punjab. There was a rise in the production of wheat from an average of 5.6 million tons between 1965 and 1970 to 8.5 million tons in 1975-6, and in rice from 1.7 to 2.6 million tons. Similar measures were taken for cotton and sugar cane.

The PPP Government also undertook other measures to revitalize the rural economy. There was a huge increase in financial allocations to the agricultural sector, from Rs 212 to Rs 1,336 million, and institutional credit rose from Rs 160 million to Rs 1,800 million between 1971-2 and 1976-7. Plant protection coverage was raised by 250% to 10 million acres, fertilizer use rose from 308,000 to 650,000 nutrient tons, tractor imports increased threefold to 12,000, and tube-wells from 88,000 to 145,000. Mubashir Hasan played a key role in these substantial achievements in economic uplift, and eventually even die-hard bureaucrats unsympathetic to his leftist views worked with dedication for him and ended by admiring their Minister.

ZAB is accused of developing agriculture at the cost of industry. To some extent this allegation is not unjustified. His feudal background, and the fact that he had been opposed by the industrialists, led to bad relations with the business community. Another influence contributing to this was the prejudice, amounting to obduracy in the 1974 textile crisis, of J. A. Rahim and Mubashir Hasan, though they were more antagonistic to the feudal class. Still, in several sectors of industry, growth was significant, as shown in the table below.[17]

| | 1970-1 | 1976-7 | % Increase |
	(In thousands of tons)		per annum
Fertilizers	375	824	20
Vegetable Ghee	136	326	23
Sugar	519	736	7
Mild Steel Products	196	270	6
Cement	2,702	3,071*	2.5

*Because of the power shortages; it had reached 3,320 in 1974-5, that is a 6% per annum increase.

To illustrate further, the Government initiated projects to double cement and oil-refining capacity, and quadruple fertilizer production; and even sugar-refining and textile-spinning capacities, mainly in the private sector, were increased by over one-third.

The earlier nationalization of industries and life insurance had indeed scared big business. Rumours that they were rushing money abroad resulted in the passports of many members of the 'twenty-two families' being impounded. Between them, they owned sixty per cent of the nation's industrial assets and eighty per cent of the insurance business. Most passports were returned following ZAB's meeting with businessmen at Rawalpindi on 5 March 1972.

The massive, some maintained excessive, devaluation of the rupee in May 1972 took the official rate from Rs 4.75 to Rs 11 per US $1. Devaluation was a condition imposed by the World Bank and the IMF for rescheduling debts and sanctioning further loans. Only the extent of devaluation was debated in Pakistan, and decided with the advice of officials. It had a devastating effect. Large loans in foreign exchange taken by the industrial community now had to be repaid in hugely increased sums of rupees. The Ayub Khan era had witnessed considerable industrial development, but this had partly been achieved through the import of machinery at low-cost dollars and the availablity of cheap raw materials, which amounted to a subsidy. There followed numerous defaults from which several important development finance institutions never recovered. Nor did domestic prices.

ZAB attempted in vain to allay the apprehensions of businessmen, and to explain this devaluation, when he met them again on 23 May 1972 in Karachi. He pointed out that 'our political and social fabric would explode unless certain basic reforms were undertaken'. After giving a summary of his measures, including land reforms, described as 'the most basic', he added: 'We accept that private enterprise has a role to play in the economic progress of Pakistan but we must rid the economy of concentration of economic power in the hands of a few to end the exploitation of the many.'[18] His concluding words, 'We can now work in harmony and cooperation and end every kind of exploitation', might as well have been saved for another audience. The industrialists left the meeting totally demoralized.

Businessmen follow money; deprived of their industries, they pursued trade instead, with profitable results. Their declining confidence in the industrial sector, and the Government's reaction in blaming them for failing to perform, resulted in further nationalization. The first assault, on 2 September 1973, was on the vegetable oil industry in order 'to maintain at reasonable prices supplies essential to the lives of the community while safeguarding the interests of the small investors in the industry'.[19] It was also the first measure under Article 253(1)(b) of the 1973 Constitution,[20] and made clear that the new Constitution afforded no protection against nationalization. Equally important, it mainly affected medium-sized businesses.

The Government also nationalized the export trade in cotton and rice, two principal commodities. The newly-established Cotton Export Corporation was for the first time able to ensure a fair price to the growers and a reasonable price to the manufacturers of yarn and textiles through Government procurement prices and regulation of exports. This was a beneficial measure, particularly as Pakistan, uniquely, exported cotton in all three forms—raw, yarn and finished textiles. Previously, about a hundred parties had handled these large cotton exports, and the businessmen said the Government could not cope; but we did.

The final blow to big business came on the first day of 1974. They were antagonized as much by the ill-conceived meeting held by ZAB the previous evening, when he asked them to co-operate in economic development, as by the magnitude of the measures taken in the early hours of the New Year. He nationalized banks, petroleum products and maritime shipping.[21] Here again, only Pakistani and not foreign companies were affected. From then onwards, not only was there little question of large-scale industries being established by big businessmen, but they repaid what they termed ZAB's 'deceit' by working wholly against him.

The nationalization of petroleum distribution proved successful and profitable, but shipping did not. The banks provided new opportunities through a change in their 'in-house' lending practices, which had previously benefited mainly the 'twenty-two families', now allowing many smaller businesses to secure loans. This was also a time when many overseas Pakistanis were contemplating returning with their savings because of the recession and unemployment in Western countries. Pakistan could have experienced a large rise in medium and small-scale industries which would have created new jobs quickly to alleviate our growing unemployment problems. These hopes were dashed by the nationalization of flour mills, rice husking mills and cotton ginning factories on 17 July 1976.[22]

This was both an economic and a political blunder. ZAB had consulted only bureaucrats—the Secretary-General of Finance, the Food and Agriculture Secretary, and the Cabinet Secretary—none of whom disagreed; Feroze Qaiser, who had in March 1976 ceased to be Special Assistant for Economic Affairs, was brought in at a late stage, though he opposed the measure. His replacement, Syed Shahid Hussain, a Vice President in the World Bank, was not aware of this major proposal but subsequently had to defend it. The plan was executed mainly by the FSF under Masood Mahmood. The Cabinet was presented with this *fait accompli* simultaneously with its public announcement; all we could do was to point out its pitfalls. Only a few of us did so, while the rest were full of praise for this step. ZAB did not appreciate the difficulties involved in administering huge numbers of small units, despite the earlier ill-conceived ventures in the fields of health and education.

ZAB's opponents maintain that this action was designed to gain a firm grip on agriculturists before the elections, since their products would have to be purchased by the new state-owned organizations established for this purpose. In fact, this was not the motive. Above all, he wanted to establish his peerless position: having outperformed the rightists on the Qadiani issue,[23] he now sought to outdo the left within the Party, quite apart from trying to strengthen his support among the smaller farmers. How the Prime Minister failed to anticipate the problems and ensuing disaster is difficult to comprehend, particularly as he himself came from a rice-growing area. It was absurd to nationalize the small rice mills, some of which were literally located in the backyards of houses; these were later handed back, and some other anomalies corrected. But the main damage was done, because now even small businessmen were hesitant to invest. Worse, the entire *bazaar* became hostile and played a significant part in the 1977 post-election agitation against ZAB.

J. A. Rahim, Mubashir Hasan and Sheikh Rashid were often blamed, even by ZAB, for antagonizing the business class, which they did at times through unnecessary rhetoric. The need for nationalization in early 1972 has already been examined, as, indeed, have the merits of some of the other economic measures. But on this occasion, neither Rahim nor Mubashir was in government, and Sheikh Rashid was not consulted; ZAB alone was responsible.

With a disgruntled and reluctant private sector, the principal burden of industrial development fell on the public sector. The PPP Government put the country on course for its main phase of industrialization and import substitution, that is, from the packaging of consumer items to the production of more basic and heavy engineering goods. In this, the Steel Mills, the Heavy Foundry and Forge at Taxila, and large investment in such vital sectors as fertilizers, cement and petro-chemicals played an important part. Although several of these projects were of long gestation, the reorganization of the public sector soon began to pay dividends. Moreover, a comparison of actual imports, on the basis of regression, shows that substantial import substitution was achieved from 1974 onwards, more than trebling to over Rs 3,000 million in 1976-7 from less than Rs. 1,000 million in 1970-1. The share of capital goods of the total imports, which was 42% in 1971-2, was reduced to 34% by 1977-8, while the share of manufactured goods in the country's exports over the same period rose from 28% to 50%.[24]

The Government maximized the use of domestic material and human resources by insisting that Pakistani companies be the prime contractors, wherever possible, with foreign firms as their subcontractors. This trend has, regrettably, been reversed by successor governments. Moreover, in order to benefit from employment opportunities in the newly-rich Gulf States, the Government eased the issue of passports, which permitted many skilled and unskilled workers as well as managers to earn foreign exchange abroad. Their remittances home brought welcome benefits to their families in Pakistan suffering from high inflation and unemployment. Where the Government failed was in not filling this gap by improving educational facilities, especially technical training; the continued neglect of this, and of literacy in general, over the past two-and-a-half decades has prevented proper growth.

The performance of the 'taken over' industries, as expected, gave rise to much criticism. Undoubtedly there were inefficiencies and deficiencies, especially in the earlier period, but their over-all operations were praiseworthy, particularly in view of the difficult circumstances in which we were functioning. On 18 June 1976, during the debate in the National Assembly on the Budget, I spoke at length in support of the major industries, which came under the Ministry of Production. In view of the hostile propaganda against the public sector both then and now, it is instructive to look at some of the figures which I quoted:[25]

...despite [world] recession, production as compared to the previous year increased by over four per cent...over the year 1974-5 which was an almost record year when we increased production by twenty-one per cent..., or put in another way, a seventy per cent increase over the base year 1972-3 is indeed no mean achievement.

Referring to a major contribution in the form of taxes, I pointed out that in 1975-6 the units under the Ministry were to pay a total of Rs 133 crores by way of corporate tax, sales tax, excise duty and other taxes as compared to Rs 121 crores in 1974-5, Rs 66 crores in 1973-4, and Rs 39 crores in 1972-3, whereas prior to nationalization the amount paid was less than Rs 20 crores. Similarly with gas and electricity, before nationalization the private sector paid hardly anything, whereas our payments had increased from Rs 11 crores in 1972-3 to Rs 28 crores in 1975-6. Regarding profits, I pointed out that in 1974-5 we made Rs 27 crores, while in 1975-6, despite power shortages and other difficulties, the profits were Rs 15.5 crores: 'We have achieved this without retrenchment of labour to which the private sector has resorted. We have done this after paying in full all electricity bills and all taxes. We have done this without cutting production, although at times paying high inventory carrying charges.' In terms of production too I gave some revealing and specific details. In the Heavy Mechanical Complex at Taxila and in the Pakistan Machine Tool Factory, Karachi, production had increased by ninety per cent in 1975-6 and seventy per cent in the previous year; and utilization of capacity in the former was sixty-six per cent in 1975-6 as opposed to thirty-nine per cent in 1974-5. The production capacity utilization in the National Refinery, for example, had been 120%, and 126% in the last quarter, of the designed capacity, which was a record. The State Cement Corporation was operating at ninety per cent capacity; when some foreign experts visited some of our plants they were surprised that factories which were over forty years old could operate at such high capacity. I also laid emphasis on the indigenization programme, the encouragement extended to domestic engineering, and how we would be manufacturing two sugar mills a year to achieve self-reliance.

ZAB was mainly supportive of these public sector endeavours but not above interference. In one proposed French scheme, which I viewed as unnecessary duplication of an existing project, the Prime Minister insisted on it being pushed through, and set up a Cabinet Sub-Committee for this purpose. The two other Ministers present had been directed to follow his instructions but I said I would not take the project on in my Ministry of Production. Present at this meeting was also the Chairman of the Defence Production Board, retired Major-General Ihsan-ul-Haq Malik, who came up to me after the meeting, although I had not met him previously, and said it was a pleasure to meet at least one Minister who could take a strong stand. The project did not materialize.

ZAB's attitude was demonstrated in two other instances. When he visited the Steel Mills for the first time after the financing arrangements with Soviet Union were finalized on 30 December 1974, he was delighted at the rapid progress made in twelve months. On 12 March 1976, the Chief Minister of Sindh, who had accompanied us, wrote to the Prime Minister saying that the Sindh Government proposed to name the site 'Zulfiqarabad', despite ZAB's earlier instructions not to name anything after him while head of government. The Prime Minister was irritated at his name being spelt with a 'q' and not a 'k', about which he was very particular, and wrote in the margin, 'Learn to spell my name...before you name anything after me'.[26] Yet it was apparent from the note of 4 April which he sent to me with the file that he wanted to accept the proposal:

> ...after inspecting the Steel Mill on the 30th of December 1975, I said to you and Rafi Raza that I was so impressed...that if ever I was tempted to break the rule which I have imposed...the sole exception would be in the case of the Steel Mill...I made those remarks in a pensive mood. Since you have been associated with me for a long time you took it as a hint but in good grace. I did not see the same reaction in Rafi Raza's face. I did not have to hint...I have the power to ORDER...I am marking the file to Rafi Raza as he is also concerned as Minister for Production and also because he did not react like a *chamcha* when I spoke. Moreover, for this very reason it is safer to get his opinion! (ZAB's own underlining and capitals; *chamcha* is a term commonly used for a stooge or sycophant.)

I replied on 8 April, fully recognizing ZAB's efforts in the past under Ayub Khan and his continued persistence after 1971 to establish this industrial landmark. However, I felt that there was no occasion for its re-naming now and that it should wait till the Mills came into preliminary operation in 1978. The Sindh Government went ahead with his approval. After this, ZAB's self-imposed ban gave way completely.

The second incident was less amusing. The US Embassy had expressed dissatisfaction that no major contracts had been awarded to American parties by the Ministry of Production. On one project the Ambassador was particularly insistent; although ZAB had earlier agreed that I should resist American pressure, he then changed his mind and asked me to give the project to them. However, I maintained that the only possible course would be to reopen the tender for fresh bids from both parties, the contract to be awarded to the best bid within the stipulated time. In fact, one European Ambassador pointedly remarked that, alternatively, we should save ourselves and others the time and cost involved in competitive bidding. When it still did not go the Americans' way, acrimonious correspondence and words were exchanged with the Ambassador. The Production Secretary stressed that it was not unusual for a bidder who was an interested party to try to twist the facts and give them a complexion which suited his convenience, and everyone in the Ministry and Corporation concerned maintained that the Americans were in the wrong. Nevertheless, in February 1976, ZAB asked the Foreign Secretary to get in touch with the Ambassador, who said that the incident spelt the end of his relationship with me. Against this, ZAB commented in writing to me, 'This is not good. We should put it right again. It should be back on the rails.' I rejected this suggestion,[27] saying I would rather resign as Minister. The Ambassador and I did not speak to each other again.

In fact, throughout ZAB's period in government, the US asserted considerable influence over economic issues, mainly through the IMF and the World Bank. Those were the days before formal structural adjustment programmes were introduced, but some senior Pakistani official, mainly the Secretary-General of the Ministry of Finance, A. G. N. Kazi, would visit Washington regularly and return with directions concerning what subsidies should be eliminated or reduced, and how the prices of certain basic commodities and utilities should be increased. Several of these moves were resisted by Mubashir Hasan while he was Finance Minister. In fact, he was the first

Third World Finance Minister to raise the issue of improved terms of trade, rather than aid, while addressing a Special Session of the UN concerning economic matters in April 1974. The task of putting the brakes on these measures also fell to Feroze Qaiser while he was Special Assistant for Economic Affairs and subsequently, and, where specific issues related to the Ministry of Production, I had to take up the cudgels. ZAB was aware of the adverse political effects of higher prices and often supported our views, even though he did not want to offend the US, the World Bank and the IMF. With the passage of years, I am increasingly convinced that Mubashir was right in saying that the prime interest of the Western donor countries is to ensure that the Third World continues to service debts in order to keep the Western banking and commercial system going, rather than to foster the well-being of the recipient nations.

The PPP Government has been wrongly criticized for its lack of success in the economic sector. Its record, and the facts, refute most of these allegations. Many failings in the public sector occurred after the fall of ZAB. This was particularly the case with banking, because of the practice of 'writing off' loans. The PPP Government was determined to make the public sector succeed for the sake of its own credibility and to prove that its decisions were correct. The Zia regime had no such stake or goal. On the contrary, it sought to show the failure of ZAB's policies, but still lacked the courage to reverse the main nationalization decisions of the PPP. ZAB's Government overcame many odds and, apart from at times unnecessarily antagonizing the private industrial sector, pursued a successful economic policy, spreading the wealth of the nation throughout the country and, in particular, the hitherto backward areas.

Recent governments in Pakistan, in order to please Western countries and the IMF-World Bank combine, have rushed to undertake a reversal of the earlier PPP Government's measures through massive and ill-prepared privatization and deregulation, concepts whose wisdom and success have in several instances been questioned even in the West. Whereas ZAB could be excused for haste in the early 1970s, there is no justification for the recent hurried actions. In a Third World country such as Pakistan, without any safety net of social security or a proper tax-collection system in place, such policies will only recreate and increase disparities, further enriching the affluent at the expense of the poor and those in underdeveloped areas. It should always be borne in mind that the main point of economic growth must remain the improvement of living conditions and human welfare. In a country like Pakistan the position has further deteriorated with the entwinement of wealth and political power, the pursuit of both becoming supreme. This has made the attainment of a just, egalitarian and democratic society more difficult and distant.

Religion and Sectarianism

We have earlier examined the issues involved in determining the role of Islam in the State of Pakistan, particularly in the context of constitution-making, which the 1973 Constitution appeared to have settled. ZAB's failure to arrive at any working political arrangement with the Opposition forced many parties, such as the NAP, into open hostility. The rightist, religious parties, mainly comprising the Jamaat-i-Islami, JUI and JUP, were equally inimical. However, having agreed to the Constitution, they could not readily take issue with the PPP Government on related matters of religion. The Islamic Summit in February 1974 proved a catalyst in bringing Muslims together, not only world-wide but also within Pakistan. Religion ceased to be a source of major division or political agitation until the disturbances over the Qadiani issue in May 1974.

The problem relating to Qadianis[28] was not new. It had given rise to serious rioting in March 1953, resulting in Martial Law being imposed temporarily in Lahore for the first time. Following that declaration of Martial Law, the Punjab Disturbances Court of Inquiry was established, comprising two eminent judges, Muhammad Munir and M. R. Kayani.[29] This

Court considered vital questions relating to Islam, and the definition of a Muslim, which demand elucidation.

A fundamental principle in Islam is *Khatm-e-Nabuwwat,* the finality of the Holy Prophet Muhammad (peace be upon him): no prophet could follow him. Mirza Ghulam Ahmad of Qadian, a small town in the Indian part of the Punjab, founded the Ahmadiyya movement in 1889. Earlier, in March 1882, he had claimed that in a revelation *(ilham)* God had entrusted him with a special mission to revive the pristine values of Islam. His view of the *Khatm-e-Nabuwwat* was that after the death of the Holy Prophet *(Nabi)* of Islam (pbuh), no prophet would appear with a new *Sharia* (Code of Islam), although this did not preclude the appearance of another prophet without a *Sharia.* He proclaimed himself a *zilli nabi* (shadow of the Prophet).

This was a critical issue. There were two other doctrinal differences of importance to Muslims. The first concerned the crucifixion of Jesus Christ, whom Islam recognizes as a prophet conceived by virgin birth, but not as the son of God, as do Christians. Muslims hold that Jesus did not die on the cross; he was lifted by God to heaven, where he lives and will appear on the Day of Resurrection.[30] Qadianis differ in that they believe Jesus was rescued from the cross before his death, and died later in Kashmir, where he was buried. They hold that another person with his attributes was to appear subsequently, and that this person was Mirza Ghulam Ahmad.[31] The second doctrinal issue related to *Jihad,* which the Qadianis viewed as a duty not limited to the sword,[32] and this gave rise early on to differences over how to oppose British rule.

A distinguished Qadiani, Chaudhri Muhammad Zafrullah Khan, was the first Foreign Minister of Pakistan. The explanation he gave of his beliefs while speaking on 'Islam as a Live Religion' before a public gathering at Jehangir Park, Karachi, on 18 May 1952, is noteworthy. He stressed the superiority and finality of Islam as a world religion. According to him, Mirza Ghulam Ahmad was a person commissioned by God for *tajdid-e-din,* that is, for reforming or renovating the religious norms of Islam, which had been distorted through the years, with a view to preserving its purity; Ahmadiyyat was implanted by God himself, and the plant had taken root to provide a guarantee for the preservation of Islam in fulfilment of the promise contained in the Holy Quran, he asserted.[33] This caused riots. By 1953 there was growing pressure to declare Qadianis non-Muslims and to remove them from responsible policy-making positions, especially Zafrullah Khan.

These disturbances, particularly in Lahore, were fanned by the newspapers and by the Provincial Government of the Punjab. One of the few sane voices at the time was that of Hamid Nizami, editor of the influential *Nawa-e-Waqt*.[34] Mian Mumtaz Daultana, who was then Chief Minister of the Punjab, had stated on 30 August 1952: 'The question of declaring Mirzais as a minority is a constitutional question. Our Constitution has not so far been framed and the Constituent Assembly has not so far taken any decision in regard to the distinction to be observed between the majority community and the minorities'.[35] The Court of Inquiry reported that it had put to the *ulema* the question, 'What is Islam and who is a *momin* or a Muslim?' The report went on to state, 'But we cannot refrain from saying here that it was a matter of infinite regret to us that the *ulema*, whose first duty should be to have settled views on this subject, were hopelessly disagreed among themselves'.[36]

Martial Law and the findings of the Court of Inquiry dampened this issue by the time the 1956 Constitution was passed. Earlier, in January 1953, thirty-three leading *ulema* had suggested an amendment to the Report of the Basic Principles Committee[37] requiring Qadianis to be included among the minorities, for whom separate seats in the Assembly should be reserved by separate election.

We have examined the Islamic provisions in the three Constitutions of 1956, 1962 and 1973.[38] The Qadiani issue was not seriously raised during the discussions which immediately preceded them. It was not until April 1974, when a minor incident occurred at Rabwah, the new Qadiani headquarters after their move from India, that trouble started again, mainly in the Punjab. A failure to tackle the situation appropriately at the time, or to give an accurate account of it, allowed the disturbances to grow and rumour to magnify. Initially, ZAB was not inclined to deal with the issue severely. He had enjoyed close relations with the Qadiani leadership and their support in the 1970 elections. More recently, however, he was disturbed by reports that they were considering transferring to retired Air Marshal Asghar Khan the support which they had previously given to the PPP.

In the course of the anti-Qadiani troubles, a new development in the form of India's 'peaceful' nuclear explosion on 18 May 1974 contributed to change ZAB's position. Within a few weeks, on 13 June, he announced that the Qadiani issue would be referred to the National Assembly for resolution. The timing of the announcement requires some comment. Ten days previously he had urged the Opposition in the National Assembly not to incite the public, stating that there was no 'genuine controversy'. The following day, 4 June, the Speaker of the Assembly ruled out a debate on the Qadiani issue, observing that 'minorities' had already been 'defined' in the Constitution. Although I was not in government, ZAB had also asked me to meet Haneef Ramay, then the Punjab Chief Minister, to 'sort out' the matter.[39] Ramay was generally regarded as a mild person, an intellectual, who acted according to ZAB's dictates; we discussed appropriate administrative measures. Then came ZAB's sudden announcement. He felt a need to counter concern over India's nuclear bomb. While there had been a certain amount of agitation on the Qadiani issue, it was not of such a serious nature as to require defusing in this way. The announcement served as a temporary expedient to distract opinion and avert further agitation. Lost sight of was the fundamental principle of whether religious issues can or should be settled in a political forum. At the time, however, the announcement was generally acclaimed.

In the prevailing atmosphere, the conclusion of the National Assembly was virtually foregone: it unanimously declared the Qadianis to be non-Muslims. The Assembly proceedings were held in camera and still remain secret. A Constitutional Amendment was adopted to incorporate the decision in the basic law of the country. Doctrine and belief apart, some legislators later regretted privately that this decision had to be taken by the Assembly. ZAB was also ambivalent about its desirability, and was concerned over this fusion of religion and politics. He was conscious that even under Prime Minister Khwaja Nazimuddin, who was deeply religious and a man of integrity, though generally considered weak, the Central Government had written to the Punjab Government on 27 February 1953: 'The Ahmedis or indeed any section of people cannot be declared a minority against their wishes. It is not part of the functions of Government to coerce any group into becoming a minority community.'[40] Now it had happened under ZAB.

Publicly he took great pride in this achievement. When the PPP Manifesto for the March 1977 elections was being prepared, he insisted on placing it as the first specific item[41] in the chapter dealing with the Party's achievements:

> Resolved the ninety-year-old Qadiani issue by clearly defining in the Constitution that a person who does not believe in the absolute and unqualified finality of the Prophethood of Muhammad (peace be upon him) is not a Muslim.[42]

The Manifesto also gave prominence to his other achievements in this field: the removal of restrictions on Hajis which enabled about 300,000 Pakistanis to perform Haj during this period; the change in the name of the Red Cross to the Red Crescent Society; the organization of the first International Seerat Conference and arranging visits by the Imams of Ka'aba Sharif and Masjid-e-Nabvi; and the declaration of Friday instead of Sunday as the weekly holiday from July 1977. The contribution of Kausar Niazi was considerable in these examples of Islamization, though not on the Qadiani decision.

Later, following the 1977 elections, ZAB introduced a ban on alcohol and gambling in an attempt to appease the agitation against him which, by then, had taken the form of a demand for *Nizam-e-Mustafa,* that is, a system of governance under the Holy Prophet (pbuh). Try as he might, he could not establish his personal credibility as a proponent of Islam. He had given up his earlier views that the State and religion should be kept apart, only to find the new ground on which he chose to tread was firmly occupied by the religious parties.

ZAB might well have heeded the words of the Court of Inquiry on the circumstances leading to Martial Law on 6 March 1953:

> (4) That nobody realized the implications of the demands, and if any one did so, he was not, out of fear of unpopularity or loss of political support, willing to explain these implications to the public. (5) That the demands were presented in such a plausible form that in view of the emphasis that had come to be laid on anything that could even be remotely related to Islam or the Islamic State, nobody dared oppose them, not even the Central Government...[43]

He had always claimed to be secular-minded, seeking in this to follow the Quaid-i-Azam. However, his actions in government were based on pragmatism, or at times on deeply-rooted feudal beliefs. The result was that religion aggressively asserted itself directly in the realm of statecraft, from which it had largely been removed by the 1970 elections. In one aggravated form or another, this has since continued.

The PPP and Preparations for the 1977 Elections

By the time ZAB came to power, the Pakistan People's Party had grown in strength. Despite his preoccupations as President in the early period, he held regular meetings of the Central Committee and stayed in contact with Party members. It was part of my function as Special Assistant to the President to act on his behalf and liaise with the Provinces. No decision or appointment of significance was made in respect of the Punjab or Sindh without my first consulting Mustafa Khar or Mumtaz Bhutto. There was similar interplay at the Federal level, particularly with Mubashir Hasan, who was Finance Minister and in charge of other important economic ministries. These arrangements worked with precision. The Party as such had developed little formal organization, but with all its leading members in the Government it did not initially matter if this new machinery dominated.

In the early months, ZAB was eager to put the PPP's imprint on all matters of governmental concern. First, he introduced a PPP uniform, 'like in China'. This was not, as generally believed, to give himself and his Partymen a military appearance. The real reasons were that he did not like the *sherwani* (long coat), which he considered 'Hindustani', he thought most Pakistanis wore ill-cut suits, and he regarded *shalwar kameez* (baggy pyjamas and shirt) as sloppy in appearance and reflective of a lazy mind. Ironically, he was subsequently credited with bringing *shalwar kameez*, or this *'awami libas'* (people's apparel), into public life. Having decided to standardize, he added distinctive features with different colour braid round the neck-band and down the trouser-side—gold for himself as President and later Prime Minister, silver for the Provincial Governors and Chief Ministers, blue for Federal Ministers and so on down the line. When I remarked that the stripe on the trousers made us look like band-boys, ZAB did not agree. However, when the uniform was first worn on a foreign trip, at a banquet in Moscow in March 1972, I noticed the stripes were missing from his trousers. After that, the stripes ceased to be part of the uniform. Secondly, he changed the emblem on his Presidential standard, letter-heads and other items, although it had been in place since 1948. He replaced the original emblem with a gold sword on a black background. The sword, *Zulfiqar-e-Ali*, was the Party election symbol. When he told me to arrange its design, I counselled against changing the insignia, but he insisted on establishing his own traditions.

However, just as his attitude towards the Opposition changed,[44] so did it towards the Party, or at least its leadership. Within one year, disillusion with ZAB's conduct began to set in. The increasing alienation in the ranks of the PPP leadership led to several resignations. In October 1972, Law Minister

Mahmud Ali Qasuri left the Government;[45] his absence was felt by the intelligentsia and laywers of Lahore, though he was not missed by the rank and file of the Party. He was subsequently expelled from the Party in February 1973. More significant was the resignation of Mairaj Mohammad Khan as a result of left-right differences. Mairaj was intellectually and temperamentally unsuited to the Government of which he had become a part, yet his talent could have been usefully channelled by ZAB, particularly in dealing with students and labour. He found the 'leftist antics' of Mairaj annoying, and allowed Maulana Kausar Niazi to attack Mairaj and his friend Tariq Aziz. They would complain to me, but my attempts to defuse the tension were unsuccessful. The effect of Mairaj's resignation of 13 October was muted by its acceptance on 20 October, the same day as the announcement of the Constitutional Accord. Mairaj eventually left the Party and became a vehement opponent, being subsequently jailed and badly treated.[46] In him and Tariq Aziz the Party lost two two outstanding speakers; it also suffered politically as Mairaj had been General Secretary, Karachi.

In Sindh, trouble with A.H. Jatoi and Darya Khan Khoso arose over the continuation of Martial Law.[47] Soon, Ali Ahmed Talpur, the elder brother of Governor Rasul Bux Talpur who was the Party President, turned hostile on being given no office, contrary to ZAB's earlier indications to him. As a result, Rasul Bux Talpur resigned abruptly as Governor on 14 February 1973. However, the impact of this was defused by the immediate appointment as Governor of Raana Liaquat Ali Khan, the widow of the first Prime Minister who had been assassinated in 1951. Rasul Bux Talpur, like Mairaj, was more suited to opposition than government, but the Party could not readily replace this long-serving office-bearer who had played an important role in the 1970 elections.

Of more significance was the ousting of Mumtaz Bhutto as Chief Minister of Sindh. Throughout, ZAB and he had a difficult relationship. As Special Assistant I sometimes managed to smooth this out, but after I left government in July 1973, tension increased, and Mumtaz followed a few months later. As a result, the Party's position in Sindh deteriorated.

It was no better in the Punjab, where there were two main causes of dissent—the differences over 'scientific socialism', and the struggle for power and office which centred on Mustafa Khar. ZAB encouraged these divisions. Khurshid Hasan Mir and Kausar Niazi were the main protagonists on the first issue, which led to Mir's resignation as Federal Minister later in 1974. While Mairaj, Qasuri and Talpur joined opposing forces outside the Party, Mir maintained his attack from within.

The tussle around Mustafa Khar had serious consequences. Khar had correctly anticipated problems. The day I became Special Assistant, Khar told me that I would be the closest to the President and requested me to look after his interests while he was Governor in Lahore. He particularly mentioned one colleague who he knew was determined to work against him. True, ZAB was noted to be *kaan ka katcha,* accepting the word of the person who had last spoken to him, and at the same time enjoying the complaints of one colleague about another. Still, I was surprised that Khar should share his concern with me as we were not particularly friendly at the time.

Khar had undeniably become high-handed and, as ZAB put it, 'too big for his boots', but his main fault lay in promoting the interests of the Punjab vigorously. Instead of trying to set this right by talking to Khar directly, ZAB in feudal tradition pitted other colleagues against him. Following his departure as Governor for the second time, Khar sought to contest a by-election in Lahore for the Provincial Assembly in October 1975. In a meeting at the Governor's House, Lahore, where several PPP stalwarts were present, I was alone in advocating that it was his right to be given a PPP ticket, which would also avoid confrontation. In the end, all means and the full force of the Federal and Provincial Governments were used to ensure his defeat by a weak PPP candidate. Several second-rank members left with Khar and the contest revealed ZAB's vulnerability in the Punjab; this became critically apparent eighteen months later, after the March 1977 general elections.

Hayat Sherpao fared no differently in the Frontier. A few months before his death, he seriously considered leaving the Party altogether. He only changed his mind on the persuasion of myself and other friends from the Frontier, and when Mubashir Hasan agreed to become Secretary-General of the Party. A few months later, in February 1975, he was assassinated. Of all those around ZAB, Sherpao's personal devotion had been the greatest, and his subsequent disillusionment was consequently the most profound.

However, the sacking of J. A. Rahim as Senior Federal Minister on the night of 2 July 1974, and the nature of his subsequent treatment, was the most serious indictment of ZAB's conduct. On leaving a lengthy meeting on Balochistan, he was infuriated to learn that Rahim had refused to wait any longer for a large dinner for politicians, military and civilian officers, and had walked out, commenting dismissively on the 'Wadera of Larkana'. He announced Rahim's sacking at the dinner and, at the same time, to my surprise, he appointed me in his place. After dinner, when I pointed out that I had not been consulted first, he replied: 'I have made the announcement in front of everyone, and there is no question of going back on it.'

Rahim was my neighbour in Islamabad. When I got home I found that he and his son had been beaten up and taken to a nearby police station. I brought them back to their house after two in the morning and, with difficulty, secured the help of a doctor. The next evening ZAB insisted that the beating-up had resulted from one of the messengers sent with the dismissal note being fired at by Rahim. I urged that if the attack on Rahim had arisen from some error there was every reason for ZAB to apologize to his friend and former colleague. Sadly, this was not to be and they never again communicated directly. Mubashir and I arranged for Rahim to go abroad to keep him out of further trouble, but on his return he continued to criticize ZAB, and was subjected to renewed ill-treatment. Clearly, this was intended to serve as a lesson and to instil fear in others.

Mubashir Hasan was the next to leave government in October 1974. His disillusionment had come early: he first tendered his resignation in November 1972 and then again in August 1973. However, ZAB's arguments prevailed with him on those two occasions. This time, he decided to stay on in Egypt, where his brother was posted, until the Cabinet reshuffle was in place. His differences with ZAB had increased over the past two years. ZAB had tired of his doctrinaire, leftist views, while Mubashir as Finance Minister found it difficult to cope with ZAB's increasing demands on behalf of friends. Many feudals, even educated ones, seem to have difficulty in accepting the separation of state and personal properties and funds. Mubashir was greatly disturbed by ZAB's disregard for the original principles and purposes of the PPP. With his departure, within less than three years of the establishment of the PPP Government, all the original six—Rahim, Mubashir, Khar, Mairaj, Mumtaz and Sherpao—had either left or been alienated. In each case, ZAB was largely responsible. The camaraderie, a main feature of the early PPP years, ceased. Mir Afzal Khan and I both warned of the danger of this state of affairs, pointing out that he was isolating himself, replacing the affection his colleagues had felt for him with awe, if not fear.

This isolation did not arise only from arrogance or his inability to brook criticism, as opponents maintain. It also resulted from his feudal background and a sense of insecurity. Much earlier, when I was his Special Assistant, he had told me after a Cabinet meeting that he wanted to hear my frank views, but that I should give them to him in private, not in front of others, who might otherwise be encouraged to be too independent and 'contrary'. He wanted to maintain an aura of infallibility. Subsequently, when I mentioned this conversation to Mubashir Hasan, he said he too had decided to speak with ZAB only in private in order to avoid ugly consequences.

Another serious shortcoming, which eventually cost ZAB and the Party dear, was his notion of loyalty, which flowed only one-way, towards and not from him. He felt everyone owed their position to him personally. True, most of us would not have attained high office but for ZAB, yet the contribution of the original nucleus of six and several others, particularly in the earlier years, was equally undeniable. By nature suspicious, he sought to have 'dirt' available against his Ministers and leading Party members and, in early 1976, assigned to his officials the task of preparing secret dossiers about them, to be used in case of need. Through Interior Minister Qayyum Khan and Yusuf Khattak, Mir Afzal Khan and several of us learnt about this. Suspicion was reciprocated.

ZAB's reliance on various intelligence and security agencies, and the importance he gave them, increased. Previously, the two main ones answerable to the civilian Federal Government had been the ISI and the DIB. He created the FSF, the Federal Investigating Agency (FIA) and the Airports Security Force; in addition, he engaged Said Ahmed Khan as his Chief Security Officer, and, later, Rao Rashid as his Special Secretary, while he appointed retired General Tikka Khan as his Special Assistant for National Security. This only increased oppression, not efficiency. In late 1975, he appointed a Special Committee to review intelligence arrangements, with myself as chairman and retired Major-General Syed Ghawas, then Frontier Governor and a former head of the ISI, and Mian Anwar Ali, Ayub Khan's and ZAB's first DIB, as members. One of our main conclusions was that the various agencies had developed a 'First Information Report' (FIR) mentality: their sole aim seemed to be to report all matters first and directly to the Prime Minister. In this way they tried to show their efficiency to please him. The review also pointed out that there was no coordination or exchange of information or views among the various agencies. The system needed complete revamping, but nothing was done.

Even earlier, soon after I had ceased to be Special Assistant in July 1973, control over Party matters shifted into bureaucratic hands. On 22 November 1973, ZAB wrote to Afzal Said Khan, Secretary to the Prime Minister, saying that he had 'set up a machinery under your command to control Party matters' and instructed him to 'put this Party Secretariat to full use'. Following the abrupt exit of J.A. Rahim, some of us felt we should attempt to revive and revitalize the Party when it came to appointing a successor as Secretary-General. Hayat Sherpao and I pressed a reluctant Mubashir Hasan. He had already decided against continuing as Federal Minister, and felt that the Secretary-Generalship was a meaningless office since ZAB did not want the Party reorganized. He only agreed when I stressed that Sherpao might quit if he refused. At the Central Committee meeting held for this purpose, three names were put forward, including Mubashir Hasan's, which was not to ZAB's liking. Supported by most of us, including Mustafa Khar and Mumtaz Bhutto, he was appointed. He was greatly respected by the Party rank and file. Unfortunately, he was right in anticipating that he would not be allowed to function effectively. In fact, since the Secretary-General was not of his choice, the Chairman did not call another meeting of the Central Commitee as such for as long as Mubashir Hasan held office. Instead, 'high-level' meetings of the Party were held. ZAB also reconstituted the 'Principles Committee', the most important body in the Party, which had earlier included Rahim and Haneef Ramay. The new Committee comprised ZAB, Mubashir Hasan, Sheikh Rashid and myself. However, this too failed to be convened by him.

The state of the Party was described in a detailed seventeen-page memorandum, *Where do we go from here?*, which Mubashir Hasan sent to ZAB on 17 September 1974. He pointed to 'the decline and deterioration of the Pakistan Peoples Party...What dreams of December 1967 and December 1971 do not stand shattered?' Mubashir emphasized that, unless the situation was corrected, 'it is a sure invitation to military intervention and disaster'. ZAB sent copies to four of us, saying he wanted a joint meeting after we had studied it.[48] But no meeting was held. Rather than reorganize the Party, he decided to reshape his Cabinet. On 24 October 1974 Mumtaz Bhutto was brought back into government as Communications Minister, and some good new Ministers like Mir Afzal Khan were appointed. A mini-reallocation of portfolios became necessary just prior to the Cabinet meeting because, until the last minute, ZAB refused to believe that Mubashir Hasan would not withdraw his resignation.[49] It was inconceivable to him as a politician that anyone would willingly relinquish power. He was as baffled by Mubashir's resignation as by my own in June 1973.

Although the reshuffle of the Cabinet reinforced the Government, it was no substitute for strengthening the Party. Another of the Party's weaknesses was its lack of a proper constitutional structure. Efforts to prepare a constitution for the PPP, which in this case ZAB wanted, were frustrated at each stage by such questions as the authority of the provincial parties, and the position of Assembly members in the framework. As a result, there were no elections within the Party, nor for that matter have any been held since. All office-bearers were either appointed or approved by Chairman Bhutto. This problem was further compounded by the fact that no local bodies elections were held in the Provinces, even though under the Accord of 6 March 1972 they were envisaged on a date to 'be fixed as early as possible after the Provincial Assemblies are convened'.[50] In a note to ZAB concerning activating the Party which I wrote on 21 December 1976, I said that, looking back over the past five years, 'one of the greatest errors' was not having local bodies elections in April 1972. No attempt was made to establish democracy at the grass roots level; control continued to be imposed from the top. The corollary of this was that all discontent and criticism was also directed at the top.

On the question of reorganizing the Party, Mubashir Hasan wrote to ZAB on 22 December 1974 pointing out that 'many big landlords and other undesirable elements' had found their place in the Party at the expense of 'dedicated and sincere' workers. This was a vital issue which in part accounted for the disintegration of the original PPP team. After coming to power, ZAB was never quite at ease with those who had helped form the Party. Instead of reforming the Party and getting rid of those who had turned corrupt, he entertained complaints against his own workers. He turned to those with whom he was most at home, his own class of feudals. I was unhappy with this development; I also disapproved of the Government's increasing use of violence and force in politics, the 'personal reasons' for which I had earlier resigned as Special Assistant. At that time,

President Bhutto simply could not understand my resignation, saying, 'Why, you are virtually the Prime Minister.' Indeed, as his Special Assistant and chief of staff I enjoyed far greater power and authority than I did subsequently when for a while I simultaneously held the portfolios of Production, Industries, Commerce and Town Planning.

A basic issue arose about the qualification for membership of the Party. Mubashir Hasan as Secretary-General wanted to retain the ideological orientation of the Party by restricting it to those who at least subscribed to the original cause. ZAB insisted on as wide a membership as possible. In his letter to Mubashir of 25 March 1976 he said, 'I feel that the objective conditions for enrolment have since changed. If, with elections coming, we have restricted enrolment, this would be tantamount to rejecting the vote of those whom we do not allow to be members of the Party'.[51]

ZAB felt that the days of an ideologically-oriented Party were over; even in China, after the 'Gang of Four', efforts were under way to end the narrow ideological content of the Communist Party by seeking wider public support. Similarly, the Nazi Party in Germany, originally national-socialist, had become a party with an all-German following. Not only his opponents, but ZAB too, had a penchant for comparisons with Hitler: he would urge Kausar Niazi as Information Minister to outdo Joseph Goebbels, Mumtaz Bhutto in Sindh to be a Heinrich Himmler, and myself as Production Minister to be another Albert Speer.

As the Party organization declined, a coterie of bureaucrats and advisers around ZAB gained his ear. Instead of public rallies, which had previously been organized by the Party and had proved to be a major asset, he now held *kutcheries* (courts) in various Divisions. These *kutcheries* were organized by government functionaries and those invited were invariably hand-picked, non-critical people. These advisers also often helped choose local office-bearers of the Party. In some areas a combination of local Party and bureaucracy proved an insurmountable barrier for poor people seeking justice or a remedy for their grievances. Such practices had been common before the PPP Government, and their reintroduction showed how far the Party had moved from its early position. ZAB's new attitude could also be seen in his choice of Governors: by 1976 all four were from the former ruling class or the army, with the Amir of Bahawalpur in the Punjab, the Nawab of Junagarh in Sindh, a retired Major-General in the Frontier and the Khan of Kalat in Balochistan. However, despite everything, his personal popularity and appeal, although diminished, remained considerable.

Of course, as in most developing countries, and even in several developed ones, tales of corruption were rife, but even opponents of the PPP now admit that corruption was on the whole small-scale compared to what occurred under subsequent governments. ZAB made no real effort to stamp out corruption. When, as early as June 1972, with the support of Chief Minister Mumtaz Bhutto, I drew attention to allegations of corruption in the Sindh Government, ZAB paid no heed. On the contrary, officials who had provided the files were ordered to be transferred. Later, he was highly indignant when I sent him a note on 6 August 1975 saying among other things:

> The Prime Minister has been pleased to refer to 'our many conversations on corruption'. Quite frankly, I don't know what more I can say. I can only reiterate that corruption cannot be tackled merely through legislation, committees or administrative measures. What is needed is action, and against the big fish. Unless this is done the whole exercise will be in vain, and would only widen the credibility gap.
>
> Some months have passed since the Prime Minister's press conference announcing a 'battle royal' against corruption. People are becoming increasingly sceptical about whether any action will be taken…
>
> After the Prime Minister's announcement, people say that if the Prime Minister knows the problem (of which formerly they said he may not know the extent), then why does the Government take no action?…

This is how the question is put when asked in a logical or polite manner. Otherwise, the language used is somewhat less polite.[52]

As with everything else in the PPP Government, ZAB was the man in charge, and several allegations of corruption centred around him. He became very conscious of this and, following one discussion on military purchases, pointed out to Mubashir Hasan and myself that we had been exchanging 'knowing glances', as if accusing him. By the time the election campaign started, specific charges were being directed against him and the PPP, particularly the Provincial Governments, and these played an important part in his fall. However, even General Zia's Martial Law could not unearth evidence, so further comment would be inappropriate in this narrative. ZAB's own attitude to financial rectitude can be illustrated by a conversation I had with him in Larkana. Qazi Fazlullah, a Sindhi former Minister, requested a small favour which ZAB asked me as Minister for Production to fulfil. After Fazlullah had left, ZAB commented: 'You always maintain that honesty in Third World politicians is an attribute which the people most admire and respect, and quote Nasser as an example. Well, you must have gathered from this talk that, for all his other faults, Fazlullah was financially straight. Is he more admired and respected than the other Sindhi politicians who were dishonest?'

By the year 1977, when general elections were held, ZAB was convinced that he was the sole asset and vote-catcher of the Party, which was otherwise a liability—a view with which the officials around him eagerly concurred, and promoted. With the selection of feudal or well-to-do candidates, able to muster a certain amount of local support, ZAB could win the elections virtually single-handed. To blame these advisers, and not him, is indeed to reverse responsibility; nevertheless, the role they played should not be underestimated. They included his Special Secretary, Rao Rashid, the Secretary to the Prime Minister, Afzal Said Khan, the DIB, his Military Secretary, Said Ahmed Khan, and Masood Mahmood of FSF fame. Vaqar Ahmed, who was the Cabinet and Establishment Secretary, mainly dealt with administrative matters but had assumed such importance that, at a dinner when a Federal Minister complained about him, ZAB dismissively pointed out that Ministers were dispensible but not Vaqar Ahmed.[53] The FSF, which had a poor reputation, also came to be regarded highly by ZAB, who appointed a committee in January 1975 to consider a charter for its 'scientific and correct motivation'. One person even proposed that the FSF should take a personal oath of allegiance to ZAB. I toned the proposals down to such an extent that, at least to my knowledge, they were never finalized or implemented.[54]

The extent of bureaucratic dominance can be gauged by the fact that, by 1976, many communications from the Prime Minister were being channelled through Rao Rashid. He was already Secretary to the newly-established National Intelligence Board (NIB), on which some of us served, when, on 29 June 1976, he informed us that ZAB had 'set up a Special Committee consisting of a limited number of Ministers of the Federal Government to advise him on matters of grave national importance'. The charter of duties of the Committee which he enclosed indicated how some bureaucrats pandered to ZAB's desire to concentrate all powers in himself. This Committee was combined with the NIB by October 1976.

With the loss of the old guard, the Party had become an adjunct of the Government at the Centre and in the Provinces; its original identity and vitality had died. It had ZAB at the pinnacle and little below him; the official administration assumed the role of the Party. What complicated matters further was that the bureaucracy itself was not behind ZAB following the wholesale dismissal of over 1300 civil servants in his early Martial Law period, and the subsequent radical administrative reforms of 20 August 1973 which particularly affected the superior services. Although these reforms were good in principle—abolishing the former 'classes' by merging the services into a unified structure, and permitting entry to older, professionally qualified persons—in practice, political considerations came into play with several appointments, especially at a later stage. Moreover, the FSF, which had been created as a substitute for the army to deal with civil disturbances, had become both very unpopular and, worse, ineffective except when terrorizing political opponents. As a result, ZAB entered the election year with a disillusioned PPP and questionable support from the bureaucracy, on which he relied.

Nevertheless, as with all matters, he planned carefully and well ahead for the elections. The first formal meeting on elections was held as early as February 1976, followed by others which were attended by some Federal Ministers and officials, by the Provincial Chief Ministers and Presidents of the Party. The Secretary-General, Mubashir Hasan, was deliberately excluded because by then he and ZAB had little in common. ZAB urged everyone to prepare meticulously, and designated me the chief organizer, the person to be contacted for decisions if ZAB was not available. Apart from my other responsibilities, I was again to be his *de facto* chief of staff. His insistence on detail can be seen in his note to me of 18 February, a few days after the first meeting, in which he commented that he had 'stupidly' forgotten to mention the importance of getting 'numerous copies of every voters' list'.

Mubashir Hasan advised me that others would do the dirty work while I was supposedly in charge of the campaign. I believe it was for this reason that he as Secretary-General both distanced himself and was deliberately excluded from participating in the preparations. A few months later, when I mentioned the assignment to Syed Babar Ali[55] as we were flying to Multan to inspect a fertilizer project under construction, he asked, 'Can't you get out of it?' I knew it was a thankless task, but I informed both him and Mubashir that I had already told the Prime Minister that I would not continue as a Federal Minister after the elections, and in the circumstances it was difficult now to refuse his request to oversee the PPP campaign.

The timing of the elections had been under consideration since the successful outcome of the Islamic Summit and the recognition of Bangladesh. Now that the West Wing legally constituted Pakistan, fresh elections appeared necessary. In November 1974, Mubashir Hasan, Sherpao and I discussed this with ZAB. We proposed that elections should be held in February or March 1975. ZAB was surprised at the suggestion of this early date, which he incorrectly thought reflected our desire to capitalize on the Islamic Summit and the Qadiani issue. In fact, Sherpao's reasons for the 1975 date was the need to clear his position in the Frontier through elections, while Mubashir Hasan wanted the Party to return to its original roots. By prior arrangement, I, as the spokesman, pointed out the need for a fresh mandate and stressed that March 1975 would be a good time from the economic standpoint. I said that very high inflation, caused by spiralling oil prices and long-gestation projects like the Steel Mills, would hit us later in 1975. Moreover, looking ahead, 1980 would be an appropriate time for the next elections, since by then the economy should be on an upturn. He appreciated the arguments but not the prospect of reducing his term as Prime Minister. He discussed the question of the timing of elections with others, including Kausar Niazi, Mustafa Jatoi and Hafeez Pirzada, but basically he kept his own counsel on this subject.

By early June 1976, several Opposition parties had published their manifestos; first the Tehrik-i-Istiklal (TIP) and the National Democratic Party (NDP), followed by the draft proposals of the United Democratic Front, which was designed to accommodate several parties. ZAB forwarded all these to me, along with numerous other documents, directives and notes. By this time, Ghafoor Ahmed of the Jamaat-i-Islami had put forward preconditions for participating in any elections. In my note to the Prime Minister of 29 June, I recommended that these should, in the main, be accepted at the appropriate time.

From August onwards I received confidential monthly directions from ZAB setting out the programme for the current and ensuing months. These related to the announcement of various reforms, the celebrations to be observed each week by such bodies as farmers, lawyers, students and even the Armed Forces, and the 'Future Programme' which sometimes covered nearly sixty points. They ranged from such subjects as foreign policy to public transport and even child care centres.

The preparations for the elections were mainly done in a block of rooms in the Prime Minister's Secretariat. The main participants in the meetings which I held from October 1976 were Maulana Kausar Niazi, Sardar Muhammad Hayat Tamman, Yusuf Buch, Rao Abdul Rashid, Pir Ali Muhammad Rashdi and Hamid Jalal. It was a curious mix, selected by ZAB. We considered the 'weeks' to be celebrated, and discussed some of the issues to be covered in the months ahead. We particularly felt that a week was too long and would be counter-productive, and instead suggested one day. Somebody present must have informed ZAB that Rashdi had compared this programme with Ayub Khan's; ZAB wrote on my note to him of 9 October, 'Only a bastard can compare these weeks with Ayub's decade of reforms.' We were right—the weeks celebrated were not successful.

I started work on the Party Election Manifesto and, in the course of collating material for the chapter on 'Promises Fulfilled' in the Manifesto, could not but be impressed by the numerous achievements of the PPP Government in virtually every field, including defence.[56] Kausar Niazi was meanwhile attending to the media and publicity, while other officials, the most active of whom was Rao Rashid, were at work in the same Secretariat under the Prime Minister's direct supervision.

Maulana Kausar Niazi, in *Zulfiqar Ali Bhutto of Pakistan— The Last Days,* recorded that there was a ' "mini-Cabinet" of the Prime Minister which conceived "Operation Victory" for the 1977 elections'. He also stated that ZAB placed 'total reliance on Rao Abdul Rashid, beside Afzal Saeed Khan, Vaqar Ahmad, Saeed Ahmad Khan, Masood Mahmood, Muhammad Hayat Tamman (Adviser for Public Affairs), Akram Sheikh (Director Intelligence Bureau)', who together with the Chief Secretaries of the four Provinces, and 'Major-General Imtiaz (Military Secretary to PM) and Hamid Jalal (Additional Secretary to PM)', constituted this mini-Cabinet.[57] Kausar Niazi went on to state, 'Rafi Raza as incharge of the election campaign very courageously tried to reject their electoral strategy'.[58]

In fact this was admitted in one case even by General Zia's *White Paper on the Conduct of the General Elections in March 1977,* which was otherwise a blanket condemnation of all ZAB's actions, and those of his associates, relating to the elections. Referring to a note from his Military Secretary about, *inter alia,* immobilizing opposition candidates, on which ZAB in his own handwriting had asked me to issue implementing directives, the White Paper said I 'shot it down', reproducing my handwritten comment of 19 February 1977: 'Spoken to the PM—not to be done as the methods are not acceptable—File.'[59]

In the meantime, the Opposition parties too had been active. Battle lines were so clearly drawn that they even refused to attend a banquet held by ZAB in honour of King Khalid of Saudi Arabia in October 1976. They stepped up their vilification of ZAB and the Government. It seemed that their hostility to him had deprived the Opposition of any positive thinking. We in turn decided, on 27 October 1976, to prepare a list of persons within the PPP who would counter allegations made by each of the Opposition leaders. There was no let-up.

In ZAB's words, we had done 'our homework', despite all our shortcomings. He was now set for general elections in 1977. On one point, however, he did not yield, in spite of my attempts to persuade him otherwise: he refused to ease confrontation with the Opposition. In this, the Opposition's hatred and suspicion of him played no small part. In the election year of 1977, the country was, as usual, the victim.

NOTES

1. Source: *Report of the 1972 Census,* published by GOP. Other sources show a total area of 310,403 square miles.
2. *The Language Accord,* Department of Films and Publications, GOP, Karachi, 1972, p. 1.
3. Ibid., p. 15.
4. *Jasarat,* Lahore, 13 March 1978.
5. *Morning News,* Karachi, 30 January 1973.
6. *The Sun,* Karachi, 31 January 1973.

7. Z.A. Bhutto, *Speeches and Statements, January 1, 1973-March 31, 1973,* Department of Films and Publications, GOP, 1973, pp. 90-1.

8. ZAB asked me to withdraw my resignation of 1 June 1973. When I refused, he accepted, by letter No. D-830-PS (President)/73 of 7 June. My resignation was to become effective on 30 June 1973 but, on the urging of Mubashir Hasan, supported by Sheikh Rashid, that if I did not want to serve in the executive I should at least be a legislator, ZAB made me a Senator from Sindh. It then took effect on my becoming a Senator, *vide* Notification No. 119/6/73-Min. of 19 July 1973.

9. *National Assembly of Pakistan Debates,* 14 February 1974, Vol. I, No. 21.

10. *Morning News,* Karachi, 4 March 1973.

11. *Dawn,* Karachi, 16 July 1974.

12. *Morning News,* Karachi, 11 December 1977.

13. *See also,* Chapter 6, p. 177 and Chapter 7, p. 219.

14. As narrated by Mustafa Khar to the author in 1978.

15. Z.A. Bhutto, *Speeches and Statements—1 April 1973 to 13 August 1973,* Department of Films and Publications GOP, 1973, pp. 134-7.

16. Source: *Economic Survey,* Statistical Supplement, 1992-3, GOP, Finance Division, Islamabad, Table 2.1 at p. 37.

17. Ibid., Tables 4.14 and 4.15 at pp. 107, 108 and 110.

18. *President's Address to Businessmen,* Department of Films and Publications, GOP, Karachi, 1972.

19. Hydrogenated Vegetable Oil Industry (Control and Development) Act, 1973, *Gazette of Pakistan,* Extraordinary, Part I, 2 September 1973.

20. *See also,* Chapter 6, p. 183.

21. (a) Banks (Nationalization) Act, 1974, *Gazette of Pakistan,* Extraordinary, Part I, 11 March 1974; (b) Marketing of Petroleum Products (Federal Control) Act, 1974, ibid., 8 March 1974; and (c) Pakistan Maritime Shipping (Regulation and Control) Act, 1974, ibid., 11 March 1974.

22. (a) Flour Milling Control and Development Ordinance, XXIV of 1976, *Gazette of Pakistan,* Extraordinary, Part I, 17 July 1976; (b) Rice Milling Control and Development Ordinance, XXV of 1976, ibid., and (c) Cotton Ginning Control and Development Ordinance, XXVI of 1975, ibid.

23. For the Qadiani issue *see,* pp. 291-6.

24. *Import Substitution Policy and Investment in Pakistan* by Dr. Muhammad Aslam Sial, former Deputy Economic Adviser, GOP, IBF, May 1989, p. 49.

25. *National Assembly of Pakistan Debates,* Official Report, 18 June 1976.

26. In fact, the correct transliteration is 'q', 'k' being an anglicized form. For further details about the Steel Mills, *see also,* Chapter 7, pp. 199 and 249-50.

27. Mubashir Hasan, a former Finance Minister, referred to this incident in his statement which was reported in several newspapers, including, *The*

Frontier Post, Peshawar, *The Muslim,* Islamabad and *Dawn,* Karachi, of 13 August 1990: 'I am aware that Mr. Raza had successfully resisted all pressures including those of the then US Ambassador (Henry Byroade) to place orders of plants on a basis other than that of commercial and technical merit and thus saved hundreds of millions of dollars'.

28. Qadianis are also referred to as Ahmadis and Mirzais.

29. The Punjab Disturbances Court of Inquiry Report of 10 April 1954 was published by the Government of Punjab in 1954.

30. Holy Quran; Sura IV, an-Nisa, Ayat 157-9 (Chapter IV, On Women, Verse 157-9).

31. The Punjab Disturbances Court of Inquiry Report, p. 189.

32. Ibid., p. 192. The Qadianis formed a force of about a thousand men to fight in Kashmir in 1948.

33. Ibid., p. 75.

34. Ibid., pp. 106-9.

35. Ibid., p. 98.

36. Ibid., p. 205. A person who discharges all his obligations to Almighty Allah and society is generally regarded as a *momin.*

37. *See,* Chapter 6, pp. 166-7 regarding the Basic Principles Committee.

38. *See,* Chapter 6, pp. 172-3, 182-3 and 185.

39. During the year I was not in the Government, I only twice sought interviews with ZAB, though I met him otherwise on several occasions; the first concerned Balochistan, and the second time was over the Qadiani issue, when we had this talk.

40. The Punjab Disturbances Court of Inquiry Report, p. 147.

41. Relevant notes, and draft Manifesto with ZAB's insertions thereon, are with the author.

42. *Manifesto,* Pakistan People's Party, January 1977, p. 13.

43. The Punjab Disturbances Court of Inquiry Report, p. 235.

44. *See also,* pp. 273-4.

45. *See,* Chapter 6, p. 175.

46. During a demonstration Mairaj Mohammad Khan was beaten up; as a result, his eyesight was impaired.

47. *See,* Chapter 5, pp. 154.

48. Copy of this memorandum by Mubashir Hasan, containing handwritten comments by ZAB on 26 September 1974, is with the author.

49. In fact, despite telling ZAB that Mubashir Hasan, who had spoken to the author of his decision a month earlier, would not change his mind, ZAB asked the author to telephone Mubashir in Egypt for confirmation in his presence.

50. Para 11 of the 6 March Accord; for further details about this Accord, *see,* Chapter 5, p. 153.

51. Various documents and correspondence pertaining to the Party, including

the draft PPP constitution, the issue of membership, the Manifesto, are with the author; several have ZAB's own handwritten comments.

52. *The White Paper on the Conduct of the General Elections in March 1977*, published by GOP, Rawalpindi, July 1978, makes many allegations against ZAB, the PPP generally, and the author, but conveniently omits such connected notes which counter its defamatory allegations.

53. Narrated by Mumtaz Bhutto to the author; the incident occurred during a dinner at Hafeez Pirzada's house at which Mumtaz, Aziz Ahmed and other Ministers were present.

54. The committee comprised the author, Yusuf Buch and Kamal Azfar; ZAB was not pleased that I disagreed with, and moderated, the proposals, which would have made the FSF more akin to the Nazi SS. Some reports of this committee appeared in the White Paper on *The Performance of the Bhutto Regime, Vol. III, Misuse of the Instruments of State Power,* GOP, Islamabad, January 1979.

55. Syed Babar Ali is a leading businessman in Lahore and, at that time, was Chairman of the National Fertilizer Corporation, a major public sector enterprise.

56. Both V.A. Jafarey, who was then Planning Secretary, and Ghulam Ishaq Khan, who had been appointed Secretary-General of the Defence Ministry, met the author several times in this connection in December 1976. They too expressed amazement at the magnitude of development. Ishaq Khan later became President of Pakistan from 1988 to 1993. *See also,* Chapter 9, pp. 328-9 concerning the Manifesto.

57. Kausar Niazi, *Zulfiqar Ali Bhutto—The Last Days,* Vikas, New Delhi, p. 43.

58. Ibid., p. 72.

59. *The White Paper on the Conduct of the General Elections in March 1977,* p. A-300 Annexure 85.

FATEFUL 1977 GENERAL ELECTIONS

Final Run-up to the Elections

No one at the start of 1977, least of all ZAB, envisaged that the months ahead would prove fatal for him and affect the course of Pakistan's history for the next eleven years. Although no longer the popular leader of the past, ZAB felt he was fully and firmly entrenched—the man in charge. The Opposition was suppressed, in disarray or in jail. The army under a seemingly pliable Chief of Staff, General Ziaul Haq, appeared to be willing to carry out all orders emanating from its political master.

Looking ahead to the elections, he ended the year 1976 by settling for the time being what he considered the last of the potentially explosive election issues in the Punjab and Sindh, namely, the apportionment of the River Indus waters. This question had remained unresolved since Independence. President Yahya Khan had appointed a Judicial Commission under a Supreme Court Judge, Fazle Akbar, but ZAB had decided not to implement its findings and, instead, to establish a new Commission to settle the matter. The 1973 Constitution had conferred on the Council of Common Interests exclusive jurisdiction to consider complaints regarding 'water from any natural source of supply' by the Federal or the Provincial Governments.[1] The Governors' Conference which ZAB held on 31 December 1976 was therefore, at the last minute, designated a meeting of this Council to fulfil constitutional requirements. It appointed a new Commission comprising the Chief Justices of Pakistan and the four provincial High Courts to recommend water apportionment. The reference to the Council was intended to prevent this issue from coming to the fore during the elections, particularly in the Punjab. He considered it a masterstroke.

His next step was to ask the Chief Election Commissioner (CEC) to announce on 2 January that, 'we are ready to conduct elections in the country'.[2] Many Pakistanis had felt that, using as a pretext the recent Proclamation of Emergency in India and that Government's decision of 5 November 1976 to extend the life of the Indian Parliament by a year, ZAB would likewise delay elections. The Government's credibility was low and some still considered the CEC's statement a ruse but, on the whole, it had a reassuring effect.

This was followed by ZAB setting in motion the reforms planned earlier. First, on 4 January, came the labour reforms. Not of major consequence, they mainly emphasized the existing improvements and gave some increase in workers' salaries. Then, on his birthday, 5 January, came land reforms. Having dealt severely with the industrialists, and being a big landlord himself, his real test lay in agricultural reforms. Under his instructions, these had been prepared in strict secrecy by a handful of officials, and were intended as a 'personal gift' from him to the people on his birthday.

The land reforms[3] provided for a reduction in the ceiling on holdings of irrigated land from 150 to 100 acres and of unirrigated acreage from 300 to 200; compensation was to be through redeemable bonds over ten years. ZAB thought this would surprise the leftists in the PPP; having 'out-flanked' the rightist parties on the Qadiani issue, he now wanted to do the same to the left. He was not pleased when I pointed out that Sheikh Rashid had, in fact, proposed to me that a 25-acre ceiling be stipulated in the Party's Election Manifesto.

Of far greater significance were the simultaneous proposals concerning taxation: the outmoded system of land revenue collection was to be replaced by the introduction, for the first time, of tax on income from agriculture. In order to protect and appease the smaller agriculturists, they were to be exempt from tax on income from twenty-five acres or less of irrigated land. Equally important, it was proposed to reduce income tax from a maximum of 60% to 50%, and super-tax on corporations from 30% to 20%. Discussions on tax on agricultural income had continued over two decades, as well as on the need to widen the tax base and streamline the over-all system. By lowering tax rates and including agricultural income, he tried to make the payment of tax more acceptable to urbanites, both businessmen and the salaried classes, a particularly difficult task in developing countries, where taxpayers feel there is no return benefit. It was a bold measure for a politician who depended on the dominant landlord class for political support in the Assemblies.[4]

The following day the Government announced an increase in pensions for civil and military personnel. In addition, the Iqbal Centenary celebrations commenced with considerable fanfare in Lahore. Both these moves were designed mainly to appeal to the majority Province, the Punjab.

When I returned from Karachi at the end of December after a few days' leave to attend my niece's wedding, ZAB told me he was going to announce, on 7 January, the elections for 7 March. He asked me to keep it confidential, adding that, apart from myself, he had informed only Hafeez Pirzada and, to my surprise, US Ambassador Byroade. I suggested a fortnight's delay in announcing the election date in order to allow time for finalizing the Manifesto and publicity arrangements, but, as ZAB had already assured Byroade of the date and said he would convey his answer on the nuclear issue after the elections, he considered that any delay might cause misunderstanding. Why he mentioned the date to Byroade, if indeed he did, and what he intended to convey, is open to question. I did not see the wisdom or necessity of his action then, nor do I now.

As in past years, I was invited to Larkana for his birthday, this time to review developments and election preparations. Although himself a product of popular support in 1970, he always felt that elections in Pakistan were harbingers of ill fortune. In the few elections that had taken place, the results were invariably a landslide for one side, which produced chaos followed by subsequent corrective action. Elections damaged authority with no commensurate benefit, he maintained. The country had witnessed this in 1954 in East Pakistan, the 1964 election had brought disrepute to Ayub Khan, and the 1970 election had led to dismemberment. Nevertheless, a fresh mandate was clearly overdue and necessary for tackling the important issues and problems facing the country. Moreover, an early election would deny the Opposition sufficient time to unite or gain momentum.

In personal terms too ZAB had little to say that was favourable about elections. His father had ended his political career after losing in Larkana, which he attributed to family rivalry. ZAB had such little confidence in the election process that in the indirect Basic Democracy elections of 1962, he had secured the post of Ambassador to Saudi Arabia for his opponent, Abdul Fatah Memon, to ensure no contest, though he was certain of winning as Ayub Khan's strongman. Even in the 1970 general elections he was cautious.

His present anxieties were aggravated by the decision of the Allahabad High Court, which had disqualified Prime Minister Indira Gandhi for an election offence; this had compelled her to declare an Emergency pending appeal to the Supreme Court. He asked me to examine his own position, and whether, as Prime Minister, he could continue to use official facilities such as helicopters in the campaign. He was particularly concerned whether the cost of such facilities would be taken into account in the limit of expenses allowed in his own constituency campaign. He was, as always, anxious about the Courts. I assured him there was no need to worry, since as Prime Minister he could use official transport. As for Larkana, I suggested that, after filing his nomination papers, he should merely inform his constituents how well he had served them, and explain that he would not return till after the elections as he had campaigning and work to do elsewhere. He was certain to be elected. He appeared satisfied with my answer.

At this point one commonly held misconception should be dispelled, relating to the belief that elections were held one and a half years earlier than required. Article 271(1) of the Constitution provided that the first National Assembly, 'unless sooner dissolved, shall continue until the fourteenth day of August one thousand nine hundred and seventy seven'. It was only under a Proclamation of Emergency that Article 234 (6) provided, 'Parliament may by law extend the term of the National Assembly for a period not exceeding one year'. Although an Emergency was in force, any extension would have required an extraordinary measure. Indira Gandhi's similar step had resulted in severe criticism. ZAB proudly proclaimed that, for the first time, we were ahead of India, especially as Indira Gandhi soon followed in announcing March elections.

Having determined the date for the elections, he addressed the National Assembly on 7 January 1977. In a long speech, he first talked about his achievements of the past five years in rebuilding the country, his contribution to Islam, and the 1973 Constitution. Expanding on economic progress, he dramatically displayed a bottle of liquid, claiming that high-quality oil had just been struck at Dhodak. Then, with quiet deliberation, he announced general elections for 7 March, with elections to the four Provincial Assemblies to follow three days later. He went on to state that the Chief Election Commissioner, whose term of office was to expire shortly, would be given a three-year extension 'to inspire confidence', and to ensure fair elections. The theatrical speech was effective, but nothing compared to the dénouement ahead.

The Elections

ZAB insisted on a prompt demonstration of the PPP's effectiveness as a political machine. He sought to impress the public with our readiness, and the absence of it in the Opposition. Thus, on 9 January, PPP Parliamentary and Appellate Boards were set up to consider the award of tickets. In contrast, most of the Opposition parties then in the UDF met in Lahore only to adjourn with the announcement that they would jointly contest against the PPP.

Suddenly, on 11 January, indecision in the Opposition camp was replaced by unity of purpose. They formed the Pakistan National Alliance (PNA) and agreed to contest the elections jointly through candidates chosen by Parliamentary Boards. The PNA adopted the 'Plough' as a common election symbol, and established a committee to prepare the PNA's organizational structure and election manifesto. This committee included Ghafoor Ahmed, Nawabzada Nasrullah Khan and Malik Wazir Ali, with Rafiq Ahmad Bajwa as Convener, and was authorized to decide the question of the PNA leadership. In the meantime, they were to launch their campaign with a public meeting in Karachi on 23 January, to be followed with large rallies at Hyderabad and Nawabshah, and then in Peshawar on 30 January.

During the past several months ZAB had set various agencies to work to prevent such a combination among the Opposition parties. The speed with which the PNA was formed, and our lack of any prior knowledge of it, took him aback. When he upbraided these agencies, he was reassured by Rao Rashid, Masood Mahmood and others that the leadership issue among the religious parties, to say nothing of Asghar Khan's ambitions, would prove a fatal stumbling block.

Despite these developments, the public was still certain that the PPP would be victorious, and that ZAB could not lose. There was a stampede for Party tickets, joined even by Ahmed Raza Kasuri, who had earlier accused ZAB of murdering his father.[5] Landlords, feudals, tribal chiefs and various notables, including Shaukat Hayat Khan, announced they were joining the PPP with their 'thousands of followers'. If even half their claims were to be accepted, the ranks of the Party would have swelled to twice the total population of Pakistan. The Prime Minister could not conceal his delight, and was irritated when I pointed out that it was like the late rush to join the Convention Muslim League under Ayub Khan, which had only resulted in an amorphous mass. Many of the PPP old guard, of whom

Mubashir Hasan, Mumtaz Bhutto and Sheikh Rashid were the most vociferous, protested about this loss of the Party's identity.

The swelling ranks of the PPP confirmed ZAB in his view that the Muslim League (Qayyum) were no longer required as coalition partners. Some PPP colleagues, including Mumtaz Bhutto, Mir Afzal Khan, Kausar Niazi and I, felt that it would be advantageous to have the support of this party, with its traditional base in the NWFP. The PPP Government in that Province was weak and unpopular, and needed this alliance to counter Maulana Mufti Mahmood's JUI and the former NAP. Qayyum Khan had earlier refused to merge his party with the PPP. A curious combination of views among progressive elements in the PPP who thought Qayyum was reactionary, and those bent on pleasing ZAB, urged that Qayyum Khan be ditched. When it came to the division of seats in the Frontier Province, ZAB's advisers such as Hafeez Pirzada made it impossible for Qayyum Khan to accept the poor terms offered. On 12 January, he and Yusuf Khattak were formally asked to resign from the Cabinet, thus leaving the PPP alone in the field against all the other political parties. This was a mistake. ZAB himself thus achieved what the Opposition had sought, the isolation of the PPP. He could have succeeded in going it alone only if, as Mubashir Hasan had advised, he had restored the PPP to its original character, instead of letting it operate as just another broad-based party.

Within the PPP itself there were serious differences over the allocation of tickets. The Prime Minister's screening process was conducted through non-Party officials, mainly intelligence and security agencies and his staff. It soon became evident that the flood of new entrants would be accommodated. He was by now convinced that only he mattered; he told Mubashir Hasan before the campaign, 'I don't need anyone in the Party, Doctor Sahib—you just arrange public meetings and I will speak'.[6] The award of Party tickets to newcomers continued to meet with resistance; in order to overcome this, ZAB designated the Appellate Board to act also as a review body. This decision of

15 January was merely cosmetic, with little change in ZAB's choice of candidates.

The award of tickets in 1977 involved an almost complete reversal of the PPP's 1970 position. The feudal class, which had been largely eliminated earlier, or at least made dependent on the Party, was now fully rehabilitated. Several individuals whom ZAB had attacked vehemently in the 1970 elections were accommodated both in Sindh and in the Punjab, where a person who had reputedly led an attack on his cavalcade near Multan was given a Provincial Assembly ticket. Earlier, Mubashir Hasan had decided not to contest, and a day before the tickets were finally awarded Mumtaz Bhutto left Islamabad in disgust. Party workers and supporters felt alienated and aggrieved, the adverse consequences of which became apparent after the elections.

These differences within the PPP were not publicized. Instead, praise of the Government's achievements continued unabated in the controlled media. On 15 January, a *White Paper on Kashmir,* painstakingly prepared by Yusuf Buch, chronicled ZAB's role in what was considered a vote-catching issue in the Punjab. With so much else happening, it elicited no interest, despite the whole day of official publicity which ZAB had ordered.

In contrast, and to ZAB's complete surprise, the PNA continued united, announcing, on 16 January, the appointment of Maulana Mufti Mahmud as its President, with Nawabzada Nasrullah Khan as Vice-President and the JUP's Rafiq Ahmed Bajwa as the Secretary-General. The Pir of Pagaro was accommodated as the head of the Central Parliamentary Board, with the other party leaders as its members. The PNA demonstrated solidarity with the imprisoned Baloch leaders and opposition to the military operation in Balochistan by deciding to boycott the election in that Province, mainly at the instance of Sherbaz Mazari of the NDP. This PNA leadership and organizational structure was an excellent arrangement which, together with the formation of the Alliance itself, proved to be the first important development in the election campaign.

In spite of his efforts, all the political parties, apart from Qayyum Khan's Muslim League, stood united against the PPP.

Rao Rashid, reporting on these developments, comforted ZAB by maintaining that Asghar Khan was finished, knowing full well that ZAB considered the retired Air Force Chief a threat. This pleased ZAB, who forwarded the file to me. I returned it with a note saying I disagreed and that we should discuss the matter in person.

The formal announcement of the names of the candidates for the National Assembly was made by the PPP on 17 January and by the PNA the following day. The next step was for the Election Commission to allot symbols for the contesting parties. Hafeez Pirzada, representing the PPP before the Commission, rightly opposed a single symbol for the PNA as its component units had retained their separate identities. Subsequently, for reasons to be explained shortly, a single symbol was permitted.

Three further developments of significance occurred at this time, which later had considerable impact. On 18 January, Indira Gandhi announced Indian elections for 24 March 1977, although she had only recently postponed them. Her personal defeat and the rout of her Congress Party, following soon after our 7 March elections, were in sharp contrast to what happened in Pakistan. Secondly, on 20 January, Jimmy Carter became President and ZAB lost his main contact in the American administration, Henry Kissinger, 'the person who understood' him. In his inaugural address, Carter pledged to seek a ban on nuclear arms. Of the greatest consequence, that very day, several newspapers banner headlined, 'The Undisputed Leader', 'The Supreme Leader' and 'The Great Leader', with large identical photographs of the Prime Minister. ZAB had been elected unopposed from Larkana.

Not only the Prime Minister, but five other PPP candidates in Sindh and three in Balochistan were returned without contest. In all, eighteen PPP candidates out of a National Assembly of two hundred enjoyed similar 'victories' because in seven more constituencies the nomination papers of their rivals were rejected. Asghar Khan declared he did 'not recognize the unopposed election' of ZAB.[7] The Opposition parties, who had little that was positive to offer so far to the electorate, were presented with a grand opportunity; they used it to the full in condemning ZAB at the first public meeting of their campaign on 23 January at Karachi. Predictably, the four Chief Ministers followed the Prime Minister by coming in unopposed to the Provincial Assemblies. The uncontested victories of these PPP candidates discredited the entire elections before the campaign was even under way.

I met ZAB on his return from Larkana and we talked into the early hours of the morning. I said his unopposed election was astonishing; no one could accept that the PNA candidate had simply failed to show up. The error was further compounded by the publicity given to the 'Undisputed Leader', as if it were a presidential election. He tetchily asked why, if I was surprised at his unopposed election, I did not inquire how my friend Mumtaz was similarly elected from Larkana: 'Surely I am better known than Mumtaz?' Later I did inquire, and my worst fears were confirmed.[8] At the time, I said a contest was necessary even if it involved some delay, because it was vital to establish the credibility of the Government and the elections. Apart from the delay, which he did not want, he felt that allowing a contest now would, in fact, entail an admission that his opponent, Maulana Jan Mohammad Abbasi, had been kidnapped, as the PNA claimed, and might implicate him. After discussing this in detail, he directed me to sort out the matter with the CEC.

We then discussed Rao Rashid's note on the PNA, and he inquired why I disagreed. His officials had failed to convey the truth of the situation, which he himself could not fully understand, namely, that the harsh treatment of the Opposition over a period of time had resulted in pent-up emotions and deep distrust and hatred of ZAB. They realized that only by combining could they combat the Prime Minister, who would otherwise continue to rule unchecked. Individually they would be finished; collectively they had their last chance to make a stand. ZAB was not prepared to accept that this formed the basis of the Opposition's actions and unity, and looked for other reasons.

We then turned to the leadership question. To me, Asghar Khan had acted wisely in acquiring an excellent platform within the PNA; if the campaign succeeded, he would gain credit, and if it failed he was not the leader to blame. Also, he had finally laid to rest accusations about his ambitions by not securing any office in the PNA. Although I did not say so at the time, Asghar Khan's lack of success in politics has always remained an enigma to me. He had all the right attributes, but perhaps he was too honest and blunt for Pakistan's wheeling-dealing politics; he had to rely for public support on Maulana Noorani's JUP. ZAB too must have had similar thoughts, because he wanted to ensure that Asghar Khan, who was contesting from four constituencies, should not successfully emulate his own performance in 1970. I said it made little difference whether Asghar Khan won more than one seat, but any interference ordered or allowed in any constituency would have serious consequences: 'manipulation of results has its own dynamics and cannot be controlled'. Unless ZAB ensured the credibility of the elections, he would be denied the fruits of his victory. He himself had created expectations of democracy which could no longer be thwarted.

On the leadership issue, he remarked that Asghar Khan did not have the political brains to think so far ahead. In that case, I replied, he must have been wisely advised. At this, he considered the possibility of an American role which, since his dismissal by Ayub Khan, had always worried ZAB. I said his intelligence agencies should ascertain the situation. However, as far as the US was concerned, it was clear they would not favour ZAB or any leader of a Muslim nation with nuclear ambitions; this should have been apparent to anyone, despite any assurance to the contrary. The best way to combat the Americans on this issue was through national consensus.

I referred to our earlier talk, when he had told me the election date and about his conversation with Byroade, at which time ZAB had refused to take the Opposition into confidence over the nuclear programme on the grounds that there were no clearly defined leaders. Now the PNA leadership was settled, I pointed out, so a national consensus could be achieved on this matter. ZAB, however, felt the Opposition would make unacceptable demands concerning the elections, and might make the nuclear question a campaign issue. I maintained that, on the contrary, the PNA would appreciate being taken into confidence. In the event, it was not raised during the elections; but neither was there any discussion of it with the Opposition.

Before leaving, I tried once again to talk about Maulana Abbasi being allowed to contest. ZAB said that I was an idealist, but 'our politics are not ideal'. He repeated what he had often said in the past, that no elections were ever fair in Pakistan: 'Do you think Ayub defeated Miss Jinnah fairly? Do you think it was this unfair election that finally overcame Ayub? No, it was Yahya and I. We even had a code arranged for the movement against Ayub, "Ceylon Tea Party", and that was how Ayub was toppled. That is the reality of politics in Pakistan.'[9] Rahim in his dark days had said this, Mubashir had hinted at it, Khar would smile at its mention, ZAB's opponents were of this view, but now I had heard it from him. Coming on top of his unopposed election, I had nothing more to say. He too realized that he had disclosed too much, and concluded our meeting, saying, 'We are both tired'. If, as his doctor reportedly conveyed to ZAB, I looked unwell as I left, I had every reason to be, and certainly felt it. Apart from anything else, our talk confirmed my decision to quit government after the elections.[10] Subsequently, ZAB gave an entirely different colour to our talk.[11] However, I have set it out as it occurred because of its importance, not only to the elections and the final outcome, but to me personally.

This talk did, however, save me from the unenviable task of tackling the CEC over the Larkana seat because, by the time I met Justice Sajjad Ahmed Jan, ZAB had decided to speak directly to him. The CEC was a considerate, judicious man, deeply worried about the spate of complaints following the unopposed elections. Previously, ZAB had dealt with him first through Sardar Tamman and then through Hafeez Pirzada, who was now busy in his own constituency. I had barely known the

CEC but, once he learnt I was leaving government, we became friendly over the next two months, and I supported his proposals to ZAB. He expressed regret about seeking an extension himself, but, for the present, we were both at the centre of grave events and trying to salvage some semblance of credibility for the elections. Like most other people, he accepted that anything which threatened the Prime Minister would affect the nation as a whole, including the Election Commission, and understood the consequences.

In dealing with the case against ZAB and other complaints, the CEC tried to show even-handedness. The Election Commission announced on 24 January that all complaints concerning uncontested seats should be submitted by 30 January and, to avoid any possibility of subsequent election petitions being filed in such cases, took the unprecedented step of issuing special notices calling for complaints before formally notifying the names of candidates declared unopposed. He accepted several appeals against the rejection of nomination papers, including that of Jan Mohammad Abbasi for NA-160 Nawabshah IV, but not for the Larkana seat, and those of Chaudhri Zahur Illahi and Haneef Ramay. Moreover, in order to 'balance' the favourable outcome in ZAB's case, the CEC obtained a concession on the issue of the 'Plough' symbol. On 27 January, ZAB amended the Election Rules to permit a common symbol for 'any political party or a combination of any two or more political parties who have agreed to put up joint candidates for elections'.[12]

As criticism mounted concerning these uncontested seats, he described as 'absolutely illogical' the charge that he had secured a few seats uncontested when he had called a nationwide election.[13] Finally, after duly processing Abbasi's petition, the Election Commission on 11 February upheld ZAB's unopposed election. It could not thereafter be challenged legally, but the real damage had already been done.

Meanwhile, the campaign was underway as ZAB had emphatically announced that the elections would not be postponed 'for any reasons'.[14] At the same time, on 24 January, he presented the PPP Election Manifesto before the assembled

Press. The Manifesto offered very little in the way of new promises because he had insisted on introducing his reforms and changes before the elections, seeking mainly to consolidate his position in the next five years. It contained three pages under the heading, 'What We Inherited', followed by twenty-nine pages of 'Promises Fulfilled' and then twenty-six pages under the caption, 'The Next Five Years'. To give it more punch, I had prepared a Foreword which set out its underlying policy: 'to carry forward the task we have undertaken, to build on our achievements'. It went on to state, 'This Manifesto is inspired by the idealism with which we launched the movement against dictatorship and the capitalist system; and balanced by realism derived from experience'.[15] It was with amusement that I read the comments of some newsmen that only the foreword showed the hand of the Prime Minister.

By now the PNA campaign had made progress, with the public appearing to favour the Alliance. They were impressed by its galaxy of leaders addressing large rallies and meetings, and Nasim Wali Khan, the wife of the incarcerated leader of the former NAP, emerged as a prominent new personality. The main thrust of the Alliance was the unfairness of the elections, for which ZAB had provided the ammunition. Asghar Khan declared that the PNA 'had already won', and no other outcome would be accepted.[16] At Gujranwala and other places in the Punjab, the public came out in sizeable numbers.

The campaign of the PPP was equally hard-hitting, with ZAB describing the PNA as a 'gang of nine aiming to restore exploitation' when he addressed a public meeting in Multan on 4 February. By mid-February, the PPP had gained the upper hand. On 20 February, his public meeting at Lahore was a resounding success. Representatives from all four Provinces spoke. I arranged for Aftab Sherpao, Hayat's younger brother, who had recently joined the Party on release from the army, to address this, his first major rally. Mubashir Hasan, Mumtaz Bhutto and I were talking together to the side of the main dais. I commented that this was a turning point—'God has given him another chance'. At the same time, we joked that, typically,

ZAB, even while speaking to the gathering, would be watching the three of us from the corner of his eye and wondering what we were brewing, so we decided to separate.

H. K. Burki, a pro-PPP journalist and political analyst, wrote that the tide had now turned.[17] Six days later he forecast that the PPP would win 130 seats.[18]

ZAB did not, by this time, want to hear the truth about the situation; he only welcomed favourable forecasts. While Mir Afzal Khan, ZAB and I were in Peshawar in the second week of February, we discussed the likely outcome, and he asked Mir Afzal and myself to predict the results in terms of percentages of seats before giving his views. The figures we gave were:

Province	Mir Afzal Khan	Rafi Raza	ZAB
NWFP	70%	55%	70%
Sindh	85%	70-75%	85%
Punjab	60%	55-60%	70-75%

Following the meeting, I asked Mir Afzal Khan why he had given such a high figure for the Frontier when just previously he had told me he expected 60-65%. He replied that there was little point in being accurate when it would only annoy and not have any beneficial effect. He was right. A few days later, in Islamabad, Yusuf Buch inquired whether anything had gone wrong between ZAB and me because, after our Peshawar meeting, ZAB had telephoned him to ask, 'What makes Rafi Raza think he knows everything about elections?' In Peshawar, ZAB had merely heard me without question.

I had prepared a fairly tough schedule of public meetings for ZAB, but the demands for his appearances increased the number of rallies. He gave his best, in no way stinting himself despite a full load as Chief Executive. He carried the main brunt of the PPP campaign, though Kausar Niazi also toured extensively. Mumtaz Bhutto was in demand in Sindh and, surprisingly, also in the Punjab. But what was missing in this campaign was the contribution of a large number of Partymen like Mairaj

Mohammad, Tariq Aziz, Hayat Sherpao, Aslam Gurdaspuri, Rasul Bux Talpur and Haneef Ramay, who had been inspiring speakers in 1970.

The last full-scale meeting to review the position before the elections took place in Lahore on 4 March. It was attended by senior PPP members, this time including Secretary-General Mubashir Hasan, civil and military intelligence agencies, and other officials concerned. We analysed the constituencies in the four Provinces. Everyone agreed that the PPP would win; the differences were over the extent of victory. A lone voice was that of Mubashir Hasan, who did not accept 'these bureaucratic calculations' made constituency-wise; he felt that the mood overall should be considered on a countrywide basis and that the PPP, despite its shortcomings, would win a significant victory because the Opposition had nothing to offer. Prior to the meeting, ZAB had considered security arrangements and whether we should announce on TV that the army would patrol in major cities like Lahore; I pointed out that 'no one believes that box any longer', and proposed a less obvious newspaper release.

The climax of the campaign came in the form of a large procession in Lahore led by ZAB, to be concluded with an address to the public. Before the rally I urged him, once again, to ensure fair balloting, particularly in view of disturbing rumours current in Lahore that the officials had been instructed to secure a PPP victory. For the first time, he readily agreed: 'I will call all the Commissioners in the Punjab and tell them personally that the elections must be absolutely fair.' When I returned to Islamabad, I met the Rawalpindi Commissioner at the airport; he had been summoned by the Prime Minister to Lahore.

Our next serious talk was on election day, well into the early hours of the morning. I had been in the 'Control Room' in the PM's Secretariat, from where we, in fact, controlled nothing but only received complaints and reports. Initially, the complaints were from PPP members—maybe they had assumed that everything would proceed smoothly for the Party in government.

By the afternoon, serious offences were being reported, especially from Karachi. The Prime Minister came to be briefed during the course of the day. Allegations and counter-allegations being part of elections in Pakistan, we did not pay them undue heed. In the evening we gathered for the results. General Tikka Khan, Yusuf Buch and a few others were present, and, while we were listening to the outcome, the Prime Minister joined us.

Results came unexpectedly early from some constituencies, while in urban seats the counting, or at least reporting of results, was unusually slow, which was later explained by the Election Commission. As time passed and mainly PPP victories were being announced, I became perturbed. In the Prime Minister's presence, I telephoned Syed Ijlal Haider Zaidi, then Director-General of Radio Pakistan, to inquire why such few PNA gains had been announced. He replied that the results were being broadcast in the order they came in, and there did not appear to be more PNA successes. Contrary to later allegations, the 'Control Room' in the PM's Secretariat did not receive the results before they were transmitted by Radio Pakistan.

When Zahur Illahi's defeat from his second constituency was announced I must have shown some surprise, because the Prime Minister asked me to accompany him to his house, which was about two hundred yards away. As we stood outside in the fresh air, he asked, 'Do you think I rigged those results?' I said the total rout of the PNA was very surprising. We walked slowly, talking in this vein. It was well after midnight. We went to his room, from where he called each Commissioner in the Punjab and asked what was happening, adding by way of confirmation, 'Did I not tell you to be impartial and fair?' In between, he called the Director, Intelligence Bureau, who, much to his annoyance, was asleep, and Benazir called from England, congratulating her parents. He also spoke to some other people including the DG, ISI.

It was very late by the time we walked back to the 'Control Room'. There were no more results in favour of the PNA. We should have celebrated, but instead there was silence. ZAB and

I then went to the room I kept in his Secretariat, where we had a long talk. I was worried by the margin of victory—136 for the PPP against 38 for the PNA, and all but eight of the 116 seats in the Punjab. He was very restrained and we talked quietly. After a while he sent for Yusuf Buch, but not General Tikka Khan or anyone else.

The victory had exceeded everyone's expectations. I mentioned three possible explanations. It could be a genuine, unanticipated landslide, in which case there was no problem. Secondly, there appeared to have been a five per cent swing in several seats, causing the error in our calculations. If this were neither the result of a landslide nor the handiwork of officialdom, we should consider whether any organized political party or foreign power was capable of manipulating such a swing. We concluded that no organization had the capacity to achieve this with such efficiency and secrecy. Thirdly, the ballot could have been rigged—in which case, I maintained, 'All hell will break loose.'[19] In the past, ZAB had always been dismissive about the consequences of unfair elections, but this time he was not, and we discussed the issue seriously. He was tired, drawn and clearly dejected when he left.

Although the final results became irrelevant, certain features should be underlined, some contradicting the allegations contained in General Zia's White Paper about rigging based on unduly high turn-out figures. With the heated, at times desperate, contest between the PPP and PNA, a high turnout was understandable which, indeed, made the 'uncontested' seats all the more questionable. The final results of the contested seats are given in the table on the next page.

The voter turnout in the 1977 elections, as compared to 1970, was marginally higher in the Punjab at 67% as against 63%, in Sindh at 62.5% as against about 60%, and in the Frontier there was a small increase to 47.4% from 46.8%, while in Balochistan there was a decline to 29.8% from 39% because of the PNA boycott in that Province. While the percentage of votes in favour of the PPP did increase in 1977, it must be borne in

RESULTS OF THE CONTESTED SEATS IN 1977

Province/Area	PPP	PNA	ML(Q)	Independents
The Punjab (including one Islamabad seat)	108	8	—	—
Sindh	17	11	—	—
NWFP	8	17	1	—
Balochistan	3	—	—	—
Tribal Areas (restricted franchise for tribals only)	—	—	—	8
Total	136	36	1	8

mind that there were very few candidates contesting as independents compared with 1970, because most important individuals had joined either the PPP or the PNA, particularly in the Punjab and Sindh.

While the high turnout can be explained in the other Provinces, it is not possible to account for, nor to accept, the figures for Balochistan following the effective PNA boycott. The Province-wise voting for the PPP and the other political parties in the 1970 and 1977 elections is given in the table on the next page.

Post-Election Events

The day following the elections, the Opposition parties declared they would not accept the 'rigged' results, demanding re-elections under neutral control and the resignation of both the Prime Minister and the CEC. They also announced a boycott

PROVINCE-WISE VOTING RESULTS IN 1977

Province	Party	% of votes 1970	1977
The Punjab	PPP	41.66	60.12
	Other parties*	35.38	34.75**
Sindh	PPP	44.95	63.00
	Other parties*	34.94	30.00**
NWFP	PPP	14.28	37.00
	Other parties*	78.70	40.00**
Balochistan	PPP	1.88	43.00
	Other parties*	48.38	33.00

* Excluding independent candidates
** Mainly the PNA in 1977

of the elections to the Provincial Assemblies scheduled for 10 March. They had seized several areas in Karachi on election day, and this was followed by disturbances in other cities. There was a marked absence of celebration or resistance to the Opposition on the part of the PPP.

ZAB had regained his composure by the time he informed a press conference in the evening that the results were as he had expected. He had met the American Ambassador in the morning.[20] The next day, he appealed publicly for a large turn-out for the elections to the Provincial Assemblies, emphasizing that a vote for the PPP was a vote for democracy, and would frustrate the PNA's boycott. He assured the people about their security on voting day, and that there would be no law and order problem. He had turned down my suggestion to postpone these polls until we had learnt what had happened in the general elections. He felt that this would be tantamount to admitting wrongdoing, and that a large vote on 10 March would provide the answer.

As it transpired, the turnout was so poor that at some voting centres there were more law-enforcement personnel present than voters. The military, who were patrolling, also witnessed this. In Rawalpindi, Partymen were genuinely aggrieved that they had not been consulted over the issue of tickets, and at the time of voting in the national elections they often found their votes already cast; now they were not prepared to come forward. Even TV cameras in search of long queues found embarrassingly few. To make matters worse, hugely inflated figures of the vote count were announced. That evening I reiterated to ZAB that I wished to be released from his Government, but he asked me to wait till the new Cabinet was sworn in.

Whoever manipulated the results of the Provincial elections, or advised ZAB that such figures were essential to confirm the voting pattern three days earlier, did a great disservice. When I made inquiries, no satisfactory answers were given. It would have been infinitely preferable to have given the correct figures, and to have explained the low turnout in terms of a 'walkover' resulting from the boycott and public concern about disturbances. The published figures confirmed the common perception that the general election results had been manipulated.

In our subsequent discussions, I maintained that re-elections were the only answer. Several Ministers opposed this view and, not surprisingly, those whose success was most questionable did not want fresh elections. In Mubashir Hasan's words, I had become 'a minority of one'. Only Lt.-Gen. Jilani, DG, ISI, put forward views similar to mine.

On 12 March, ZAB addressed the nation for eighty minutes on both radio and television, pointing out that he was ready for dialogue, and that 'everything' could be discussed and remedies found 'except for the National Assembly elections, which were a settled matter and would not be held again'. He also maintained that there was no foreign intervention.[21]

By now it had become apparent there were serious allegations of rigging in some forty constituencies, a figure largely confirmed by the PNA complaints; in Karachi, the PNA had also frustrated fair elections. The damage caused by the 'unopposed elections' had barely been tempered when these incidents virtually invalidated the entire elections. All the hard work that ZAB and several others among us had put into the preparations for the elections had gone for nothing because he had sought to be the 'Undisputed Leader', and had failed to ensure that the people would be the ultimate arbiter through the ballot box.

I made a final effort to convince him that it was necessary to secure a mandate which was universally accepted. This time he put forward specific objections. His concern was that, if the 7 March elections were considered void, a second election would curtail the duration of the constitutional safeguards for the Prime Minister relating to no-confidence motions. Also, he did not want 'to go down in history as a rigger of elections'. I assured him that the safeguards were up to ten years or the second election, 'whichever occurs later', under Article 96, and so remained unaffected. On his main worry, I said he would regain his former popularity if he told the public that he had ordered fair elections, but some candidates, including Federal Ministers, and officials had acted to the contrary, and he was consequently ordering fresh elections. One bold announcement would avoid protracted negotiations with the PNA and being gradually forced into concessions. After examining each aspect of my arguments, I felt he had almost agreed. However, when he went to Lahore soon after, he told the Press that he would not hold re-elections which would entail being called a 'rigger of elections'. The views of other advisers had prevailed.

More surprising to me was the report on 16 March that Bajwa had met the Prime Minister, and that he had been removed as Vice President of the JUP and Secretary-General of the PNA for doing so without the Alliance's authorization. This determined attitude on the part of the PNA seemed to have further firmed the Prime Minister's mind. Maulana Mufti Mahmood rejected the offer of a dialogue, and repeated the demand for the resignations of the Prime Minister and the CEC, calling for fresh elections under the supervision of the Judiciary and the army.

As ZAB had decided against re-elections, he felt that he should end the prevailing uncertainty, which only showed weakness on the part of the Government, by convening the National Assembly as soon as possible. He discussed this with Mir Afzal Khan and myself, and we were told to settle the matter with the CEC. The next day, 17 March, the Assembly was summoned to meet on 26 March. At the same time the CEC sought summary powers to deal speedily with cases where glaring malpractices had occurred. The Prime Minister, and a few Ministers whose seats would have been in jeopardy, argued this would be like a sword of Damocles hanging over PPP MNAs; the measure would only upset our supporters without satisfying the Opposition. After some discussion about the need to satisfy the public while ensuring that there would be no problems with the Larkana seat, an Ordinance was issued on 21 March which made a 'poll in any constituency' liable to be set aside by the Election Commission 'after such summary inquiry' as necessary for 'reason of grave illegalities or violation of the provisions of this Act or the Rules'.[22] 'Poll' meant that only contested elections could be challenged, not ZAB's 'unopposed' one.

Efforts at dialogue with the PNA continued with ZAB's reply on 19 March to Mufti Mahmood's letter of 17 March, which had demanded the Prime Minister's resignation and fresh elections. He repeated his offer of talks, but without pre-conditions, comparing his 'moderation and restraint' with the PNA's position of 'inciting violence', creating 'antipathy to the country's political stability' and dislocating the 'social and economic life' which 'caused prices to rise higher'. He pointed out that the PNA allegation that the 'countrywide preplanned rigging of general elections' had reduced them 'to a complete farce' had been contradicted by the votes received by the Opposition. He also justified the arrest of those who had 'flagrantly violated the law, incited people to violence, burnt and looted properties and killed innocent people'.

In the PPP camp we continued to discuss how the Prime Minister should deal with the PNA. Some maintained that all the leaders should be arrested and the movement crushed with a strong hand; thereafter, the Prime Minister could deal with the situation as he thought proper. Others, including myself, were against the arrest of those leaders who had not directly incited violence, because there would be no democracy if all the Opposition leaders were in jail. ZAB compromised, arresting some leaders but leaving the President of the PNA and others free.

Maulana Mufti Mahmood replied the following day that, since most of the leaders were in jail, their immediate release was essential to convene a meeting of the 'Alliance Heads' to consider the Prime Minister's letter. We discussed whether this should be permitted, but no clear policy emerged and 'ad hocism', a charge that ZAB hurled at his predecessors, prevailed. Most believed that the agitation would subside, and that clever ZAB would overcome the crisis and continue in charge. He had always had the reputation of being *kaan ka katcha,* and now his staff had a field-day. With my known impending withdrawal, the Prime Minister's main ministerial advisers were Hafeez Pirzada, who had become Finance Minister, Yahya Bakhtiar, Kausar Niazi and Sindh Chief Minister Jatoi. ZAB wavered between a political approach, strong administrative action, and then inaction. He was still in control but seemed to have lost his way.

The PNA leaders were released for their General Council meeting, after which they issued a long, scathing reply on 24 March, reiterating earlier accusations and rejecting the Prime Minister's claims, while insisting on his resignation and fresh elections. The letter was released to the Press before it reached the Prime Minister. ZAB was even more angered by a PNA letter written simultaneously to President Fazal Ellahi Chaudhry calling upon 'the Head of State...to order fresh elections to the National and Provincial Assemblies under an administration and through a machinery that ensures that these elections' be conducted 'as prescribed in the Constitution'.[23] In fact, the President neither could nor would act against the Prime Minister, but at least it was an attempt to follow a constitutional path, unlike the appeal to the Armed Forces by retired Air Marshal Asghar Khan: 'You should by now have realized that military action in East Pakistan was a conspiracy in which the present Prime Minister played a Machiavellian role.' He stressed that the elections had been blatantly rigged, and concluded with a call to arms: 'As men of honour it is your responsibility to do your duty, and the call of duty in these trying circumstances is not the blind obedience of unlawful commands. There comes a time in the lives of nations when each man has to ask himself whether he is doing the right thing. For you that time has come. Answer this call and save Pakistan. God be with you.'[24]

The military, however, did not respond to this appeal and the stalemate continued, with agitation against the Government and the arrest of many Opposition leaders and supporters. Maulana Kausar Niazi, knowing the rightist activists, maintained that either the Government's intelligence was very poor or there was deliberate sabotage, because the main agitators were not being arrested.

With the meeting of the National Assembly and the re-election of the Prime Minister, the Constitution provided for the termination of the previous Cabinet. I then formally wrote to ZAB requesting that I should not be included in the new Cabinet.[25] The night before it was to be sworn in, he invited to dinner those to be appointed Ministers. He asked me to come a little earlier, along with the Cabinet Secretary, and sought my views on the portfolios to be allocated. The Cabinet Secretary said it might not be wise to have a Sindhi as Finance Minister. but the Prime Minister replied that it would be acceptable as Aziz Ahmed, a Punjabi, was being promoted to Foreign Minister. I merely recommended that Farooq Leghari be made Production Minister,[26] instead of Minister for Tourism, and some reallocation of portfolios followed. I wanted to see my hard work in this Ministry bear fruit. I can find no explanation for the Prime Minister seeking my views on his new Cabinet at this stage, unless it was to indicate that the position remained open to me despite my earlier refusal. In the drawing room where the Ministers-to-be were assembled, Vaqar Ahmed read out the portfolios assigned. The Prime Minister said I had earlier asked not to be included for personal reasons, and thanked me for my hard work. Over the dinner that followed, the conversation was surprisingly light considering the gravity of the situation. The new Ministers were delighted with their assignments; indeed, many aspirants had earlier requested me to propose their names for appointment as Ministers. The only Minister who was not happy to continue in the Cabinet was Mir Afzal Khan, but he felt that the departure of a second Minister besides myself would be unacceptable to ZAB.[27] The swearing-in was held on 30 March.

The following day it was back to normal work with a Cabinet Sub-Committee meeting on pay for officers of the Government and in the public sector. On receiving a notice to attend, I telephoned the Cabinet Secretary pointing out it must have been sent in error, only to learn that this had been directed by the Prime Minister. At the meeting, Ghulam Ishaq Khan was sitting between Aziz Ahmed and myself. I congratulated Aziz Ahmed on his appointment as Foreign Minister, as did Ghulam Ishaq, who then turned to me and said, 'I think I should also congratulate you'. Such was the prestige of the Government. The one disturbing event that day was the summary setting aside of the election of Hafeezullah Cheema, who had been Railways Minister in the preceding PPP Cabinet, which alarmed some MNAs.

The Prime Minister next set about establishing PPP Governments in the Provinces; the four Assemblies were summoned on different dates so he could be present personally on the day each Chief Minister was elected, and he asked me to be in Peshawar and Lahore. We anticipated trouble, but nothing untoward occurred in Peshawar. The position in Lahore was different when the Assembly met on 9 April. We ensured extra security and, in the course of our talk about these measures, Aziz Ahmed, the new Foreign Minister, mentioned that General Zia had, without Foreign Office clearance, given a dinner for US Ambassador Byroade which was attended by several senior officials. ZAB was more concerned by this reference to what might have been considered questionable loyalty on the part of the Army Chief in front of those present at the meeting, especially in the context of the US, than he was by Zia's breach of rules.

The Punjab Assembly met in the midst of serious disturbances, even though the PNA efforts to break the security cordon did not succeed. There were disturbances also in other parts of the Punjab, including Gujranwala, Gujrat, Faisalabad, Sahiwal and Sargodha. The Government announced that 8 people had been killed and 77 injured, whereas according to military estimates, 30 had been killed and 250 wounded in Lahore alone.[28] In the prevailing hostile atmosphere, rumours of FSF and police brutalities spread. In any event, the fact of such strong resistance to ZAB in a Province where he had recently claimed an overwhelming victory confirmed the general impression that the election results were not correct. The troubles in Lahore not only gave new momentum to the PNA agitation but also unsettled ZAB. By this time, the demand for *Nizam-e-Mustafa* had come to the forefront and gained a following; the religious parties had taken over the agitation. The next twelve days were to witness particularly significant events.

The same evening ZAB called a meeting of several Federal Ministers, the Chief Minister and senior Party members of the Province. I was the only non-Punjabi, non-Minister present. Concern was writ large on everyone's face, and discussion centred on the 'Islamic' demands of the PNA, which had been triggered by the 'Westernization' introduced by the PPP Government, and its extravagance and corruption. In some ways the movement, although of short duration in Pakistan, was the precursor of similar developments in Iran and other parts of the Muslim World. It fed on general decadence and high inflation, which people felt were imported from the West. Even earlier, in the course of the campaign, it had given credence to the impossible call for the return of prices to pre-PPP 1971 levels; the public refused to accept that the main impetus for inflation came from the increase in the price of Middle East oil.

It was a curious meeting, but what followed was even stranger and more surprising. At the meeting itself, right-wing Party members such as Kausar Niazi argued that it was desirable to concede to the 'Islamic' demands, including prohibition, and those of the left like Sheikh Rashid agreed that this was necessary and inevitable. The Prime Minister sat silent, listening. He then got up, asked the others to wait, and invited me to accompany him. We went out onto the lawn of the Governor's House, where he talked astonishingly openly about his views.

He said we had worked very closely together and I understood his thinking. After listening to the others, he had realized that their perceptions were quite different. He did not want to end his days like Ayub Khan, who had at first staked his reputation on the Presidential and Basic Democracy system, but by the end had been ready to bargain it all away at the Round Table Conference, which achieved nothing. The concessions had proved futile, and Ayub's end had followed. ZAB had 'always stood for modern ideas' and to concede these principles now would make him 'no different from Ayub'. He would rather quit than accept what people like Maulana Maudoodi preached: 'The rightists can never be appeased; their demands will keep escalating and I know I could not accept them in the ultimate analysis'. It was 'axiomatic', his favourite expression, that there could be no agreement between the PPP and Jamaat-i-Islami on basic statecraft. He paused for my comments, but proceeded to soliloquize when nothing was forthcoming from me. For the first time he actually questioned whether politics in Pakistan was worthwhile; in return for almost killing himself with work, he and his family had received only abuse, and even his young daughter, Benazir, had not been spared. I was leaving so maybe he too should quit. This surprised me. Was he playing the devil's advocate again or did he want a straight reply? I said I was of no significance, and my departure was of little consequence; unlike me, politics meant everything to him. He again talked about quitting rather than compromising on the issue of *Nizam-e-Mustafa;* then he said, 'I can talk like this only to you', as if explaining why he was so talking. We discussed fresh elections, and this time he looked genuinely sad and said he was tired and could not readily face them again so soon. He regretted not having given more serious thought to my earlier advice, and so it went on. A certain embarrassment crept in as if, in return for this baring of his soul, I was expected to say something in return, like agreeing to continue in Government. But I had made up my mind, and merely said I would stay back to help in the case of elections, but not as a Minister. To avoid being drawn into the subject, I pointed out that the others should not be kept waiting, and we returned.

The gathering inside dispersed. There had been speculation about our talk, for we had been outside for a long time; several Ministers inquired whether the Prime Minister had asked me to return to Government. To their disbelief, I said no, but they did not appreciate that pride alone would have prevented such a direct request by the Prime Minister after my earlier refusal. When I told Mubashir Hasan about this talk, he was adamant that I should leave Pakistan: 'You are not in Government, your continued presence will only give rise to jealousy and problems.'

I was staying in the Fertilizer Corporation guest house, where Ministers including Mumtaz Bhutto, Mir Afzal and Kausar Niazi, and others like Yusuf Buch, would drop in to discuss the situation. The question of the Prime Minister resigning also came up in conversation. When I told Mumtaz about the Prime Minister mentioning this, he said the Prime Minister knew his own mind best and silence on my part was correct.[29] Indeed, it was difficult to visualize him quitting, convinced as he was that the destiny of Pakistan was linked with his own fate.

The Prime Minister remained in Lahore for over a week, and I was regularly summoned by him. The Opposition agitation did not cease despite many people being jailed. Some PPP members from the Punjab decided to leave the Party, and one MPA's house was attacked by the FSF to ensure no more resignations. It had become the PNA versus the Prime Minister.

I met Mubashir Hasan several times in Lahore, and on each occasion he expressed disgust with the state of affairs. He mentioned sending his resignation as Secretary-General on 28 March, but agreed first to have a straight talk with the Chairman before its publication. In the late afternoon of 11 April, a very disturbed Prime Minister told me that Mubashir had just spent several hours with him, and had attacked corruption among his relatives and servants, compelling him to ring up the chairman of a state corporation for confirmation. ZAB, with tears in his eyes, said such an attack was unwarranted. He described Mubashir's political solution: to oust the feudal landlords who had joined the Party; to get rid of the coterie of bureaucrats around him, and the FSF; to take to task the corrupt, however close they might be and whether Ministers, relatives, friends or staff; and not to concede any quarter to the advocates of *Nizam-e-Mustafa*. Above all, Mubashir had called for a revival of the original impetus and force of the Party to take on its opponents, saying that massive reforms would give the people a stake to fight for, and that a revolutionary change was needed to meet the crisis. ZAB narrated what he termed 'these unjust demands', pointing out that such radical changes could not be introduced at this critical juncture. He asked why all his old colleagues in the PPP had quit. There was not much I could say when he was already so disturbed except, tritely, that times had moved on from those early days—both he and the PPP had changed.

The next day, Mubashir Hasan confirmed his discussion with ZAB, providing me with specific allegations of corruption. He felt strongly that the Party, which was also his creation, had been betrayed and was in need of shock revival treatment. Earlier, on several occasions, he had suggested similar remedies, but his proposals had gone unheeded.[30] To my knowledge, this was the first time he had criticized the Chairman directly, face to face, since 'the crisis did not permit otherwise'. Previously he had put his objections in writing. He said he was going abroad and again urged me to leave, which I said I would do once things were more settled. He called me naive, and expounded on the feudal mind and how loyalty was one-way; feudals wanted people under them and I had always remained independent; they liked to be asked for, and to dispense, favours, and I should curb my independence and ask for a favour. When I replied that I had nothing to request, Mubashir shook his head, saying I still did not understand; the request was not to fill my need but ZAB's—he would be happy that, finally, I had asked for a favour. If I did not, ZAB's annoyance would only increase after my departure. As I later learnt, his advice was as sound as was his comprehension of ZAB's feudal mentality. Mubashir was and remains a man of integrity—one of the few leaders in the PPP who did not change or compromise in or out of office.

Of the original nucleus of the PPP, only Mumtaz Bhutto continued with ZAB, and that too for reasons of family. Mumtaz and I commented on how ZAB again turned to us at this time, even though I was not then a Minister, rather than to his security agencies—although, unfortunately, by then the mess had already been created. He seemed at times almost helpless. He had little new to say or offer, and appeared to be drifting without direction. He had lost his early PPP friends and associates; the public and Party, or what remained of it, did not rally to his support; and he must have realized the consequences of his own over-dependence on force. In fact, Mumtaz had during the past months predicted to me that the way ZAB was proceeding would result in a military *coup*.

Personal reminiscences, though important for a proper understanding of ZAB and events at the time, should not divert too much from the onrush of political developments. For instance, the news of Mubashir Hasan's resignation as Secretary-General of the PPP appeared in the Press, and Yahya Bakhtiar put forward a formula under which fresh elections to the Provincial Assemblies would be held and, in the event that the PNA won, general elections would follow. Over dinner, ZAB told us that the proposal did not have his prior approval, but he agreed to adopt it. To me, nothing short of early new elections were the answer, though most others refused or failed to see this.

While in Lahore, apart from discussing the 'Islamic' demands, we also considered the call for the revocation of the Proclamation of Emergency. The Attorney-General had sent the Prime Minister a detailed note on its legal and political implications, emphasizing that it would prove disastrous if the Emergency were lifted in the existing conditions. The Prime Minister sent me the note, with a request that I consider it seriously and give my views. After dealing with the various issues raised, I stated in my reply that:

> I had mentioned that one of the major factors for complaint during the past few years had been the general lack of *insaaf* (justice), and that an individual's *izzat* (honour/respect) is no longer safe. In a nutshell that there has been too much arbitrariness, and not enough fairness, on the part of Government. With this the Prime Minister agreed...if the political impasse is to be overcome then undoubtedly they (political detainees) will have to be released...
>
> As to the political implications, the...Attorney General has said that Pakistan did not need Emergency as much in 1975 and 1976 as it is needed today...However, the situation is also totally different. The post election events, particularly those since April 9th, have given rise to and require a whole new dimension in thinking.

There were reports of restiveness in the army and rumours of some resignations by Pakistani Ambassadors abroad, reminiscent of the Bengalis resigning in protest under Yahya Khan. On 16 April it was confirmed that Gul Hasan and Rahim Khan, two former Service Chiefs, had resigned as Ambassadors. Mir Afzal Khan and I returned together to Islamabad. On the way to the airport, we stopped at Mubashir's house to say goodbye, but he had already gone abroad. Then, later on 17 April, there was another surprise. In a national broadcast and TV appearance, ZAB announced that *Shariat* Law would be introduced in six months, and there was to be an immediate ban on alcohol, horse-racing and gambling. He had two days earlier, equally surprisingly, visited Maulana Maudoodi's house to secure support. Coming a week after our talk in Lahore, it was difficult to comprehend what compelled him to make this announcement. During the TV broadcast he lit a cigar and said with a smile, 'Fortunately, smoking cigars has not been banned'; only then did a flicker of vintage ZAB appear. At the same time, he also formally put forward Bakhtiar's plan; 'If the PNA won a majority of the aggregate of the four Assemblies', fresh polls would follow for the National Assembly.

ZAB had thereby tried to appease the religious parties, but without satisfying the movement against him. He had also accepted the principle of re-elections, first through the Provincial Assemblies, without addressing the basic demand for the rejection of the 7 March results and fresh elections.

Maulana Maudoodi maintained that the only solution lay in fresh elections. The Bakhtiar formula was rejected. Agitation continued. ZAB returned to Rawalpindi, and, when invited to call on him, I was surprised to be offered a drink despite the ban; he was sticking to red wine. He talked of the agitation. He said it would have to be put down finally by enforcing Martial Law in the three troubled cities of Karachi, Hyderabad and Lahore. I questioned the advisability and constitutionality of this—the army could aid civil power, but handing over three principal cities would mean that the political government had ceased to have control. With the army already in Balochistan, this could lead to a military take-over. ZAB felt he could maintain political supremacy over the army, but said he would further consider the matter.

I returned home dejected. Abdul Hafeez Memon, a friend and former law colleague, who was then a Sindh High Court Judge appointed to the Election Commission, came to see me. While we were talking about the situation, a crated gift arrived from the Prime Minister. Hafeez Memon was surprised; I had resigned and yet was being called to so many meetings, and now this gift had arrived. He too maintained that I would be coming back as a Minister; but I repeated that I was going to London by the end of the month.

Two days later, when I next met the Prime Minister on the question of 'limited Martial Law', as it came to be called, all issues including constitutional aspects appeared to have been settled already. The Attorney-General, Yahya Bakhtiar, and the Army Chief of the General Staff, Abdullah Malik, were the only others present, not General Ziaul Haq or General M. Shariff, the Chairman of the Joint Chiefs of Staff Committee. In the past ZAB had indicated favour and promotion by dealing directly with senior officers over the heads of their superiors, but this was not the time for such tactics.

Limited Martial Law was declared in three principal cities on 22 April. Soon after, reports came in of unease among the army officers performing Martial Law duties—seniors not wanting to pass, and juniors not wanting to obey, unpopular orders. These reports were fanned by the PNA. In order to avoid the impression that he was seeking a military solution, on 27 April ZAB announced the Government's continued interest in a political solution. At the same time, he got the Service Chiefs to issue a joint statement of loyalty: 'We wish to make it absolutely clear that Pakistan Army, Navy and Air Force are totally united to discharge their constitutional obligations and to fulfil the pledge their officers and men have to defend the country's independence even to the peril of their lives.'[31]

I have dwelt on various discussions with ZAB up to this period because they throw some light on his thinking at the time. It was also the end of my personal knowledge of what transpired, as at this point I made my farewell call before leaving for Karachi *en route* to London on 30 April 1977. It was a sombre meeting, ending a close relationship which had dominated critical years of my life. He asked me to meet his son and daughter in England and put them in the picture about what was happening: 'You need not disguise the facts, but put it gently, not in your usual manner.' We talked about the research I proposed to do in London on the struggle for Pakistan, and he said, 'Why write about the past when you have experienced historic events? Write about them.'[32] These were almost his last words to me.

Neither ZAB nor I realized this would be our last meeting, that his removal, trial for murder, and execution lay ahead. If we did not fully appreciate what was to happen, others in Government were blind to everything. Some Federal Ministers felt that at last ZAB had achieved his goal of a 'One-Party State', controlling the Federal Parliament and all four Provincial Assemblies, while dominating the civil bureaucracy and the Armed Forces.

Negotiations with the PNA and the 5 July *Coup*

The PNA remained united and continued its agitation, forcing ZAB to the negotiating table. The PPP team comprised the Prime Minister, Hafeez Pirzada and Kausar Niazi, while the PNA was represented by Maulana Mufti Mahmood, Nawabzada Nasrullah Khan and Ghafoor Ahmed. Some of the details were, from time to time, left to Hafeez Pirzada and Ghafoor Ahmed to determine. It is clear that no final accord emerged. Controversy centres on whether a basic agreement was actually reached, which General Zia's pre-emptive *coup* was designed to prevent being signed and implemented.

There are several versions of what transpired before midnight on 4 July. A member of the PPP negotiating team, Kausar Niazi, has given a firsthand account,[33] as indeed has Ghafoor Ahmed of the PNA.[34] An important member of General Zia's military junta, Lt.-Gen. Faiz Ali Chishti, has also written his version of the events.[35] Chishti was regarded as the strongman of the Zia regime and was actually in charge of 'Operation Fair Play', the code-name for the 5 July *coup d'état.* Moreover, apart from examining ZAB's account in various affidavits filed in Court and in *If I am Assassinated,* which was reportedly written by him,[36] I have sought to verify the position from others involved at the time from the PPP and the military[37] and, in particular, from Sardar Sherbaz Mazari,[38] who was a leading member of the PNA agitation and a hardliner against ZAB. Sherbaz Mazari remains a man of principle and integrity who, to my knowledge, has turned down high office; also, he knew ZAB personally over many years, long before the decade 1967-77.

From the various versions of events leading up to the night of 4 July, the following facts emerge. The PNA planned a 'long march' on Rawalpindi for 30 April. It was prevented by a massive force of police and the FSF, with the army in support positions. That day, for the first time since the elections, ZAB directly approached the public in Rawalpindi-Islamabad by driving around and displaying a letter from the US Secretary of State, Cyrus Vance, complaining that ZAB 'should have refrained from levelling open allegations as that can only lead to worsening of relations'. It was a successful day for him, but Lt.-Gen. Chishti subsequently made the significant comment that, 'If on 30 April the deployment ensured that nothing should get into the Prime Minister's house, it could be used in the reverse order to ensure that nothing moved out of the Prime Minister's house as well'.[39]

In the course of May, the representatives of some Muslim countries tried to get a dialogue going between the PNA and the Government, noteworthy among them being the Foreign Minister of the UAE, Ahmed Khalifa Alswedi, the Foreign Minister of Libya, Ali Al-Tariqi, and the Saudi Arabian Ambassador, Sheikh Riaz Al-Khateeb. The role of the latter was the most significant both because of Saudi Arabia's importance and the fact that the PNA felt he would be less pro-ZAB than the emissaries of Sheikh Zayed and President Gadhafi. Also in May and, increasingly, in June, the military top brass began to be invited frequently to Cabinet and other meetings.

On 4 May, the PNA presented its demands in the form of a detailed draft Accord prepared by its team of lawyers; it ranged from the dissolution of all five Assemblies, the holding of general elections on 7 October and elections to the Provincial Assemblies on 10 October, the recall of the army in Balochistan, the repeal of Amendments to the Constitution, and the termination of all Special Tribunals, to such minutiae as how the election results were to be announced.[40] The PNA had rightly rejected Ministerial-level talks, because only ZAB could make the important decisions. However, as it transpired, it was the PNA negotiating team's lack of authority which ultimately caused problems.

On 11 May, ZAB belatedly expressed willingness to negotiate directly. Such dramatic gestures as visiting Maulana Mufti Mahmood and Nawabzada Nasrullah Khan in jail at Sihala yielded only temporary gains, and were soon overtaken by ZAB's shift from promised negotiations to a referendum proposal, which he made before the National Assembly on 13 May.

Despite the fact that the Pir of Pagaro rejected the idea of a referendum the following day on behalf of the PNA, the Seventh Amendment was hurried through Parliament on 16 May enabling a referendum to be held to show confidence in ZAB, with elections to follow in the event of a negative vote. It was a non-starter, rejected both by the PNA and the army; it only created further doubts in the minds of the PNA leaders about his sincerity in the negotiating process. Again, on 26 May, through the efforts of the emissaries of Muslim countries, the PNA announced their readiness to resume negotiations. By this time the country had already suffered considerably; Attorney-General Bakhtiar later admitted before the Supreme Court in the course of the hearing on the validity of 'limited Martial Law' that, from mid-March to 25 May, 241 people were killed and 1,198 wounded, several members of the security forces were also killed and wounded, 1,622 vehicles were destroyed, and hundreds of buildings, including banks, cinemas, offices and shops, had been destroyed or damaged.

Negotiations started in earnest on 3 June, with the Government accepting such demands as the release of the Opposition leaders and an end to hostile propaganda against the PNA. Both sides agreed to suspend processions and rallies. On 6 June, detailed discussions commenced on whether to have fresh polls as proposed by the PNA or re-elections only for the disputed seats. When the first was accepted the following day, it was left to Ghafoor Ahmed and Hafeez Pirzada to work out the details. There were still major differences, including whether the polling should be within three months, as demanded by the PNA, or twelve, as proposed by the PPP. Within the PNA itself there were also disagreements; its negotiating team did not press for the Prime Minister's resignation, while those outside insisted on it.

Over the next few days deadlock ensued on the guarantees which the PNA required concerning the elections. By 11 June, not only were such hardliners as Asghar Khan speaking threateningly but even Ghafoor Ahmed expressed dissatisfaction with the lack of progress. As a result, on 12 June, the Government presented its own proposals to the PNA. The differences over the election date had been resolved, but Sherbaz Mazari, Nasim Wali Khan and Asghar Khan, all three considered hardliners, pressed two of their original demands, namely, the withdrawal of troops from Balochistan and the termination of the Hyderabad Tribunal.[41] On both issues, the army, by now deeply involved, proved reluctant to concede. ZAB wanted to meet Sherbaz Mazari and Nasim Wali Khan separately but they refused. Even earlier, on 13 April, shortly after the trouble over the Punjab Assembly meeting in Lahore, he had sent Mazhar Ali, an editor with leftist views, and his wife to meet Mazari in Sukkur jail, where Maulana Noorani and Ghafoor Ahmed were also interned. The 'offer' they brought from ZAB was for the NDP to form governments in Balochistan and the NWFP: 'Why are you attaching yourself to the beards,'[42] meaning the religious parties. Whether the motive was genuine or to divide the PNA is not clear, but shortly ZAB himself met Maulana Maudoodi, and on 17 April announced 'Islamization'.

Negotiations continued and, after the meeting of 15 June between the two teams, Ghafoor Ahmed and Kausar Niazi addressed a press conference in optimistic terms about an agreement. At the same time, however, Asghar Khan told the Press separately that any agreement was worthless unless all the details were settled; both he and Sherbaz Mazari insisted on ZAB agreeing to all the thirty-two original demands of the PNA.

By this time, there had been a reconciliation between ZAB and Khar, who became Special Assistant to the Prime Minister.

More surprisingly, after telling Mufti Mahmood that he was going to Larkana for a few days' rest, he left instead on 17 June for a five-nation tour of Muslim countries. Publicly he maintained that he was going to thank their leaders for their help; privately, he told close aides that he intended to secure finances for the forthcoming elections. His unexpected departure once again raised serious misgivings about his intentions in the minds of the PNA leaders, even among those who sought a settlement. In his absence, the question of constitutional safeguards for the proposed Accord, particularly concerning the Implementation Council, resulted in a deadlock. Ghafoor Ahmed said he would not negotiate further with Hafeez Pirzada, and the PNA would only talk directly to ZAB on his return. All involved at the time agreed that two factors caused the breakdown: the lack of trust in ZAB and the hawkish attitude of Hafeez Pirzada who, Lt.-Gen. Chishti maintained, 'seemed to be the deciding authority'.[43] By the time the PNA put these demands to ZAB, its stand on the constitutional status of the Implementation Council, which was to act on any accord, had become increasingly rigid. All through the night of 1 July, until dawn the next day, most of the points were thrashed out again between the two teams. However, when the PNA negotiating team reported back to its General Council, Asghar Khan vehemently opposed the draft, and he was supported by Sherbaz Mazari and Nasim Wali Khan.

According to Kausar Niazi,[44] Sardar Qayyum Khan warned him on 3 July that some PNA leaders were in contact with the Generals and asked him to get ZAB to sign immediately, but ZAB dismissed the warning. Meanwhile, the draft had been given for examination to the respective legal advisers of the two sides before they met again on 3 July. Lt.-Gen. Jilani, DG, ISI, had also urged him to sign without further delay.[45] After some private discussions between ZAB, Hafeez Pirzada and Kausar Niazi, the PPP team rejoined the meeting with the PNA leaders and informed them that further consultations were necessary regarding the constitutional status of the Implementation Council. The matter was then considered the following day at a

Cabinet meeting at which General Ziaul Haq was also present. That night, ZAB invited Mustafa Jatoi, Mumtaz Bhutto and Hafeez Pirzada for discussions. Pirzada still maintained there was no urgency, but ZAB held a press conference at midnight to announce that the agreement would be finalized the next day. Unfortunately, the next day was not to be on ZAB's terms: at 1.30 a.m. 'Operation Fair Play' was put into action by General Zia.

The sequence of events before the *coup* has been set out as accurately as possible, but the question of whether the *coup* was deliberately timed to prevent the agreement remains to be answered. ZAB's position was that it had been, though Kausar Niazi's account of the negotiations does not entirely endorse this view.[46] At the time of the *coup*, the PNA neither maintained that an agreement had been reached nor condemned General Zia for acting precipitately. Several of them readily joined General Zia's Advisory Council and later, after 23 August 1978, were inducted into his Cabinet. When military rule was subsequently extended, some of the leaders of the PNA claimed that there had indeed been an agreement prior to the *coup*. The one PNA leader who at no time cooperated with the military was Sherbaz Mazari, so greater reliance is placed on his view.

According to Sherbaz Mazari, who mainly supports Ghafoor Ahmed's version, ZAB tried to draw out the negotiations in the belief that the PNA's unity would crack, their movement would lose momentum, and they would be unable to restart the agitation. The PNA remained united mainly because of their distrust and dislike of ZAB, though there were differing views on the right course of action to take. Some wanted to conclude the negotiations; others felt it might be better to have a brief Martial Law followed by elections which would end ZAB's rule, especially in view of Asghar Khan's assurance that the army would hold elections within ninety days. On what basis Asghar Khan said this is difficult to comprehend, particularly as, after the military *coup*, he spent the next several years in detention. A little earlier, in June, Asghar Khan had produced a document, purportedly from the Intelligence Bureau, containing a 'hit-list' of PNA leaders to be assassinated, which only increased the resolve of Mazari and others to stand firm.

Sherbaz Mazari also maintains that, until the very end, the 'two main sticking points', namely, the withdrawal of the military to their barracks in Balochistan and the trial in ordinary courts of those arraigned before the Hyderabad Tribunal, remained unresolved. General Zia had rejected them, although after the *coup*, he accepted both positions. Whether it was Zia's plan to prevent a settlement, as ZAB later suggested, or the Prime Minister had put the General forward to object, as Mazari and others believed at the time, remains unclear. Moreover, according to Sherbaz Mazari, the 'interim constitutional arrangement' prepared by the team of PNA leaders to 'sanctify' the terms of the agreement did not materialize because the negotiating team 'never had the authority'. He explained that the situation was 'like in Ayub's RTC when the Opposition could not agree'. There were also suspicions within the PNA, as it had been reported that Mufti Mahmood and Nawabzada Nasrullah had met ZAB separately. As for ZAB's claim that he had agreed to all the terms by the night of 4 July, the PNA leaders neither heard the press conference which started at midnight, nor were they informed—'That is if he really did agree.' Sherbaz Mazari summed up the reasons for the failure of the negotiations thus: 'Total distrust of Bhutto. Tell me, did anyone trust him? In any case, how can you blame the PNA when the fault for the late decision, if any, was Bhutto's.'

Lt.-Gen. Chishti's account from the army viewpoint does not differ in any substantial manner. He maintains that the *coup* was delayed to allow the PPP-PNA negotiations to be finalized on 3 July; ZAB said his answer would be conveyed on 4 July; there was no communication throughout the day, and 'at five minutes past midnight' ZAB told the Press that he would 'meet' the PNA team the next day. By then General Zia had already ordered the execution of the planned *coup*. Chishti states categorically that, if ZAB had signed the Accord earlier, there would have been no military take-over, and puts the blame squarely on ZAB for prolonging the negotiations, despite prompting by the army.[47] He questions whether ZAB would finally have settled because, even earlier, on 15 June and 3 July, similar announcements of success had been made. He denies that General Zia was part of any conspiracy against the Prime Minister.

It is now necessary to consider ZAB's allegation of an 'international conspiracy', which he first mentioned in his speech to Parliament on 28 April: 'This is not a *desi* (indigenous) conspiracy, this is an international conspiracy...a huge colossal conspiracy against the Islamic State of Pakistan.'[48] Although he did not specify the US, there was little doubt at whom his accusation was directed. This raises two important questions: first, whether the US had a hand in the formation of the PNA, its campaign and agitation; and second, the contribution of the US to the *coup* itself. All aspects need examination, including cause and effect—was it caused by a foreign hand or did the effect produced by the PNA unity and agitation allow scope for outside mischief?

On the question of a foreign hand in the formation of the PNA, its creation and unity was, in fact, mainly the consequence of ZAB's own actions. The Opposition parties had been pushed to the wall: the NAP had been banned and its leaders jailed; other opponents had been prevented from propagating their views by the cancellation of licences for newspapers and magazines and by the control of newsprint and advertisements. An Emergency and Section 144 had been imposed most of the time throughout the country, so public meetings could not be held. The levers of power and patronage remained solely in the hands of the Prime Minister, who apparently intended to perpetuate his power through elections which the Opposition were certain would not be fair.

The PPP Government's record in by-elections justified this fear. All large cities had experienced unfair elections. In Lyallpur (now Faisalabad) there had been overkill by both disqualifying Mukhtar Rana and having his sister defeated in 1972. In Karachi, a PPP by-election victory had been severely criticized, and in Hyderabad the result was ultimately set aside. The defeat of

Mustafa Khar in Lahore's constituency no. 6 in 1975 became a byword for election rigging. The by-election in Quetta was not too different. On a larger scale, the elections in Azad Jammu and Kashmir in the summer of 1975 showed to all how a highly questionable landslide victory could be won. K. H. Khurshid, who had recently joined the PPP, pleaded with me to impress upon the Prime Minister that, with Sardar Qayyum Khan removed as President of Azad Jammu and Kashmir, victory was at hand without manipulation; but my speaking to ZAB was to no avail. At the time, he rejected these pleas because Khurshid 'had his own ambitions for a united Kashmir'. Not till the very end did he recognize the importance of fair elections in March 1977.

The Opposition parties also understood that, in the 1970 general elections, the PPP had not got a majority of the votes, even in Sindh and the Punjab. They felt that they could do better this time if they united, especially as the PPP had lost ground since 1970. The PNA in 1977 was neither the first nor the last instance of unity among diverse parties, whether in elections or agitation. In 1954 in East Pakistan, the Jugto Front was established to overwhelm the Muslim League Government in the Province; the Combined Opposition Parties (COP) was formed to challenge Ayub Khan in the 1964 presidential elections; and the agitation in 1968 and early 1969 saw all parties at work against Ayub Khan, in both East and West Pakistan. There had also been the earlier, though less successful, UDF, and then the PNA against ZAB. Subsequently, after the ZAB period, some PNA elements and the PPP combined in the 1983 Movement for the Restoration of Democracy (MRD) against the Zia regime, and various parties joined against the PPP in August 1990.

In fact, the formation of the PNA in January 1977 was not an overnight occurrence. Serious negotiations had been under way by the end of October 1976, but were unsuccessful at the time because Asghar Khan and Noorani demanded sixty per cent of the total seats. Sherbaz Mazari, who was Convener of the All Parties Conference called for this purpose, maintains that basic

arrangements had been settled well before the formal announcement of the PNA's formation. According to him, there were no
differences on the leadership issue when he put forward the
name of Mufti Mahmood; the other 'candidate' under consideration was the Pir of Pagaro, not Asghar Khan. ZAB's
intelligence agencies were clearly ill-informed.

There is no evidence to show that the PNA was established
with foreign backing. The component parties of the PNA had
each in their own way struggled for democracy in the past, and
the accusation that they were motivated from abroad is both
unjustified and unfair. The fact that their opposition resulted in
Martial Law is also not peculiar to 1977. In fact, it has always
been the case that neither the Government nor the Opposition
has known when to concede or call a halt, allowing the military
to intervene directly or indirectly. This happened with the
Qadiani troubles in Lahore and the introduction of limited
Martial Law in 1953, the dismissal of the Jugto Front Government in East Pakistan and the installation of Iskander Mirza as
Governor in 1954, the agitation against Ayub Khan resulting in
Yahya Khan's Martial Law in March 1969, with ZAB, and,
more recently, with the simultaneous 'resignations' of both
President Ghulam Ishaq Khan and Prime Minister Nawaz Sharif
in July 1993.

However, despite the assertions of Lt.-Gen. Chishti, there
were several indications that the army was in contact with some
PNA leaders. Quite apart from the certainty with which Asghar
Khan gave assurances on their behalf, there were persistent
rumours that some agencies had a hand in the troubles. Plainly,
there was merit in the argument that the army leadership had
not reconciled themselves to a secondary role in a civilian set-
up.

The army had enjoyed power, indirectly or directly, from
1954 until the PPP Government in December 1971. The Armed
Forces were not only involved in defence, which is inextricably
linked to foreign affairs, but were also concerned with internal
security and even planning and the budget, of which they
consumed a large percentage. Prior to 1972 there had been talk

of a constitutional role for the Armed Forces. While considering the proposed Constitution before the Accord of October 1972, ZAB and I discussed whether, for a period of five to ten years, we should have a Council for National Security to consider such issues as war and emergency, in the form of an expanded Council of Common Interests with the inclusion of the Service Chiefs. It would have involved the Provinces as well as the military in important decisions in a constitutional forum, thus avoiding an 'us and them' situation and, hopefully, any future Martial Law. After considering the idea, ZAB rejected it; when he raised the same proposal with General Zia after the post-election crisis, it was too late.

As we have already noted, ZAB involved the army in political issues from 1973 onwards. They were brought into Balochistan and Dir, and he introduced limited Martial Law in three principal cities in April 1977. He included the army leadership in political discussions and assessments, and in such matters as the proposed referendum. They witnessed the decline in his popularity and his growing dependence on them. They were, no doubt, also concerned about rumours of changes in their top echelons. By early July it was evident that General Zia should strike without the delay of even one day. Speed was essential; as Lt.-Gen. Chishti maintains, 'there were good chances of the leakage' of the *coup* plans, which would have proved fatal for General Zia.[49] However, one extra day would have settled the question which has been much debated ever since, namely, whether an accord could have been finalized on 5 July, thus obviating the need for a *coup*. The general situation in the country had remained reasonably calm since 26 May, and there was no necessity for immediate action other than concern that the plans for the *coup* might not remain secret. Further delay by General Zia might have meant a missed opportunity, and might also have allowed ZAB scope for further 'tricks'. Zia, like ZAB, had to act in the 'national interest', that much-abused phrase. Ironically, the same arguments advanced by Attorney-General Bakhtiar in May and June before the Lahore High Court and the Supreme Court to establish the role of the army were used to advantage by the military junta following the *coup*.

That General Ziaul Haq was a man with ambition and independent ideas became apparent from the manner in which he exercised power and perpetuated his rule for eleven years. His ability to promote himself was clearly underestimated by ZAB and most others from 1975 till 5 July 1977. However, whether he was US-inspired, as alleged by ZAB, is a matter that requires examination.

Lt.-Gen. Chishti, who was Ziaul Haq's close associate and colleague, suggests that 'it is possible that the CIA got hold of him when he was training in the USA'.[50] However, many senior officers of the Pakistan Army and other services received training and attended courses of varying length in the US. With most of our military hardware coming from the US until the 1960s, American influence on our Armed Forces was undeniable. It is a sad commentary on the political life of Pakistan that many leaders have had similar accusations levelled against them, including Governor-General Ghulam Mohammad, President Iskander Mirza, Prime Minister Mohammad Ali Bogra and Finance Minister M. Shoaib. Even ZAB was labelled a CIA agent by J. A. Rahim. Whatever the truth, it is commonly perceived that the US has always had an all-pervasive role in Pakistan. Certainly, in the case of all sudden changes of regimes which have occurred in Pakistan since 1953, the blessings of the US have been sought, even if they did not play a direct role. This was no doubt as true of General Zia as of other leaders, including ZAB himself.[51]

Foreign intervention is similar to corruption in one important respect—everyone is aware of its existence but, at the same time, evidence is not readily available. The accusation is invariably denied, ignored or not commented upon by the country concerned. However, there can be little doubt that the US wanted, if it did not encourage, the exit of ZAB. Even in the last days before his execution, when Mumtaz Bhutto and others approached various embassies to help save him, they ignored the US Embassy as they felt they could expect nothing from the

Americans.[52] The US did not exert any significant pressure on the Government to prevent his execution. Nevertheless, till the very end, ZAB and his family hoped for help from abroad, and concentrated their efforts on foreign support, particularly from the US.[53] The UK was no more helpful. In London, I took his son Mir Murtaza to meet a Foreign Office Minister, and subsequently called on Margaret Thatcher, then Leader of the Opposition. There was no effective response from the British Government.

While the interests of General Ziaul Haq and the US did converge in ZAB's ouster, just as the interests of the US coincided with those of Yahya Khan and ZAB in overthrowing Ayub Khan in 1969,[54] there is no evidence to support ZAB's allegation that General Zia was US-inspired. Nor is it supported by Zia's subsequent conduct of an independent foreign policy, particularly on the nuclear issue, which caused the US to cut off aid to Pakistan. The scene only changed after the Soviet invasion of Afghanistan late in December 1979.

Reverting to the question of the 'international conspiracy' itself, ZAB cited me as the person who initially brought it to his attention: 'The unity of the PNA was not a *desi* conspiracy. Rafi Raza was the first person to describe its foreign colours to me.' He was referring to our talk in late January, which I have already set out in detail.[55] In the course of this I had put forward my own understanding of the domestic and international situation. His account of our conversation bears no relation to what we actually discussed, and does a disservice to his own much-vaunted proficiency in foreign affairs. In previous Chapters, I have recalled his fears and concerns about US influence, and his policy of trying to appease, at every juncture, President Nixon, President Gerald Ford, Secretary of State Kissinger and others, including Ambassador Byroade.[56] His failure to realize that it was axiomatic for the US Administration in its own global interests to view with suspicion any independent-minded Third World leader, let alone one from a Muslim country with nuclear ambitions, belies his own intelligence.

Long before the end of January 1977, ZAB and I had discussed how the US might react to our nuclear programme; and if one were to accept the generally accepted version of his exchange with Henry Kissinger in August 1976, he should certainly have expected the worst.[57] He indulged in unnecessary rhetoric about the nuclear programme, in sharp contrast to all other countries, apart from the Nuclear Club, which either had the device or were much more advanced in their programmes than Pakistan, such as India and Israel. It is difficult to understand the reason for his rhetoric, especially when he was convinced that his own earlier removal as Foreign Minister had been at the behest of President Johnson, and in the light of what he himself called 'the Myth of Independence'. Did ZAB believe he could fool the US or carry them along with his aims? Why, as he claimed, did he tell Ambassador Byroade about the election date in late December 1976, informing him at the same time that he would convey his answer on the nuclear issue after the elections? Surely, if nothing else, his rhetoric offered an invitation for US pressure. Even taking into account his illusions of grandeur, there had to be some limits to self-delusion.

Nevertheless, one fact is beyond question: ZAB slipped badly over the elections, and he had no dearth of enemies to seize this opportunity and help bring about his downfall. Equally certainly, as should have been clear to any student of international relations, the US was inimical to his ambitions, and it would have been most unlikely for them to lose this opportunity to encourage, if not stir up, trouble for him. However, the extent of the role they actually played is difficult to determine because of the lack of direct evidence. In fact, two days before ZAB attacked the US, though without naming them, in the National Assembly on 28 April 1977, he held a meeting with the heads of his intelligence agencies, the Foreign Office and other officials, at which no one come forward with evidence against the US.[58]

In addition to ZAB's earlier errors, which have already been enumerated, real trouble for him burst into the open with the 1977 elections. The main justification put forward for the *coup* by his opponents was that he had lost the right to govern because of election rigging, on which the Zia regime issued a detailed White Paper. The purpose of the White Paper was to discredit ZAB, and to damage him in the course of his trial. It was wrong in many of its allegations, as also were ZAB's rebuttals.[59] It would, however, serve a useful purpose to consider some of its aspects.

The White Paper itself contained only flimsy charges, with little or no direct evidence against ZAB. It starts with the so-called 'Larkana Plan'. This was a poorly-written proposal which ZAB routinely forwarded to me, as he did many papers concerning the elections from February 1976 onwards. After glancing through it, I told him that it must have been prepared by some over-zealous bureaucrat,[60] as it would have involved almost all the available jeeps in Sindh being utilized in two constituencies. He laughed in agreement. If anyone cared to read this 'Plan' it would be abundantly clear that it was not written by him. I expressed these views at the time, during discussions in London about the White Paper with the journalist David Housego, who wrote in the *Financial Times* on 26 July 1978:

> One of the major pieces of evidence brought forward to demonstrate that Mr Bhutto intended massive rigging of the election is the so-called 'Larkana Plan'...He passed it to Mr Rafi Raza...
>
> But was the 'Larkana Plan' prepared *'personally'* by Mr Bhutto as the White Paper asserts, or was it one of those windy documents drawn up by a sycophantic official, to which Mr Bhutto foolishly put his name? The signature is certainly Mr Bhutto's but the language is not. Take for example...This is not the style of the arrogant Oxford-educated Mr Bhutto.
>
> Mr Rafi Raza...evidently did not take the document very seriously...
>
> Mr Bhutto created an atmosphere in which his officials felt they would gain favour by ensuring the overwhelming victory of his candidates irrespective of the means used.
>
> But the White Paper does not convincingly show that he ordered the rigging of the elections...

Among other canards, the White Paper referred to a 'Million Dollar Election Fund'. This, in fact, was non-existent, as the inquiry set up by the Zia regime itself confirmed. However, in their attempt to link this fund, which had purportedly been raised through the Ministry of Production, to the election campaign, of which I was formally in charge, the regime overlooked the fact that the two important documents on which reliance was placed for the establishment of this so-called fund, referred to at page 231 and at Annexures 241 and 242 of the White Paper, were of 30 August 1973 and 18 June 1974. I became Minister for Production in July 1974, and subsequently instituted an inquiry.[61]

The White Paper also incorrectly alleged that all election results were channelled and announced through the Prime Minister's election cell, which took appropriate decisions. In spite of having full access to all of ZAB's records, and those of the Election Commission, the Zia regime's extensive efforts produced little tangible evidence against him. Still, as with corruption and foreign intervention, it could be argued that actual evidence of election rigging is difficult to find. The subject thus needs to be looked at in other ways.

Let us start with the 'unopposed elections'. At a time when the PNA were determined to challenge ZAB, it is hard to understand how they could have permitted nearly ten per cent of the seats in the National Assembly to go uncontested in favour of the PPP. However, this itself did not spark off the agitation, though it was a major contributory factor to what happened on 7 March.

On that day, it is generally acknowledged, there were serious malpractices, some committed by the PNA. What needs to be considered is, first, whether the PNA would have accepted any results that showed ZAB as successful, the events of 7 March being just an excuse; and, second, whether the results were manipulated by him or were the outcome of other uncontrolled manipulation which had 'its own dynamics', as I had earlier asserted.

On the first point it must be stressed that, if the events of 7 March had merely been an excuse for the PNA action, the public would not have responded in the way it did to the

Opposition's call for agitation. After all, the people had not accepted the earlier claim by Asghar Khan that the elections had already been won by the PNA in view of the massive support shown at their rallies. The public mood was also greatly influenced by the almost complete absence of resistance to the agitation on the part of the PPP workers, who were deeply disillusioned. But for the rigging, there would have been no serious agitation to allow the army to intervene in July 1977.

The allegation that he manipulated the results gained support from his own past record, particularly in relation to the earlier by-elections and the unopposed elections. His failure to disassociate himself immediately from the malpractices of 7 March, and his insistence on proceeding with the farce of elections to the Provincial Assembly three days later, greatly strengthened the case of the Opposition. However, without any rigging and in the fairest of elections, it is not only my considered opinion, but that of others including several opponents, that ZAB would clearly have won the general elections.

It is generally believed that the main reason for the rigging was his desire for a two-thirds majority in order to amend the Constitution. His biographer, Stanley Wolpert, who was given access by ZAB's family to his papers, says, 'He wanted to be sure, however, of winning at least the two-thirds majority required for changing the Constitution to return Pakistan to a presidential system'. Wolpert also states that Leslie Wolf-Phillips of the London School of Economics and Political Science worked out 'secret codes' and flew to Rawalpindi in July to brief ZAB on his 'top secret' endeavours to change the Constitution.[62] This is not correct. Wolf-Phillips had come for two months in mid-1976 to examine constitutional issues; nothing developed along the lines suggested by Wolpert.

In my numerous discussions with ZAB concerning elections, not once did he give any indication of a desire to secure a two-thirds majority to change the Constitution. Six Constitutional Amendments had already been passed by his first Parliament, and he could at this time also have changed the parliamentary system or, for that matter, not agreed to it in 1972. But his reputation for seeking absolute power proved his undoing; not even his friends and Ministers have readily accepted what I have stated concerning this issue.

Post-*Coup* and the Courts

5 July was a black day for Pakistan. Whether ZAB failed, the Constitution was too inflexible, the Opposition was too intransigent or the military strike was pre-emptive, are questions that have been asked and not fully answered. Hopefully, this narrative serves to fill the gaps in that period and to answer several of the questions.

For the Quaid-i-Awam, or Leader of the People, it was a tragedy in personal terms. He was the only leader, whether in opposition or government, to have toured virtually every part of Pakistan. Yet, at the time of his ouster, hardly a voice was raised in his favour. Zia's unconstitutional *coup* was not condemned at home or abroad.

With the declaration of Martial Law, General Zia placed the PPP and PNA leaders under 'temporary protective custody'. ZAB was taken to Murree on 5 July and lodged at President House, where Zia met him twice. On his release, he returned to Rawalpindi on 29 July, from where he went to Larkana. During this period he swung from bluster to deep concern. General Zia had promised elections on 18 October, for which the filing of nominations commenced on 8 August. ZAB arrived that day in Lahore via Multan, the PPP having announced four days previously its readiness to participate in the elections.

ZAB once again turned to the public and they responded enthusiastically to his appeal in Lahore on 8 August. Mustafa Khar had earlier advised him not to attack Zia and the military,[63] but there was no holding him back—he was on the stage, looking into the eyes of the people and responding. The people of Lahore seemed to have rediscovered their leader or, more correctly, their leader had found his way back to them. Only ZAB was capable of achieving such astonishing swings in the public's perception of him.

General Zia had previously thought ZAB was finished. He was disturbed and angered by this resurgence of support and by ZAB's 'insulting language'. When they spoke on the telephone following the Lahore reception, Zia said he was not going to spare ZAB, who had embarrassed him, and whom he accused of being a murderer.[64] Several cases had been registered against ZAB, including the murder in 1972 of a Jamaat-i-Islami parliamentarian, Nazir Ahmad, and it became apparent that he would soon be detained. In the early hours of 3 September, he was arrested at his house in Karachi for the murder, on 10 November 1974, of Ahmed Raza Kasuri's father. When the Lahore High Court ten days later granted him bail, General Zia was taken aback. Three days afterwards, Zia had him rearrested, this time under Martial Law, which precluded his release.

ZAB was confronted with his nemesis, the Courts. They had the final say in two vital cases: the petition filed by Nusrat Bhutto challenging the validity of Martial Law, and his murder trial. The case and the trial itself have been commented upon in great detail both domestically and internationally, so little purpose is served by going over well-trodden ground. But this narrative must make some brief mention of certain important facts.

I have already noted how, in the Begum Nusrat Bhutto Case, the Supreme Court upheld the declaration of Martial Law, deciding that its earlier judgment in the Asma Jilani Case was not applicable. Moreover, the Supreme Court permitted General Ziaul Haq to amend the 1973 Constitution without any qualification, enabling him to perpetuate his rule.[65]

In the second case, the charge against ZAB was of conspiracy to murder Ahmed Raza Kasuri, an MNA. Members of the FSF had fired on Kasuri's car on the night of 10 November 1974, allegedly to kill him at ZAB's behest, but instead had killed Kasuri's father. The acting Chief Justice of the Lahore High Court, Justice Mushtaq Hussain, constituted a Bench of five Judges to hear the case, carefully omitting the Judge who had earlier granted ZAB bail. ZAB filed various petitions challenging the constitution of the Court and requesting the transfer of the case—he apprehended that he would not get a fair trial because of the bias of the acting Chief Justice. At one stage the Press and public were excluded, and he 'boycotted' the proceedings. On 2 March 1978 the Court unanimously sentenced him to death, and almost a year later, on 6 February 1979, the Supreme Court rejected his appeal with a 4-3 split decision. ZAB and others subsequently blamed his lawyers, particularly for protracting the hearing, as two Supreme Court Judges who were favourably inclined to ZAB had ceased to be on the Bench at the time of the judgment. The lawyers, in turn, maintained that the decision to make the case a political rather than a straight murder trial was taken by him. This proved to have been a wrong move, as did the decision that, in the appeal, no plea would be made for a reduction of sentence as that might have involved acceptance of the charge. He was similarly mistaken in personally addressing the Supreme Court for four days, and in admitting that he had received a full and fair hearing.

The question of a reduction of sentence was only argued in the subsequent Review Petition, when it was not allowed, even by Justice Dorab Patel, one of the three earlier dissenting Judges. However, Justice Patel helped to secure the inclusion of the following observation in the main decision of 24 March 1979:

> Although we have not found it possible in law to review the sentence of death on the grounds urged by Mr Yahya Bakhtiar, yet these are relevant for consideration by the executive authorities in the exercise of prerogative of mercy.[66]

The judgment of the Lahore High Court, on which the Supreme Court heard the Appeal, was very detailed. Yet there was no worthwhile corroboration of ZAB's actual role in the affair, as there was against the other accused. The main evidence was given by the former head of the FSF, Masood Mahmood, who had turned approver and become the principal witness for the State against him. Three Judges of the Supreme Court rejected the evidence, using such strong expressions as 'utterly unnatural and improbable', 'patently absurd', 'carefully tailored and pre-orchestrated', 'ridiculous', and 'tantamount to doing violence to basic human intelligence'. Under Islamic law, which does not recognize an approver, such evidence would in any event be worthless. Under the laws of Pakistan, largely based on Common Law, not only is its legality highly questionable, but its acceptance constitutes a danger for any innocent person.

Clearly there was considerable activity, both in the Court and outside. According to Lt.-Gen. Chishti, the confessions of the co-accused who had been found guilty with ZAB had been obtained by the inducement that they would not be hanged.[67] The three Supreme Court Judges, who had held him and one other accused person not guilty, had convicted the remaining three co-accused mainly on the basis of their 'voluntary confessions'. If indeed any inducement had been held out, then the confessions were not 'voluntary' and could not be used against these accused. It follows that, if the very foundation on which the superstructure of the case against ZAB was built, namely, the 'voluntary confessions', were removed, then indubitably there would be no case against him. On the other hand, General Khalid Mahmud Arif, who was General Zia's Chief of Staff at the time, has stated that Chishti's narration of events is not correct.[68]

Over the past year I have discussed the events of this period with Sharifuddin Pirzada, who was the Attorney-General at the time. He has, at my request, recently set out in a brief signed note (copy at Appendix B) some facts relating to the question of inducement and the mercy petition. It supports Chishti's version. It is important to appreciate that when General Zia telephoned Pirzada in the evening of 27 March, the General did not repudiate what Justice Mushtaq had conveyed to Sharifuddin Pirzada about the assurance held out to the co-accused that they would not be hanged. Instead, General Zia mentioned that the Supreme Court would issue a clarification concerning Justice Safdar Shah's statement to the media that the observations of the Supreme Court could not be disregarded by the executive in deciding the question of implementing the death sentence. On 29 March 1979, the Supreme Court took the unprecedented step of issuing a formal clarification to the effect that Justice Safdar Shah had reflected his personal views and not those of the other Judges—this, after the Court itself had ceased to function for the purposes of the case following its final decision on 24 March.

Sharifuddin Pirzada also suggested that General Zia had arranged to send him abroad, probably to avoid discussion on the issue. When the mercy petition was considered, Law Minister A. K. Brohi gave the opinion that General Zia undoubtedly wanted, as stated in Pirzada's note: 'Mercy, remission, or commutation is negation of justice, and justice is not only to be done to the killer who is surviving because of legal formalities, but is also to be done to the deceased who cannot be heard but whose soul looks for justice—the revenge—death for death—and that in fact is the humanitarian consideration.'

This dexterous management of the case against ZAB was almost inevitable in the light of the commonly held view that General Ziaul Haq foresaw one grave, which would contain either his or ZAB's body. He deeply feared ZAB and did not trust him alive, even in the exile abroad proposed by some countries. He rejected all mercy petitions from ZAB's family and supporters, and paid no heed to appeals from foreign leaders to spare his life on humanitarian or other considerations. In the early hours of 4 April 1979, ZAB was hanged. His execution took place within a few hundred yards of the official residence from where he had ruled the country, first as President and then as Prime Minister, for five-and-a-half years.

As for the elections promised by General Zia's 'Operation Fair Play', they were put off on 1 October 1977 pending the completion of the 'process of accountability', and several postponements followed. It was not until after Zia's death in August 1988 that the first proper party-based elections took place, in November 1988, under acting President Ghulam Ishaq Khan. These brought to power, as Prime Minister, ZAB's daughter, Benazir Bhutto.

The circumstances of ZAB's arrest, trial and execution, and the lack of popular agitation to save him, have created numerous accusations and counter-accusations even between his widow Nusrat and his cousin Mumtaz.[69] Wolpert refers to ZAB writing that his lawyer, Bakhtiar, 'had capitulated, broken down, and joined the conspiracy... But why is everyone...doing it...to dig my grave'.[70] This does injustice to Bakhtiar, who was loyal and devoted, and worked ceaselessly without remuneration. Many other such accusations were made by ZAB from jail. I believe the best explanation for this was given to me by our mutual friend, H.K. Burki: 'You must understand how lonely, isolated and humiliated Bhutto must have felt in jail; surely no one should accept all he said in those circumstances.'[71]

Despite his faults, in the final analysis he did not deserve the humiliation of his last days and his legally flawed execution. Critics have argued that they were the logical outcome of the violence and divisiveness of his period in office. In fact, the treatment meted out to him ignored his real achievements and, far from quashing dissent and ending the troubles facing the country, led to even greater division and turmoil. As I stated to the BBC on 5 April 1979:

> The execution of Mr Zulfikar Ali Bhutto is...a tragedy for Pakistan...(and)...will produce a trauma in the body politic which will drive a deep wedge into a country already sadly divided.
>
> ...Mr Bhutto is dead but his memory and inspiration will live on. He gave Pakistan political awareness. He gave hope and dignity to the poor and exploited. He put women on the path to emancipation. This can never be forgotten in Pakistan. Internationally, he gave a new direction to the Muslim and Third World, helping it assert its rightful place in the brotherhood of nations.
>
> His name will prove a rallying point for all those to whom the name of Pakistan is dear.

NOTES

1. Article 155 of the 1973 Constitution, clauses (1) and (6).
2. *The Pakistan Times,* Lahore, 3 January 1977.
3. Land Reforms Act, II of 1977, published in the *Gazette of Pakistan, Extraordinary,* Part I, of 9 January 1977.
4. General Zia subsequently aborted these reforms despite being an all-powerful Martial Law dictator.
5. Later, at ZAB's trial and in subsequent statements, Ahmed Raza Kasuri maintained that this was a tactical move for self-protection.
6. As narrated by Mubashir Hasan to the author.
7. *The Pakistan Times,* Lahore, 21 January 1977.
8. Mumtaz Bhutto narrated to the author the exact sequence of events; he said he would make it public at the appropriate time.
9. ZAB said the code was settled with Lt.-Gen. S.G.G.M. Peerzada. *See also,* Chapter 1, pp. 15-6, but the author remains of the view that ZAB did not frustrate the Round Table Conference called by Ayub Khan, as stated at p. 16.
10. The author had already informed ZAB of his intention to leave government after the elections. In October 1976 the author had accepted the offer of a Visiting Fellowship at the Centre for International Studies, the London School of Economics and Political Science, for the academic year 1977-8.
11. In *If I Am Assassinated,* p. 107, ZAB referred to our talk and alleged, in relation to the PNA plan, that, 'Rafi Raza showed me its blueprint and also dynamite with which to explode it'. I was not in possession of any such plan, so there was no question of showing it to ZAB or anyone else. Others have also suggested that I knew about the plan but refused to disclose it. In view of the treatment meted out to J. A. Rahim and others, it is unlikely that I would have been allowed to survive if I had refused to disclose information I admitted to possessing. The absurdity of such allegations is underlined by the fact that I continued to be consulted on all important issues until my departure for London at the end of April 1977. ZAB also reportedly claimed that I warned him in our talk that unless he gave up the nuclear reprocessing plant, it would be his undoing. This too is untrue. I talked about the nuclear issue only in relation to the need to discuss the matter with the Opposition in order to achieve a national consensus. My warning that ZAB might be denied the fruits of his victory related solely to the consequences of failing to ensure free and fair elections. I gave ZAB my own views, as I always did, and conveyed no message from anyone about the US role.

    Hafeez Pirzada's claim to have participated in any talk with ZAB and me on this subject, and about my not disclosing these matters to ZAB before the announcement of the elections, is false. He was not present at any time during these discussions nor, for that matter, did ZAB mention Pirzada's presence, although he refers to his doctor meeting me outside after this talk. My previous rebuttal of Pirzada's concoction is reproduced in *The Muslim,* Islamabad, 25 June 1985, *Daily News,* Karachi, 26 June 1985, and other newspapers of those dates.

12. *The Pakistan Times,* Lahore, 28 January 1977. Kausar Niazi, without knowing the real reason, said that it was 'the biggest flaw' to permit the PNA a common symbol, *Zulfiqar Ali Bhutto of Pakistan,* Vikas, New Delhi, 1992, pp. 65-6; however, he also refers to ZAB stating in the course of later negotiations with the PNA, 'Then I'll withdraw the symbol of the Plough', ibid., p. 222.

13. *The Pakistan Times,* Lahore, 3 February 1977.

14. *The Pakistan Times,* Lahore, 25 January 1977.

15. *Manifesto,* Pakistan People's Party, Khursheed Printers, Islamabad, January 1977, p. 4.

16. *The Pakistan Times,* Lahore, 11 February 1977.

17. *The Pakistan Times,* Lahore, 21 February 1977.

18. *The Pakistan Times,* Lahore, 27 February 1977.

19. Subsequently, others confirmed that my conviction concerning the adverse consequences was a major cause of ZAB's suspicions that I knew more than I was revealing: 'How could Rafi Raza have been so certain unless he knew?' ZAB failed to see that I was voicing the view of any sane political analyst—of whom there were sadly few around him at the end.

20. Stanley Wolpert narrates the impressions of the American Ambassador, p. 279.

21. *The Pakistan Times,* Lahore, 13 March 1977.

22. Representation of the People (Second Amendment) Ordinance, XV of 1977, published in the *Gazette of Pakistan, Extraordinary,* Part I, of 21 March 1977.

23. The full text of the PNA letter is set out in Ghafoor Ahmed's *Phir Martial Law Aa Gaya,* (Again There Came Martial Law), Jang Publishers, Lahore, 1988, p. 20 of the Appendix in English.

24. The full text can be seen in Kausar Niazi's *Zulfiqar Ali Bhutto of Pakistan,* pp. 227-30.

25. ZAB asked me to write a letter similar to the one I had sent him on 1 June 1973 giving 'personal reasons' for my resignation.

26. Sardar Farooq Ahmed Leghari became President of Pakistan in 1993.

27. Mir Afzal Khan mentioned his talk with the author in his letter to ZAB at the time he declined a PPP ticket to contest in General Zia's promised elections to be held in October 1977.

28. Lt.-Gen. Faiz Ahmad Chishti, *Betrayals of Another Kind—Islam, Democracy and The Army in Pakistan,* Tricolour Books, Delhi, 1989, p. 42.

29. A few weeks after I had left Pakistan at the end of April 1977, ZAB asked Mumtaz Bhutto about this 'meeting' in the guest house concerning his resignation. Who informed the Prime Minister, or whether the room was bugged, is not known.
30. *See also*, Chapter 8, p. 303-5 concerning Mubashir Hasan and the PPP. Lt.-Gen. M. G. Jilani told me later, in 1990, that ZAB had on several occasions, while thinking aloud when he was distressed, said that Mubashir Hasan might well have been right.
31. *The Pakistan Times,* Lahore, 28 April 1977.
32. I explained the problems regarding writing about my period in government. He gave me special permission, saying, 'You have my permission; you know all the facts, write about them'. Although I met Mir Murtaza, I was unable to see Benazir as in May 1977 I had to go to Iran, where my brother-in-law had just died.
33. Kausar Niazi, *Zulfiqar Ali Bhutto of Pakistan—The Last Days, see,* Note 12 above.
34. Ghafoor Ahmed, *Phir Martial Law Aa Gaya* (Again There Came Martial Law), *see,* Note 23 above.
35. Lt.-Gen. Faiz Ali Chishti, *Betrayals of Another Kind, see,* Note 28 above.
36. Zulfikar Ali Bhutto, *If I am Assassinated,* Vikas, New Delhi, 1979. This was reportedly based on an affidavit which ZAB submitted in his murder case before the Supreme Court, which did not accept it on the record. Some doubts have been cast on its authorship.
37. Among others, Mumtaz Bhutto, Mir Afzal Khan, Mustafa Jatoi, Mustafa Khar, Kausar Niazi, Lt.-Gen. Ghulam Jilani and General M. Shariff; several did not want me to name them.
38. The author had several discussions with Sardar Sherbaz Mazari, and showed him in draft form the relevant parts of this Chapter, specifically asking for comments on this section. He expressed agreement.
39. Chishti, op. cit., p. 64.
40. For the text of the draft, *see,* Kausar Niazi, op. cit., pp. 181-7.
41. *See,* Chapter 8, pp. 271-2 and p. 276 above on these two issues.
42. As narrated by Sardar Sherbaz Mazari to the author; similarly, with other references to Sherbaz Mazari in this section.
43. Chishti, op. cit. pp. 56, 58 and 60; also Kausar Niazi, op. cit., pp. 204-7, 208 and 240.
44. Kausar Niazi, op. cit. pp. 217-18.
45. Lt.-Gen. Ghulam Jilani confirmed to the author in 1990 that two days before the *coup* he said this to ZAB in the presence of Rao Rashid, then DIB.
46. There are, however, several inaccuracies in Kausar Niazi's account. When I pointed them out, he merely replied that he had praised me, and left the matter there.

47. Chishti, op. cit., pp. 61-3.

48. *If I am Assassinated,* p. 107, quoting from the *White Paper on the Conduct of the General Elections in March 1977.*

49. Chishti, op. cit., pp. 61-2.

50. Chishti, op. cit., p. 28; but it must be borne in mind that, after leaving Ziaul Haq's Government, Chishti became very critical of him.

51. *See,* Chapter 4, p. 135, and Chapter 7, p. 238.

52. As narrated to the author by Mumtaz Bhutto.

53. General Khalid Mahmud Arif, *Working with Zia—Pakistan's Power Politics 1977-88,* OUP, Karachi, 1995, at pp. 191-4 quotes Benazir Bhutto's messages to her brother who was abroad.

54. *See,* Chapter 1, pp. 13 and 16, and p. 327.

55. *See,* pp. 325-7. ZAB's quote is at p. 107 of *If I am Assassinated.*

56. *See,* for examples, Chapter 4, pp. 127 and 135-6 and Chapter 5, pp. 147-8, Chapter 7, pp. 236-45, and Chapter 9, p. 318.

57. *See,* Chapter 7, p. 243.

58. As confirmed to the author in 1995 by Agha Shahi, the Foreign Secretary in 1977.

59. *The White Paper on the Conduct of the General Elections in March 1977,* published by GOP, Rawalpindi, July 1978; ZAB's rebuttal is contained, *inter ali,* in *If I am Assassinated.* When Mir Murtaza Bhutto organized a Jurists' Convention in London in 1979, he asked the author to sit on the main panel. In the presence of Mustafa Khar, I told Mir Murtaza that I did not want to have to contradict ZAB's affidavits lest it jeopardize his life; I would instead be in the audience, and would clarify or put forward appropriate points, which I did. ZAB was executed the day before the Convention was held.

60. The author still believes this was the case but, in order to protect the person concerned, ZAB wrote in *If I am Assassinated,* p. 69, that it was prepared by a 'devoted follower'.

61. *White Paper,* op. cit., p. 231 and Annexure 244 at p. A-701.

62. Wolpert, op. cit., p. 267.

63. Mustafa Khar told the author this in 1978.

64. Mumtaz Bhutto conveyed this to the author; Zia had felt insulted because his friend, Crown Prince Hasan of Jordan, was his guest in Pakistan at the time.

65. For the Asma Jilani Case and the Begum Nusrat Bhutto Case, *see,* Chapter 5, pp. 156-8, and Chapter 6, pp. 184-5. If the Chief Justice, S. Anwarul Haq, had not allowed this power to General Ziaul Haq, Justice Yaqub Ali could well have been reinducted as Chief Justice, a post he would have continued to hold by virtue of earlier Constitutional Amendments under ZAB, some of which General Zia had done away with.

66. Zulfikar Ali Bhutto v. State, PLD 1979 SC 741 at p. 812. Justice Dorab Patel has since narrated to the author the details of what transpired but has not authorized me to give details. These and other facts may be disclosed by him.

67. Chishti, op. cit., pp. 80-1, states that these accused were assured they would not be hanged. It has also been reported that they were told that they would be released after a few years.

68. Arif, op. cit., pp. 211-2.

69. Mumtaz Bhutto has given the author a detailed account of what happened between the time he was released from jail on 4 February 1979 and ZAB's execution. This he intends to write about in due course and, at his request, I have refrained from including this material in my narrative. *See also,* Note 8 above.

70. Wolpert, op. cit., p. 259.

71. H. K. Burki first stated this to the author in 1988, and repeated it thereafter.

CHAPTER 10

WITH HINDSIGHT

When ZAB came to power on 20 December 1971, he was supported by virtually the whole of what is now Pakistan. Everyone, including his opponents, wanted him to succeed, because this meant the survival of the dismembered country. Yet, five and a half years later, he was removed from office and spent most of his last months in solitary confinement before being executed on 4 April 1979. How and why did this happen?

From jail, he expounded at length on the reasons behind the *coup d'état* of 5 July 1977. As he saw it, everyone else was to blame; he was the lone warrior without whom national integrity and independence, including the pursuit of our nuclear programme, would have been jeopardized. Coming from a death cell, where he was fighting for his life, this self-portrait is to some extent understandable. On the other hand, his detractors were equally biased, and conceded him no virtues. One fact is undoubted: he was throughout the subject of controversy, whether over the fall of Ayub Khan and, later, East Pakistan, or the circumstances surrounding his ouster as Prime Minister, and his subsequent trial and execution.

His achievements in the period from 1967 were undeniably considerable. The creation of the PPP and, within three years, a massive electoral victory in 1970 remain unparalleled. Without ZAB, the PPP might not have come into existence, and would certainly not have become a dominant force in the politics of Pakistan to this day. He became the spokesman for the poor people of Pakistan. After the country's defeat in December 1971, he picked up the pieces and successfully rehabilitated the nation through such historic landmarks as the Simla Summit in 1972, the permanent Constitution in 1973 and the Islamic Summit in February 1974. Seen in their entirety and within the framework of their time, his earlier reforms were salutary; he helped to spread the wealth of the nation throughout the country and, most significantly, freed the poor from submission to the 'abominable status quo', another of his favourite expressions. He changed many facets of national life, mainly for the better. Introducing such reforms inevitably involved making numerous enemies; but did these enemies alone account for ZAB's sudden fall?

Let us start with his own explanations. He accused the Opposition, General Ziaul Haq and the United States of conspiring against him. We have already examined the causes that led to the unity of the Opposition parties in 1977, and their sustained agitation. Even apart from the intense hatred engendered by ZAB's own errors and personality, we have also seen that such unity within the ranks of otherwise disparate opposition parties in Pakistan was not unique in 1977, nor was the resultant military intervention. The allegations of collusion by the PNA with the army and the US have not been substantiated, nor have they been sustained by ZAB's own daughter, Benazir, who has since sought alliances with the same Opposition leaders.

We have already expanded on the role of the army in Pakistan, and there can now be no doubt about General Zia's ambitions. But it was ZAB himself who presented General Zia with an opportunity to step in when he failed, first, to ensure fair elections and, thereafter, to satisfy the Opposition. The fact that ZAB's presence no longer suited the US, and that they wanted his early exit, is also reasonably apparent. His failure to anticipate this, despite early warnings in the Press and by advisers such as myself, is difficult to understand. The US would have similarly appreciated the premature end of other Third World and Muslim leaders such as Fidel Castro, Ayatollah Khomeini and President Gadhafi, against whom they launched one form or another of physical aggression, including bombing

Gadhafi's home. However, these leaders survived because they at all times enjoyed strong domestic support.

His allegations that his colleagues were responsible for his downfall are unfair. No one can claim that those around him were without shortcomings. They had their fair share of faults, including those I have pin-pointed in him. But though many were undoubtedly unhappy with him and his behaviour, none deliberately contributed to the *coup* of 5 July 1977 or directly benefited from his fall. Several were jailed or lived in exile. Some did, however, later succumb to pressure or joined General Zia, and, if one were to accept Lt.-Gen. Chishti's version, advised against the return of ZAB to power. If any individuals let him down in July 1977, they were his own hand-picked people who, like General Ziaul Haq, were promoted over their seniors, or, like Said Ahmed Khan and Masood Mahmood, were appointed against the advice of his colleagues.

In *If I am Assassinated*, ZAB also suggested another reason for his ouster: 'I was to embark on a massive re-organization and reform programme on the strength of my fresh mandate after the elections of March 1977'.[1] In fact, no such programme was envisaged in the 1977 Manifesto, which called instead for a period of consolidation: 'In the next five years, we will build upon the base we have established, and consolidate the results of our endeavours.'[2] What ZAB viewed as consolidation, the Opposition feared would entail the strengthening of his stranglehold on all power.

In the same context, while making a comparison with President Nixon's fall, ZAB pointed out that Islamabad had 'four major power blocs: (a) The Military, (b) The Bureaucracy, (c) Big Business, and (d) The Politicians'. Unfortunately, he antagonized each of them and enabled them to capitalize on his weakened position, which subsequently led to his fall from power. This list, if anything, was too short; by 1977 he had alienated almost all other important groups both at home and abroad.

Since his explanations are partial, we must look deeper into the real reasons for his downfall. His fall can best be described in terms of Shakespearean tragedy: a man of distinction brought low and finally undone by his personal flaws, which caused or compounded his political errors and misjudgements. Behind his flamboyance and confident facade lay a deep insecurity, probably emanating from his early years. He was full of complexes and contradictions: a feudal with socialist ideas, an educated man whose language was demotic, a populist with aristocratic and autocratic attitudes who sought popular acclaim and adulation yet viewed individual human beings with regrettable contempt. He had a great belief in his own capacity, though at times he showed a surprising inability to see the reality of a situation. He was described by the Lahore High Court in 1978 as a 'compulsive liar', though he rejected this accusation. In reality, truth to him was what he wanted it to be at the time. This failing is not uncommon among politicians but, like much else, he carried it to excess.

His insecurity drove him unrelentingly to prove that he was better by trying to outdo and outperform everyone. It was partly the cause of his restless nature and his ceaseless activity, which I once described—to his annoyance—as at times resembling a top spinning on the same spot without making actual progress. It resulted in his trusting no individual or institution, making everything other than himself dispensible, including his own achievements and creations.

As a result, several factors had converged by 1977 to bring about his fall. Even his friends and colleagues accept that, in the end, with tragic consequences for himself and the country, he 'blew it'. He had deprived himself of the sheet-anchor of the 1973 Constitution and the protection of the Superior Courts, because of his insecurity and inability to accept any check on his authority. Eventually they were of no avail to him against Martial Law in 1977. His efforts to out-perform the left through his later nationalization measures turned the entire *bazaar* against him. His attempt to outdo the rightist parties boomeranged on him in the form of the demand for *Nizam-e-Mustafa* in 1977. Moreover, in trying to mix the left with the right, he fell short on both fronts. He also displayed no confidence in either the parliamentary system or fair elections, which resulted in open hostility from the Opposition.

He placed little trust in his Party and colleagues, sub-ordinating them to officialdom, and to security and intelligence agencies. Having surpassed both previous Presidents, Ayub Khan and Yahya Khan, by dismissing 1300 civil servants and introducing administrative reforms not to the liking of the superior services, he failed to realize that their support of him was at best tenuous. He had alienated the intelligentsia and students by his high-handedness and unwillingness to accept criticism. Even the peasant and urban worker, his real support base, suffered, in the absence of a proper Party, from bureau-cratic and political excesses. As a result, there was nothing to counter the agitation launched by his adversaries in March 1977.

At the same time he reactivated other forces, both domestic and international, that could work against him. To reinforce his power and position as military supremo, he brought the army back fully into politics. In the international arena, the role he chose was beyond his power to play successfully at the time. Not content with being Co-Chairman of the Islamic Conference, and at the same time nursing the nuclear programme, his desire to outperform all others with his rhetoric on the 'Islamic bomb', and to claim the leadership of the Third World, created enemies and lost friends in the comity of nations.

When he was removed the Himalayas did not weep as he had expected. The sad fact was that by 1977 his greatest talent seemed to be one for self-destruction. It is a phenomenon that repeatedly overtakes those in power, particularly in Islamabad. It had happened before ZAB and has occurred since, but, in his case, his intelligence and past experience, which could have helped him avoid these mistakes, were undermined by his insecurity. In any place and society, power and achievement can be a distancing experience, justifying the sense of always being right, of acting in the 'national interest'. This mentality was, and remains, fatal in the environment of Pakistan, which permits little freedom to opposition parties and the Press, and particularly in Islamabad, cut off as it is from the realities of the rest of the country.

Several of us had been aware of ZAB's shortcomings, even when we joined him in establishing the PPP, but we felt that his period in the wilderness following his exit as Foreign Minister, the camaraderie we had developed in the early PPP years and, above all, his own success, would help him overcome them. We recognized his intelligence and powers of leadership, and hoped that our ideas would materialize through the PPP.

Apart from his personal shortcomings, his career was determined by one further element in his political outlook—his belief in the inevitable role of the army and the use of force. In part, his early career had taught him this. Working with Ayub Khan for eight years, and then associating himself with Yahya Khan in Ayub's ouster in 1969, had ingrained in him a belief in the necessity of giving the army a role in domestic affairs. For him, the PPP was never a sufficient power base. In contending with Mujibur Rahman two years later, he again decided to accept the military path. After the defeat in December 1971, the army could have been confined to its real function, the defence of the country. However, instead of seeking democratic and political solutions to the problems in Balochistan and with the Opposition both before and during 1977, he followed the course of action with which he was by then most familiar, the use of force. Ironically, having assumed power as a democratically-elected leader after thirteen years of Martial Law and semi-military rule, he merely perpetuated the role of the army and ushered in a further eight years of Martial Law.

There are several further lessons to be drawn from ZAB's rule. Most are simple truisms which he did not accept and other leaders in Pakistan would do well to remember. No individual, however charismatic or brilliant, is a substitute for institutions. No democracy can succeed without trust in institutions, tolerance, and acceptance of opposition parties and fair criticism from a free Press and the public. Without such elementary checks there can be no real improvement in the country. Governments do not always know what is best, as they claim, in 'the national interest'. There can be no good government unless it is open and transparent to the people. In Pakistan we have had a history of Inquiries and Commissions into such vitally important issues as the assassination of the first Prime Minister and the dismemberment of the country without the public ever being informed of their findings. White Papers are only issued when they suit those in power, and often merely result in witch-hunts and the pursuit of political rivals under the guise of accountability.

Pakistan as a nation, already truncated in 1971, is coming up to its fiftieth birthday, yet we have no accurate account of what has transpired in these fifty years. We have indulged in catch-phrases such as 'Pakistan Ideology' and 'Islamic Ideology' without any attempt to give them serious content. This should be remedied by those interested in the future of the nation. Concern for national identity should be accompanied by genuine practical programmes to solve the problems of the vast majority of Pakistanis who remain without safe drinking water, sanitation, housing, health and education facilities and employment opportunities. Ideological claims alone, as we have seen in Pakistan in 1971, and more recently in the Soviet Union, cannot hold a country together. Only through concrete measures can we combat the growing cynicism and suspicion of democracy, if not its outright rejection, which have overtaken the country following the ZAB era, General Zia's rule and subsequent governments.

Our leaders should also learn one more lesson which, in my view, explains the phenomenon of why ZAB's name still lives on. He is best remembered by the poor people of Pakistan for his efforts—'at least trying'—on their behalf. Pakistan has truly become a country of public poverty and private affluence, with seemingly uncaring leaders. Because he tried to improve the lot of the common man, every effort has failed to destroy 'Bhuttoism', an unfortunate term reminiscent of Peronism and used both by his supporters and detractors. Eleven years after his ouster, his daughter, Benazir, became Prime Minister on the strength of his name. In fact, his support resurged within five weeks of his overthrow, when he went back to the people, who welcomed the return of their leader without the trappings of Prime Ministerial office, forcing General Ziaul Haq to postpone elections indefinitely and execute ZAB. The people remembered his efforts, the man who gave them a voice. They forgot and forgave his personal flaws. If they had short memories, so did ZAB who, within a few years of assuming office, forgot the power of the people which the PPP had helped to galvanize— his only real source of power.

I have here set out the facts as a witness to these events, together with my views. While we may have to wait for history to give its ultimate verdict, many in Pakistan have already pronounced theirs; not even the hangman's noose could deprive ZAB of his real distinction, the cry of the poor that *'Bhutto zinda hai!'* (Bhutto lives!) A more general epitaph for my narrative can be drawn from the Holy Quran:

Lo! herein verily is a lesson for those who have eyes.[3]

NOTES

1. *If I am Assassinated* p. 61.
2. *Manifesto,* Chapter III, 'The Next Five Years', under the caption 'Looking Ahead', p. 39. In fact, V.A. Jafarey, who was the Planning Secretary at the time, was specifically directed by ZAB to provide the author with material from the GOP's proposed Five Year Plan (1978-83) to ensure that Chapter III of the PPP Manifesto should not promise anything in excess of it. *See also,* Chapter 9, pp. 328-9.
3. *Sura III—Aley Imran: Ayat 13* (Chapter III—The Family of Imran: Verse 13.)

APPENDIX A

TEXTS OF THE SIX-POINT FORMULA AS ORIGINALLY PUBLISHED AND SUBSEQUENTLY AMENDED IN THE AWAMI LEAGUE'S MANIFESTO

Point No. 1

Original: The Constitution should provide for a Federation of Pakistan in its true sense on the basis of the Lahore Resolution, and Parliamentary form of Government with supremacy of Legislature directly elected on the basis of universal adult franchise.

Amended: The character of the government shall be federal and parliamentary, in which the election to the federal legislature and to the legislatures of the federating units shall be direct and on the basis of universal adult franchise. The representation in the federal legislature shall be on the basis of population.

Point No. 2

Original: Federal Government shall deal with only two subjects, viz: Defence and Foreign Affairs, and all other residuary subjects shall vest in the federating states.

Amended: The federal government shall be responsible only for defence and foreign affairs and, subject to the conditions provided in (3) below, currency.

Point No. 3

Original: A. Two separate but freely convertible currencies for two wings may be introduced, or

B. One currency for the whole country may be maintained. In this case effective constitutional provisions are to be made to stop flight of capital from East to West Pakistan. Separate Banking Reserve is to be made and separate fiscal and monetary policy to be adopted for East Pakistan.

Amended: There shall be two separate currencies mutually or freely convertible in each wing for each region, or in the alternative a single currency, subject to the establishment of a federal reserve system in which there will be regional federal reserve banks which shall devise measures to prevent the transfer of resources and flight of capital from one region to another.

Point No. 4

Original: The power of taxation and revenue collection shall vest in the federating units and that the Federal Centre will have no such power. The Federation will have a share in the state taxes for meeting their required expenditure. The Consolidated Federal Fund shall come out of a levy of certain percentage of all state taxes.

Amended: Fiscal policy shall be the responsibility of the federating units. The federal government shall be provided with requisite revenue resources for meeting the requirements of defence and foreign affairs, which revenue resources would be automatically appropriable by the federal government in the manner provided and on the basis of the ratio to be determined by the procedure laid down in the Constitution. Such constitutional provisions would ensure that the federal government's revenue requirements are met consistently with the objective of ensuring control over the fiscal policy by the governments of the federating units.

Point No. 5

Original: (1) There shall be two separate accounts for foreign exchange earnings of the two wings,

(2) earnings of East Pakistan shall be under the control of East Pakistan Government and that of West Pakistan under the control of West Pakistan Government.

(3) foreign exchange requirement of the Federal Government shall be met by the two wings either equally or in a ratio to be fixed,

(4) indigenous products shall move free of duty between two wings,

(5) the Constitution shall empower the unit Governments to establish trade and commercial relations with, set up trade missions in and enter into agreements with, foreign countries.

Amended: Constitutional provisions shall be made to enable separate accounts to be maintained of the foreign exchange earnings of each of the federating units, under the control of the respective governments of the federating units. The foreign exchange requirements of the federal government shall be met by the governments of the federating units on the basis of a ratio to be determined in accordance with the procedure laid down in the Constitution. The Regional governments shall have power under the Constitution to negotiate foreign trade and aid within the framework of the foreign policy of the country, which shall be the responsibility of the federal government.

Point No. 6

Original: The setting up of a militia or a para-military force for East Pakistan.

Amended: The governments of the federating units shall be empowered to maintain a militia or para-military force in order to contribute effectively towards national security.

APPENDIX B

<u>RE:</u>   <u>MERCY PETITION</u>

On 24 March 1979 the Supreme Court, while dismissing the Review Petition of Mr. Z.A. Bhutto, observed that 'grounds urged by the Counsel for the Appellant for mitigation of the death sentence were relevant for consideration by the executive authorities in the exercise of their prerogative of clemency. On 26 March Justice Safdar Shah told media correspondents that the observations of the Supreme Court could not be disregarded by the executive while deciding the question of implementing the death sentence.

On or about 27th March Justice Moulvi Mushtaq conveyed to me that he happened to meet General Iqbal at a reception, who was inclined to accept the version of Justice Safdar Shah Moulvi Sahib desired to have my views. To my query about the confessing accused Moulvi Sahib disclosed that he had indeed recently learnt that Arshad Iqbal, Rana Iftikhar Ahmed and Ghulam Mustafa were assured by the authorities that they would not be hanged and that during the hearing of the Appeal in the Supreme Court the relatives of Mian Mohammad Abbas obtained similar assurances through General Chishti. I commented that in the circumstances the confessions were not voluntary and that would affect the validity of conviction and in any case the recommendation of the Supreme Court was entitled to great weight. Moulvi Sahib said that he would apprise General Zia about my views.

In the evening, General Zia phoned me and said that Moulvi Sahib had briefed him and indicated that Supreme Court was likely to issue a clarification.

S.S. Riza²
13.6.95

Then in a cordial way General Zia conveyed that arrangements had been made in the New York hospital for my bypass surgery and that an appointment had been fixed with Dr. Rosenfeld on 2 April. I reminded the General that the doctors had advised me to have the operation in February, 1979, but he had asked me to postpone it till the last week of April. General Zia said all arrangements had now been finalized and I should leave as soon as possible. He wished me well. On 29 March 1979, in a press release issued by the Supreme Court it was stated that whatever Mr. Justice Safdar Shah had said reflected his personal views only and he had no authority to speak on behalf of other members of the bench. I left for New York on 30th March and underwent bypass surgery.

In the meantime, the Law Ministry and the Law Minister (Mr. A.K. Brohi) in their opinion on the summary to the President observed: "Allah the Almighty has clearly provided as to how the murderer is to be dealt with. He does not permit of mercy to a merciless by the judge or the President. The authority that allows merciful commutation of a sentence is merciless of the deceased, his heirs and his relatives. Mercy remission or commutation is negation of justice, and justice is not only to be done to the killer who is surviving because of legal formalities but is also to be done to the deceased who cannot be heard but whose soul looks for justice - the revenge - death for death - and that in fact is the humanitarian consideration".

On 1st April, 1979 the President rejected the mercy petition and Mr. Bhutto was hanged on 4th April, 1979.

S. S. Pirzada
13.6.95

INDEX

East Pakistan, 30, 31, 32, 40, 144,
166-70 passim, 187, 191, 260-1;
1954 elections, 25, 38, 358; 1971
war, 115-34 passim, 137; civil war,
91-104; political crisis, 6, 7, 8, 16,
35-8, 43-87 passim, 91-114 passim,
see also, Zulifikar Ali Bhutto
East Pakistan Rifles, 96
Economic Reforms Order 1972, see
nationalization under Zulfikar Ali
Bhutto
Election Commission, 63, 177, 324,
328, 332, 338, 348, 365
Estaing, Giscard d', 242, 246

F

Faiz, Faiz Ahmed, 202
Fallaci, Oriana, 207, 232
Farland, Joseph, 237
Farman Ali, Maj.-Gen., 124, 132
Federal Republic of Germany, 242,
246, Constitution, 179, 184
Ford, Gerald, 241, 242, 248, 362
France, 126, 127, 128, 242, 244, 245,
246, 247

G

Gadhafi, Muammar, 220, 228, 235,
236, 351, 379, 380
Gandapur, Haq Nawaz, 20
Gandapur, Inayatullah Khan, 275
Gandhi, Indira; 206, 207, 211, 212;
Allahabad High Court, 222, 319;
assassination, 226; US, 113, 192;
visits Europe 113; see also,
Zulfikar Ali Bhutto, India, Simla
Gauhar, Altaf, 194
General Assembly(UN), 122, 123,
125, 130
Government of India Act 1935, 153,
166, 262

Gul, General Rakhman, 61
Gul, Justice Mohammad, 155, 177,
178

H

Haksar, P. N., 207, 208, 209, 210,
211, 212, 216, 217, 218,
Hala, Makhdoom of, 31
Hamoodur Rahman Commission, 81,
158
Hanif, Rana, 144
Haq, General Ziaul: 158, 187, 188,
194; 276; 361, 362, 368-70, 384;
and Zulfikar Ali Bhutto, 162, 251,
316, 349, 355, 371, 379, 380, 385
Haroon, Mahmood, 19, 225
Hasan, Mubashir, vii, 5, 7, 20, 32,
33, 34, 146, 148, 176, 285,
289-90, 300, 301, 304-5, 331, 336,
344, 347, and Zulfikar Ali Bhutto,
4, 12, 13, 113, 120, 121, 144, 213,
281, 296, 303, 307, 309, 310, 322,
323, 327, 329, 345, 346
Hassan, Col. M. A., 77
Hazarvi, Maulana, 55, 61, 120, 153
Hazelhurst, Peter, 16, 43, 238
Henry, Paul Marc, 132
Hitler, Adolf, 115, 162, 305
Housego, David, 364
Hussain, Justice Mushtaq, 368, 391
Hussain, Kamal, 50, 94, 105
Hussain, Syed Shahid, 284
Hyderabad Conspiracy Case, 276;
Tribunal, 276, 353, 356

I

Illahi, Chaudhri Zahur, 328, 332
India, 91, 94, 96, 121, 133, 150, 173,
187, 102, 137, 195, 216-24
passim, 226, 249, 363; Pakistan,
8, 9, 52, 54, 56, 58, 93, 94, 98,